The Economic Dynamics of Modern Biotechnology

The Economic Dynamics of Modern Biotechnology

Edited by

Maureen McKelvey

Professor of Economics of Innovation, Department of Industrial Dynamics, School of Technology Management and Economics, Chalmers University of Technology, Sweden

Annika Rickne

Assistant Professor, Department of Industrial Dynamics, School of Technology Management and Economics, Chalmers University of Technology, Sweden

Jens Laage-Hellman

Associate Professor, Department of Industrial Marketing, School of Technology Management and Economics, Chalmers University of Technology, Sweden

Edward Elgar
Cheltenham, UK • Northampton, MA, USA

Published by
Edward Elgar Publishing Limited
Glensanda House
Montpellier Parade
Cheltenham
Glos GL50 1UA
UK

Edward Elgar Publishing, Inc.
136 West Street
Suite 202
Northampton
Massachusetts 01060
USA

A catalogue record for this book
is available from the British Library

Library of Congress Cataloguing in Publication Data

The economic dynamics of modern biotechnology / edited by Maureen McKelvey, Annika Rickne, Jens Laage-Hellman.
    p. cm.
  Includes index.
  1. Biotechnology industries–Europe. 2. Biotechnology industries–Technological innovations–Europe. 3. Biotechnology–Research–Europe. I. McKelvey, Maureen D. II. Rickne, Annika, 1966- III. Laage-Hellman, Jens.
  HD9999.B443E85134 2004
  338.4'76606'094–dc22

2003064854

ISBN 1 84376 519 5 (cased)

Printed and bound in Great Britain by MPG Books Ltd, Bodmin, Cornwall

Contents

Contributors

Corinne Autant-Bernard
CREUSET, University Jean Monnet, France

Johan Brink
Chalmers University of Technology, Sweden

Stefano Brusoni
SPRU, University of Sussex, UK

Steven Casper
Judge Institute of Management, University of Cambridge, UK

Rachel Cutts
SPRU, University of Sussex, UK

Linus Dahlander
Chalmers University of Technology, Sweden

Sally Gee
CRIC, University of Manchester, UK

Aldo Geuna
SPRU, University of Sussex, UK

Mark Harvey
CRIC, University of Manchester, UK

Michael M. Hopkins
SPRU, University of Sussex, UK

Hannah Kettler
University of California at San Francisco, USA

Jens Laage-Hellman
Chalmers University of Technology, Sweden

Rasmus Lund Jensen
Copenhagen Business School, Denmark

Vincent Mangematin
INRA/SERD, University Pierre Mendes France

Nadine Massard
CREUSET, University Jean Monnet, France

Maureen McKelvey
Chalmers University of Technology, Sweden

Andrew McMeekin
CRIC, University of Manchester, UK

Fiona Murray
Sloan School of Management, Massachusetts Institute of Technology, USA

Paul Nightingale
SPRU, University of Sussex, UK

Luigi Orsenigo
Bocconi University, Italy

Michel Quéré
CNRS-IDEFI, Sophia-Antipolis, France

Annika Rickne
Chalmers University of Technology, Sweden

Jacqueline Senker
SPRU, University of Sussex, UK

Keith Smith
United Nations University/INTECH, The Netherlands

Finn Valentin
Copenhagen Business School, Denmark

Preface

This book is the outcome of a delightful and remarkable learning experience. I hope that you will learn as much as the contributors who have been involved in the process.

For me, the journey to explore economic dynamics and modern biotechnology began a number of years ago, in the European and European-American communities working on neo-Schumpeterian economics, innovation processes and science and technology. Without thanking them by name, I am privileged to have met some remarkable individuals along the way. These individuals have continued to influence my conceptualization of science, technology and innovation as inherently economic activities. Some of you have been directly involved in this book, as friends and colleagues.

For the co-editors and PhD students involved in this book, the learning experience also relies on our network 'Management and Economics of Biotechnology'. This network is run by Jens, Annika and I of the School of Technology Management and Economics, Chalmers University of Technology, but our participants are distributed around Sweden. Participants bring together their experiences from firms, government agencies and academic research in natural science/bioengineering and in social sciences (including economics and business). Many thanks to each and every one of you. Even if our network discussions may not be directly mentioned or recorded here, I speak for the whole Chalmers group in thanking you for allowing detailed insights and thereby stimulating our thinking.

This book is more directly the result of an international workshop and a collaborative effort of my colleagues at Chalmers University of Technology. Jens, Annika and I together with the excellent support of Johan Brink (PhD student) organized the workshop 'The Economics and Business of Bio-Sciences and Bio-Technology' at Hällsnäs and Chalmers in September 2002. Despite the beautiful autumnal weather – in a Scandinavian forest by a lake – the workshop participants from around Europe wouldn't stop talking! Naturally, I take this as a positive sign, signalling their concentration and interest in the original empirical data and deeper theoretical issues of the workshop. Thank you all! Of the 19 workshop papers presented, we selected 10 chapters for inclusion in this book.

In the process of moving from an initial idea for a workshop into a book, Jens, Annika and I have shared the work. We have worked together in our aim not only to deliver an edited anthology of papers on time, but more than that, to form these contributions and thoughts into a book with a message. They have made valuable comments on all chapters. Moreover, that is why the book now has four co-authored chapters to form the starting-point and ending-point for this book, rather than the standard one chapter introduction to the anthology. In this process, we have benefited from comments, with thanks particularly to Magnus Holmén, Stéphane Malo, Vincent Mangematin, Keith Smith and other participants at our Chalmers May 2003 workshop 'Innovation and Entrepreneurship: IT/ Telecommunication and Biotech/Pharmaceuticals'. The usual caveats apply. Linus Dahlander (PhD student) has helped at the end of the process in delivering the manuscript to the publisher.

Luckily, we three co-editors have complementary skills and personalities. I say 'luckily', given that the two of them have had to cope with me, a co-editor, who is at least occasionally chaotic and overcommitted as well as who may change her mind about where the book is headed. But for me, it has been a pleasure to work together. The same for my other co-authors. Many thanks also to Dymphna Evans, our editor at Edward Elgar, for her support and belief in this manuscript. And I don't think any other of my book editors will ever beat your speed of response!

These ideas and lively debates in the international academic communities, network and workshop would not of themselves have led to this book. In addition, a book requires time, effort and resources. You could say that resources are necessary to make this book into a codification of some ideas and results into words, figures and diagrams – as captured in subsequent pages.

This book would not have been possible without the contributions of each author, with their respective intellectual environments and financiers. Thank you very much for participating! The list of contributors and their scientific settings gives some idea of the diversity of the community working on these issues, as it has been emerging in Europe.

Most importantly, the financiers have played an important role in the workshop as well as in the writing and editing process for the book. I would like to acknowledge the support of the following projects.

- 'Dynamics of Knowledge in IT and Bio-Science' (a faculty-wide strategic project) financed by the School of Technology Management and Economics, Chalmers University of Technology;
- 'Endowment for McKelvey chair in Economics of Innovation', Office of the President, Chalmers University of Technology;
- 'Economic Dynamics of Knowledge in Bio-Science' financed by VINNOVA; administered by IMIT;

- 'From Where Does Novelty Arise? Flexibility and Stability through Innovation' financed by the Ruben Rausing Foundation; administered by IMIT.

Without you, this book would not have been possible.

Maureen McKelvey
Wienhausen convent, Germany;
Copenhagen Airport, Denmark;
and Kärrbogärde, Sweden

PART I

Introduction

1. Introduction

Jens Laage-Hellman, Maureen McKelvey and Annika Rickne

1. INTRODUCTION TO ECONOMIC DYNAMICS

The Economic Dynamics of Modern Biotechnology is about the dynamics of knowledge – the economic processes that shape the exploration and exploitation of new knowledge in society. This book rests on the idea that innovations and the related development of new knowledge and information go hand in hand with economic transformation.[1] The economy is in flux, changing fundamentally over time, with new products, firms and activities starting up and with existing ones being significantly modified or disappearing. This constant transformation is based on learning and innovation, which are largely economic processes with economic objectives. This book explores such processes, using the empirical probe of modern biotechnology.

Analysing these types of processes requires a theoretical insight into a fundamental puzzle of the economics of the modern learning society. The puzzle concerns how and why the development of knowledge and ideas interact with market processes and with the formation of industries and firms. Using evidence from modern biotechnology, the book probes more general and abstract issues about how to conceptualize and analyse the modern economy. In doing so, the focus is on relationships between firms, markets, governments and the ongoing innovation processes.

Modern biotechnology is both a broad emerging technological area and a specific economic activity. It involves many industries, and as such the processes of innovation and economic change are interwoven in society. This has, in turn, led to a broader societal debate around modern biotechnology, a debate that raises at least four paradoxes. These four paradoxes help frame our understanding of why modern biotechnology is an interesting empirical probe, providing insight into more general and abstract theoretical issues.

First, controversies continue to abound over the negative versus positive societal impacts. On the one hand, modern biotechnology is often claimed to be crucially important to many industrial sectors, to large as well as to small firms, and to address basic human needs and societal problems. On the other hand, modern biotechnology is also the centre of controversies about the modification of nature, food safety, animal welfare, environmental protection and impacts on global poverty. Given these different views, deep chasms divide those who focus on the potential benefits based on assumptions of individual creativity and social welfare returns, from those who debate the potential problems and therefore propose solutions such as government regulation and strong social control.

Second, despite such controversies about the economic and social potentials of modern biotechnology, little truly comparative statistics or empirical evidence exists. Many studies are narrowly based, with poor comparative scope, and the definitions, methods and empirical data on modern biotechnology differ greatly between studies. These differences frequently make it impossible to compare what is going on in one country or region with another. The reasons why such differences matter is that these types of empirical and methodological decisions greatly affect which aspects of the phenomena are picked up, compared and highlighted as the major strengths and weaknesses of a firm, region or country. One can speculate what implications follow if US data mainly report firms in the range of 100–500 employees, while data from Sweden or Germany mostly include firms with up to 100 employees. Of course, one plausible reason why few truly internationally comparative studies exist is the dearth of official data and accepted definitions, although the Organisation for Economic Co-operation and Development (OECD) has been engaged in important work to develop international statistics (OECD 2001).

Third, modern biotechnology is at once fundamentally global in terms of the knowledge flows resulting from movement of skilled persons, ideas, services and products – and yet, it is simultaneously extremely local in terms of co-located actors. Not only are actors located in specific geographical spots and networks, they also seem to agglomerate in those spots and maintain such linkages over long time periods. Some regions and nations seem to have a higher concentration and density of network relationships than others – and thereby gain competitive advantage. At the same time, global mobility is also visible among individual scientists, industrial researchers and firms. There is also a mobility of ideas through scientific competition, publications, reverse engineering, imitation of government policy and so on. Due to these simultaneously global and local aspects, this book is much more about the diversity of Europe as part of global trends, rather than Europe as a specific case.[2]

Fourth, modern biotechnology has seemed for several decades to be primarily a US phenomenon with the rest of the world lagging behind. Certainly, debates have raged about whether – and for what reasons – the

USA has competitive advantage, with explanations usually being based on empirical descriptions, and often resulting in advice for government policy. In relation to firms, the emergence of a dedicated biotechnology industry has been less successful in Europe. In 2001, there were 1453 biotech companies in the USA employing 141 000 people and with revenues exceeding $25 billion. Europe had 1879 biotech companies, but they were on the average significantly smaller than their US counterparts, employing around 34 000 people and with sales totalling $7.5 billion (Ernst & Young 2002).

The US competitive advantage is thought to be in the ability to generate and use a much higher quantity and quality of basic science as well as to commercialize diverse and rapidly adapting innovation opportunities. One explanation for this may lie in how and why the national institutional structure affects incentives for individuals and organizations. On the flip side, policy advice has also been abundant, particularly about how other countries and firms can get into or win the biotech race. This advice is often along the lines of 'imitate the leader'. More recently, Europe, Asia and the Pacific are claiming to be catching up quickly in terms of scientific achievements, university technology transfer offices, firm formation, venture capital investment and so on. This fourth paradox thus focuses on the existing distance between the 'leader' and the 'laggards' but competition should be understood as dynamic, not static.

An interesting aspect of the recent analysis and policy recommendations is the clear assumption of positive impacts of modern biotechnology, so those making policy recommendations seem to be in little doubt about the first of these paradoxes. A more nuanced analysis might be in order if the enduring reliability and validity of these claims is to be tested further. It is impossible to argue that the laggards are 'behind' or have 'caught up', if the relative positions of leaders and laggards has not been carefully and systematically compared in a historical perspective. Hence, it is not clear if this assertion of the supremacy of US science, commercialization, firms and so on, reflects true differences or whether it simply reflects early empirical evidence concentrated on the US experience. Probably, the answer is a bit of both and hence choices about definitions, indicators and data matter. Following on from the third paradox, one should ask whether an existing competitive advantage can be explained through national variables or as specialized actors within global processes. The question is thereby whether the USA as such generates this high potential internally or whether the research and business contexts have allowed a dynamic and self-reinforcing specialization to emerge thereby drawing the best individuals, within the global context. Social science plays an important role in this task of questioning paradoxes and 'accepted wisdom' with impacts on future allocation on societal and government resources.[3]

These four paradoxes indicate that modern biotechnology is an area rich in controversy. These broader contemporary debates illuminate the interwoven nature of societal and economic processes in emerging technological areas in many industries, as well as characteristics of the modern learning economy. This book is set within these broader debates, but the 14 chapters focus primarily on phenomena of relevance to the economic dynamics of knowledge. This book addresses a number of general and abstract issues concerning how and why to conceptualize the modern economy. This dynamic perspective is based on an understanding of how and why the development of knowledge and ideas interact with market processes and with the formation of industries and firms.[4]

The common starting-points for this book can be summarized as four 'stylized facts' (see Chapter 3). Each chapter implicitly uses at least one of these starting-points – and often several – even though each chapter also addresses a narrower research question and thereby makes a contribution in its own right. These stylized facts reflect more fundamental dimensions of the phenomena 'economic dynamics of knowledge', which is here applied to modern biotechnology.

1. *Innovations emerge from uncertain, complex processes involving knowledge and markets.* The development of new knowledge, technologies and markets involve ongoing, complex, dynamic processes within society and these affect, in turn, the development of innovation opportunities. Such processes affect both knowledge creation and knowledge exploitation. Science, technology and innovation are driven by an internal logic of knowledge accumulation as well as by market factors such as demand, customers' preferences, profitability, investment as related to current and future returns and so on. The opening up of new innovation opportunities – and the exploitation of existing ones – affect broader economic phenomena like firm formation, industry structure and globalization. These processes are ongoing and thereby historically rooted. The institutional structure and incentives to be engaged in knowledge creation and knowledge exploitation change over time. As innovations and innovation opportunities emerge, this challenges the existing as well as forms the new.

2. *New scientific and technological areas create economic value in many different ways. The impacts of the economic value accrue within global actors as well as for geographically situated agglomerations.* Changes in knowledge bases relevant to science, technology and innovation create economic value in different ways. They can be used directly in goods and service products or can be incorporated as technological or organizational process improvements. Other ways of creating economic

value are also possible. The knowledge can be used as an intermediary good, an input into other industries and/or directly commercialized through selling licences and future options on the results of scientific research.[5] These impacts accrue in various ways, such as property rights, firm profitability, employment, productivity increases, economic growth and so on. The positive and negative impacts may accrue to global actors as well as within geographically situated agglomerations.

3. *Societal linkages exist among diverse actors. Such networks as well as societal institutions affect innovation processes.* Actors are linked to society through market relationships as well as through other societal and economic relationships and social structures, such as innovation systems, networks and institutions. This implies that there is a diversity of relationships in society, and that these structures of relationships are likely to be unevenly distributed across the population of actors. One implication of this stylized fact is that public policy may contemplate a range of action, which is much wider than compensation for 'market failure'.

4. *The firm as an organizational form plays a particularly important role in knowledge exploration and exploitation.* The firm as an organizational form plays a crucial role in exploration and exploitation of science, technology and innovation. This is so even though the firm and its future options are above argued to depend on broader processes as well as other actors, relationships and social structures. The argument is that it is significant how and why the firm (and a population of diverse firms) acts – and survives – in the face of the opportunities and challenges of innovations and markets. It matters because the firm – and population of diverse firms – will affect the rate, direction and outcome of knowledge creation and knowledge exploitation in the long run.

These four stylized facts are relevant to many types of innovation processes and to the general phenomena of economic dynamics of knowledge. In this book, these four stylized facts are common starting-points, as applied to the economic dynamics of modern biotechnology. Their relevance applies both to the empirical understanding of trends and in theoretical explanations of how and why such processes occur – and diverse actors are involved.

In approaching the dynamics of the modern learning economy, chapters in this book naturally address specific research questions, and the chapters use – and further develop – a variety of theories. As such, this book contributes to a novel and broader conceptualization, drawing on fields such as evolutionary and institutional economics, innovation studies, economic sociology, management and business economics. Theoretical explanations are necessary to answer the questions and to capture the phenomena, and thereby a variety of more specific research questions and theories is relevant to this book. This variety may be necessary to understand the phenomena, given that the phenomena are seldom the result of a singular factor, such as an ideal market, social relationships or internal scientific institutions.

The theories used throughout this book may be seen as focusing primarily on (a) innovations and developments in science and technology as related to impacts on firms and regions; (b) markets, commercialization and industries as related to dynamics introduced by innovations, and new scientific and technological knowledge; (c) networks and innovation systems as enabling and constraining diverse actors; and finally (d) the firm as a particularly crucial actor, which is accessing resources and developing dynamic capabilities. Each chapter refers to theoretical literature appropriate to their specific research question. Still, the theoretical explanations found here are related to each other, in that the fundamental issues of the economic dynamics of modern biotechnology unite the contributions.

In summary, this book seeks to provide insight into one of the most important – and yet more controversial and paradoxical – knowledge fields for society, globally. The following chapters each address a specific theoretical and empirical issue, but the book as a whole also has a message. Unravelling modern biotechnology requires an understanding of the long-term interlinkages among government, markets and development of knowledge – as affecting decision-makers in firms, governments, universities and other organizations. Hence, the book argues for a more complex understanding of the economic dynamics of modern biotechnology, which is relevant for social science research as well as decision-making in governments, universities and firms.

2.   MODERN BIOTECHNOLOGY

Modern biotechnology is a broad area of scientific and technological knowledge, which has been developing rapidly, in terms of the scope and rate of development of new ideas, techniques and tools. The development of knowledge – in combination with the economic uses – impacts on more aggregate economic variables, such as productivity, profitability and returns, firm performance and economic growth. This section gives a brief overview with regard to knowledge base and economic impacts – as an introduction to subsequent chapters (see Chapter 2).

The knowledge base continues to expand. The 'modern' in modern biotechnology refers to the post-genetic engineering era, that is after scientists had developed the knowledge, techniques and tools to intervene directly at the gene level. This obviously gives more control than previous 'biotechnology', but the logic still follows upon more traditional ways of modifying cells and biological agents (see Katz and Sattelle 1991; McKelvey 1996: Ch. 4). Humans have used 'traditional' methods for thousands of years – to breed animals, improve plants, make beer and bread and so on.

Most definitions of modern biotechnology refer back to principles of living organisms. The influential OECD provided a working definition in 2001: 'Biotechnology is the application of scientific and engineering principles to the processing of materials by biological agents to provide goods and services'.[6] Similar definitions may be found within national policy agencies, industry associations and so on – but putting these definitions into practice, for example applying them to categorize a particular piece of science, company, or product, is often more difficult.

Modern biotechnology is comprised of a broad range of knowledge fields. The OECD *ad hoc* meeting on biotechnology statistics held in Paris in 2001 further clarifies the above working definition to include the following five categories:

1. DNA (the coding): genomics, pharmaco-genetics, gene probes, DNA sequencing/synthesis/amplification, genetic engineering;
2. Proteins and molecules (the functional blocks): protein/peptide sequencing/synthesis, lipid/protein engineering, proteomics, hormones and growth factors, cell receptors/signalling/pheromones;
3. Cell and tissue culture and engineering: cell/tissue culture, tissue engineering, hybridization, cellular fusion, vaccine/immune stimulants, embryo manipulation;
4. Process biotechnology: bioreactors, fermentation, bioprocessing, bioleaching, biopulping, biobleaching, biodesulphurization, bioremediation and biofiltration;
5. Sub-cellular organisms: gene therapy, viral vectors.

Hence, a broad range of scientific and technological knowledge fields are included within the definition of modern biotechnology – and this range keeps changing, as science and technology progress over time.

Modern biotechnology also has many economic impacts. Over time, many different ideas – including knowledge, techniques, tools, biological materials and so on – become useful economically. These ideas may come from various sources, and they may be sparked through scientific research, industrial research and development (R&D), insight into customer needs and so on. Some ideas are realized and turned into goods and service products for sale on markets. Other knowledge, techniques and tools are used directly by the inventors and/or provide supplies and equipment to other firms and other industries. Not all succeed technically or economically – but some do.

This broad knowledge area has made it possible to develop new firms and industries – but also substantially affected existing firms and industries. The development of new ideas opens up innovation opportunities, in a dedicated 'biotech industry' as well as in many existing industries like pharmaceuticals, agriculture, forestry pulp and paper and so on.

The empirical evidence shows that both small dedicated biotech firms and larger firms in existing industries can turn such innovation opportunities into economic value. On the one hand, some of these innovative ideas are commercialized through the start-up of new biotech firms. For example, there may be individuals – such as scientists and professional entrepreneurs – who are using their broad contact networks to develop business ideas and build companies. Such firms, for historical reasons, typically have strong links to other actors such as universities, government agencies, venture capitalists and other firms.

On the other hand, for the incumbent firm, changes in the knowledge base and in innovation opportunities may be perceived as a peril to their organizational existence and to their routine ways of doing things. In other cases, existing firms may perceive these processes as the starting-point for reformulating their need to access new knowledge fields in order to develop strategies and access resources in order to make a profit and to survive. In such cases, the firm must thereby also realign its position within networks to access resources and relevant knowledge. These processes of adaptation occur globally as well as within specific regions and countries.

Hence, modern biotechnology involves moving phenomena, analysed in economic terms. This can be seen through a variety of indicators, such as in terms of affecting the type and number of products for sale, the processes being used, the start-up of new firms, the conditions for competing in different industries and so on. Within agriculture and the food industry, for example, heated debates over the advantages and disadvantages of genetically modified organisms (GMOs) in Europe have focused attention on the ability to modify plants and animals for very specific purposes. Differing regulations has led to differing firm strategy and sales by US and European companies in these product markets.

It is clear that major changes in knowledge, techniques and tools within a wide variety of medical, scientific and engineering disciplines result in an expanding definition. Initially in the mid- to late 1970s, modern biotechnology was mainly confined to genetic engineering techniques. In the decades since, the term has expanded to include many other things, which are as diverse as protein engineering, bioinformatics and life sciences for agro-food. Some changes in definition are related to actors' preferences, and thereby negotiation over concepts to gain resources affects the definitions commonly used by analysts and researchers. Two examples are relevant here. As firms are started up and develop business plans for new goods and services not previously found on the market, they have to decide whether to market themselves as primarily 'core biotech' firms or as firms located in other industries. The firms may label themselves in different ways, but in either case, they are likely selling biotech-related knowledge, goods and services to consumers and/or firms in existing industries like pharmaceuticals, agriculture and medical technology. Moreover, as research funding becomes abundant within 'biotechnology', researchers within various disciplines have to decide whether to change their line of inquiry and relabel them to be awarded research grants. These two examples indicate that the boundaries between rapidly developing science and products are not clearly differentiated, but instead they are negotiated among actors at specific time periods. Hence, there are reasons to suspect that the definitions will continue to change over time and that a variety of definitions will be used at any one time.

This can be exemplified in relation to the large amount of studies done on biotechnology as related to human health care, particularly the pharmaceutical industry (Drew 1999). One explanation for this focus is that much biotech R&D has traditionally been related to the pharmaceutical industry and/or that much scientific research on biotechnology has emerged from medical fields. A second reason could be that more reliable indicators and data are assembled for the pharmaceutical industry and the large pharmaceutical firms can be identified and analysed.

A third reason for this focus is that modern biotechnology has truly had – and continues to have – a significant impact on existing economic activities in pharmaceuticals, and has also opened up opportunities for new activities to emerge. 'From the very beginning, the growth of biotechnology has been driven mainly – though not exclusively – by perceived application and commercialisation opportunities in the pharmaceuticals or health care field' (Granberg and Stankiewicz 2002, p. 6). Indeed, innovation opportunities coupled with customer demands has promised a high potential for profitability in pursuing the biotech track of development. Granberg and Stankiewicz rightly point out that such promises must be seen in the light of the industry's need to increase the efficiency and productivity of the drug development process, where modern biotechnology may provide a solution to technological hindrances. Modern biotechnology has greatly transformed the knowledge base within pharmaceuticals, both in terms of scope and cognitive differentiation. This can be seen in terms of knowledge needed for different parts of the value chain, including new disciplines such as genomics, proteomics, combinatorial chemistry, bioinformatics and so on. The discovery phase involves new means and tools for target and lead identification, while in the development phase the ways to perform clinical trials change. Moreover, production may now entail, for example, cell culturing. As new commercial opportunities arise, new firms and other types of actors may emerge. The biotechnology-pharmaceutical sector is one example of the dynamic processes involving knowledge and markets.

Even so, global criticism has focused on the '10/90' gap, namely that only 10 per cent of resources go to research on diseases responsible for 90 per cent of the world's 'burden of diseases' (WHO 2001). Or, to put it the other way around, North America, Europe and Japan constitute the overwhelming majority of global demand for pharmaceuticals, given that they consume 90 per cent. This highlights the need to consider applications of modern biotechnology to human health care from a global perspective.

3.   THE SCOPE OF THIS BOOK

This section provides a 'road-map' to individual chapters and the book as a whole. A brief recapitulation of the objectives of the book is useful. The 14 chapters in this book address a series of theoretical and empirical questions, but are also united by a common message. In doing so, they provide an insight into modern biotechnology in the European case – or more specifically, an insight into diverse national and sectoral cases found in Europe in relation to global trends. Modern biotechnology is seen both as specific empirical phenomena as well as an empirical probe to explore the general and abstract issues of economic dynamics of knowledge. There are two chapters of introduction in the first Part; 11 chapters grouped into the following three Parts; and a concluding chapter in the final Part.

In terms of this book, the ongoing development of knowledge bases and commercial opportunities matters because a diversity of definitions is applied to the empirical material (see Chapter 2). Subsequent chapters follow the broad definition of 'modern biotechnology', but may relate that to other concepts, like 'life sciences'. In other words, the chapters focus on a number of somewhat different, somewhat overlapping concepts. For example, one contribution delves into the convergence of modern biotechnology with information technology (IT), whereas another chapter examines the impacts of genetic information on insurance, which have not previously been studied in the biotech context. All of them fall, however, within the broader concept of 'modern biotechnology'.

Given the common agreement and interest in the broad concept of 'modern biotechnology', this diversity among chapters is a choice made in editing this collection of work into the book. It allows authors to capture the moving boundary of the intersection of modern biotechnology with other phenomena and other activities. This relates back to our objective to understand emerging technological areas, including impacts on industries. The concept must be stringent enough to define existing boundaries, but also plastic enough to capture changes in knowledge and products over time, where the boundaries are not clearly differentiated. The chapters are also united by their awareness of definitions, methodology and data, in that most chapters present novel empirical work.

'Introduction' constitutes Part I where Chapters 1 and 2 form the common introduction to the book. The current chapter 'Introduction' by Jens Laage-Hellman, Maureen McKelvey and Annika Rickne has introduced the idea of 'economic dynamics of knowledge' as well as the broader societal debate about modern biotechnology, thereafter introducing four stylized facts and a broad definition and the book chapters. Chapter 2, 'Conceptualizing and measuring modern biotechnology', by Johan Brink, Maureen McKelvey and Keith Smith focuses on how and why measurement and methodology matter. This chapter introduces key issues related to choosing definitions, indicators and empirical data as well as the implications of such choices for drawing conclusions and implications for decision-making.

The subsequent chapters are grouped into four parts: 'Setting the scene' (Chapters 3–5), 'Challenging the existing' (Chapters 6–9), 'Forming the new' (Chapters 10–13) and 'Conclusions' (Chapter 14). Each part contains chapters that share some common characteristics and that are of relevance to that topic.

'Setting the scene', (Chapters 3, 4 and 5) paints a broad view through historical accounts and broad overviews of trends, as linked to empirical details. These three chapters are thereby useful for orienting the reader and for setting the scene of the book as a whole. They help set the scene, in the sense of providing arguments about historical trajectories and possible future ones as well as detailing diversity across time, across geographic space and across industrial versus scientific institutions. Chapter 3 provides insight into innovation processes in modern biotechnology, as organized around four stylized facts. Chapter 4 provides one possible future in the development of international trends while Chapter 5 provides a detailed analysis of the current situation in eight European countries.

Chapter 3, 'Stylized facts about innovation processes in modern biotechnology', by Jens Laage-Hellman, Maureen McKelvey and Annika Rickne makes the argument that modern biotechnology has been – and continues to be – developing at a rapid pace at the intersection of scientific and technological knowledge with innovation and business opportunities. This chapter is organized around the four stylized facts presented in Part I and draws on existing research as well as a case study of genomic companies and human biobanks.

Chapter 4, 'The post-genome era: rupture in the organization of the life science industry?', by Michel Quéré details one possible future trajectory. He argues how and why the current post-genome era (PGE) will be a structural shock in the organization of firms active in the life sciences industry. Quéré claims that under conditions where exploration as well as combination and adaptation of scientific opportunities are key, dedicated biotech firms (DBFs) will play an increasingly important role as drivers of knowledge dynamics. Further, the importance of the small firms will increase in the future, even though large firms have so far kept their advantages. So far, the large firms have had advantages in handling the complex character of knowledge as well as the necessity to coordinate complex capabilities in the pharmaceutical industry.[7] In the future, environmental conditions will continue to change and should give advantages to the smaller firms.

Chapter 5, 'An overview of biotechnology innovation in Europe: firms, demand, government policy and research', by Jacqueline Senker demonstrates the diversity of western Europe, based on evidence within three sectors and eight countries. The integrated framework to analyse these countries includes factors such as industry structure, supply, demand, financial system, industrial development and social acceptability as well as knowledge and skill formation. The comparison is based on this common set of variables, which was used to organize a thorough industry survey and summarize secondary material. This chapter argues that three prevailing factors affect the firm's propensity to innovate in different sectors and countries, namely, the existing structure of production, demand (including social acceptance) and a well-funded science base.

'Challenging the existing', (Chapters 6, 7, 8 and 9) gives an insight into the turbulence of modern biotechnology, including disruptive effects on existing businesses and economic phenomena.[8] These contributions present arguments about how and why existing actors and broader social structures attempt to cope with the challenges of modern biotechnology. These four chapters address, respectively, the challenges of modern biotechnology to existing firms and industries; firm strategy and reallocation of risk; firm strategy to cope with new information and access new competences; and emergence of new actors within an existing structure. Moreover, they use evidence from firms and industries that are not included in more 'traditional' analysis of biotech industries, namely within agriculture, genetic testing, insurance and IT. In coping with these broader societal impacts, the 'existing' actors are of course changed as part of ongoing processes. Hence, these four chapters address issues about 'Challenging the existing' because they focus mainly on how and why the 'existing' becomes renewed and fundamentally changed over time.

Chapter 6, 'Risk management and the commercialization of human genetic testing in the UK', by Michael Hopkins and Paul Nightingale, analyses corporate strategic management, using genetic testing as an example within human health care. The authors conclude that effective risk handling and reallocation of risk to third parties is key when the firm is moving into new markets where uncertainty is high. Risk management is here seen as a combination of firm strategy and public concern. The chapter is based on case studies of four UK firms providing genetic testing services.

Chapter 7, 'Networks and technology systems in science-driven fields', by Finn Valentin and Rasmus L. Jensen, examines patterns of network formation taking a microorganism widely employed in food processing (lactic acid bacteria) as its focal case. By tracking biotech patents since 1980 it is obvious that the European rate exceeds that of the USA. Competing theories on systems of innovation are examined to explain the global organizations of this field. Using formal network analysis this chapter suggests novel ways of differentiating technology systems and national systems of innovation as distinct, yet interrelated, causal mechanisms. These interrelationships have implications for understanding the responsiveness and flexibility of institutions, particularly those associated with production and dissemination of scientific knowledge.

Chapter 8, 'Future imperfect: the response of the insurance industry to the emergence of predictive genetic testing', by Stefano Brusoni, Rachel Cutts and Aldo Geuna, addresses the challenges of modern biotechnology to insurance companies. This chapter examines how firms cope with building new capabilities as a rejoinder to changes in the knowledge base and in consumer-driven industries. The survey of firms located in the UK (but often with international customers) concludes that especially large insurance firms are actively learning about the new knowledge field of genetic testing, both through internal efforts and collaboration. In addition to firm size and characteristics of the final market, the type of insurance involved affects the firm's involvement in learning as well as their strategies for integrating outside competences into the main business.

Chapter 9, 'Emergent bioinformatics and newly distributed innovation processes', by Andrew McMeekin, Mark Harvey and Sally Gee, argues that developments within bioinformatics clearly demonstrate a distributed innovation process, based on new and old elements. As such, their chapter functions as a transition between 'Challenging the existing' and 'Forming the new'. Supported by a meticulous case study of a bioinformatics firm located in various European countries and globally, this chapter analyses how new classes of economic agents emerge as a response to innovation opportunities as well as the dynamics of evolving coordination mechanisms and collaborative patterns.

'Forming the new', (Chapters 10, 11, 12 and 13) takes up issues directly related to the development of new business and economic activities around modern biotechnology. These four chapters particularly focus on forming the new – but as part of specialization and competition within global developments and/or as part of spatial bounded network patterns. Hence, these chapters analyse economic dynamics both in terms of historical trajectories as well as relative specialization within specific regions and countries. The chapters address one or both of the following: (1) the extent to which innovation processes are spatially dependent and/or emerge as long-term specialization within specific sub-fields; and (2) the effects of local (regional) networks on firm formation and performance. The four chapters in 'Forming the new' thereby follow on from 'Challenging the existing', by providing an insight into how and why the 'new' are linked to the historical trajectories.

Chapter 10, 'The dynamics of regional specialization in modern biotechnology: comparing two regions in Sweden and two regions in Australia, 1977–2001', by Johan Brink, Linus Dahlander and Maureen McKelvey, examines whether regional specialization requires a co-evolution of scientific, technological and business activities at the regional level, or whether firms are simply tied into global trends. This analysis is based on arguments to examine whether knowledge formation within specific sub-fields of modern biotechnology influences long-term trajectories of specialization at the regional level. Data are presented about scientific, technological and business activities over 25 years within two regions in Sweden and two regions in Australia. The results reveal that, relative to global trends, the regional economic transformation is a process that can be described as path-dependent in terms of specialization over time, but with points of rupture and change. The question remains how, why and when diverse types of actors may provoke those points of rupture. Interestingly, knowledge externalities in modern biotechnology between disciplines and industrial sectors implies that analysts and practitioners must understand the patterns of specialization for interrelated activities and sectors, and not just the knowledge base.

In Chapter 11 'On the spatial dimension of firm formation' Annika Rickne takes a somewhat different twist, by focusing on the process of firm formation within spatially delimited 'regions', relative to specific features of science, technology and innovation. This chapter is based on an in-depth case study analysis of biomaterials-related firms in Sweden, Ohio and

Massachusetts. It argues that the volume and profile of the firms which are established, as well as their propensity for co-location, are related to characteristics of the technological regime, the profile of the regional science base, and, most importantly, to the specificity of pre-firm activities and networks.

Chapter 12 'Examining the marketplace for ideas: how local are Europe's biotechnology clusters?', by Steven Casper and Fiona Murray, asks the question of why some clusters are more successful than others in commercializing biosciences. The chapter examines the emergence of two European clusters (Cambridge and Munich) and one American (Boston) along three dimensions: the source of technology, scientific linkages and access to labour markets. It is concluded that the European clusters are too small to support purely local cluster development. Therefore, this view has conclusions that differ from many existing policy recommendations. Rather than supporting local networking, effective policies should centre on the development of institutions that attract firms and other actors from outside the region and enhance the local actors' ability to build up geographically dispersed collaborations.

Chapter 13, 'Creation and growth of high-tech small- to medium-sized enterprises (SMEs): the role of the local environment', by Corinne Autant-Bernard, Vincent Mangematin and Nadine Massard addresses the question of whether the determinants of firm creation as well as growth of high-tech firms can be found at a regional level. Through econometric analysis of high-tech SMEs in France during the 1990s, the authors argue that new business creation is promoted by regional activity, especially by diversity (rather than quantitative potential) of scientific competences, patterns of public–private interactions and market agglomeration effects. In contrast, the growth potential of firms in the region depend by and large on internal characteristics and strategic choices of the firms.

'Conclusions' constitutes Part V. In Chapter 14, 'Reflections and ways forward', Hannah Kettler, Maureen McKelvey and Luigi Orsenigo reflect on the contributions of the book and suggest future avenues for social science research. In addition to revisiting the four paradoxes, the chapter highlights the specific issue of applying these perspectives to human health care in developing countries. Finally, the chapter reflects on the challenges of modern biotechnology to our conceptual understanding of the economic dynamics of knowledge.

NOTES

1. Economic transformation refers to evolutionary change processes at many levels, which, in turn, influence flexibility and stability of society and the economy. The relative degree of flexibility and stability – as well as reasons for different outcomes – can be analysed at various levels, such as within firms, industries, national economies and so on. Based on neo-Schumpeterian and evolutionary economics, the argument would be that economic transformation relies on qualitative (and not such quantitative) change among inputs, outputs and transformative intermediates like business organizations and technology.
2. Most chapters in the book provide evidence of Europe, or more correctly, of the diversity of Europe through a variety of evidence of firms, sectors and countries. Hence, this book provides a multitude of evidence for what is ongoing in Europe – as related to global trends – but especially as related to theoretical and conceptual issues. This implies that the book as a whole does not purport to provide one 'holistic' picture of 'the' situation in Europe – indeed, if such a thing is possible. Individual chapters here instead provide a series of 'snapshots' – or rather, moving pictures – of trends in development within Europe.
3. Social science is here used as a broad term, including both economics and business studies.
4. Neither the firm alone nor an 'innovation system' as such can explain the interactions among knowledge and markets. The theoretical structure required must include institutional settings as well as incentives, decisions and actions by a diverse set of actors.
5. The implication is that modern biotechnology involves a variety of different types of innovations.
6. The definition and the five categories were developed in *ad hoc* meetings. They can be found explicitly on the OECD's homepage under 'definition of biotechnology' and under electronic documents. The OECD (2001) document that followed on from these meetings does not provide these exact wordings, but is based on these discussions. See OECD (2002) and OECD (2003).
7. The argument is that emerging organizational structure will lead to an increasing importance of innovation networks, since cooperation gives possibilities for complementary resources and activities rather than specialization within the firm.
8. The 'challenges to the existing' can be seen through, for example, changes to traditional innovation processes, new organizational structures, the emergence of new types of actors, the need for incumbent firms in existing industries to renew their knowledge base and modify routine ways of doing things and so on.

REFERENCES

Drew, J. (1999), *In Quest of Tomorrow's Medicine: An Eminent Scientist Talks About the Pharmaceutical Industry, Biotechnology, and the Future of Drug Research*, New York: Springer Verlag.

Ernst & Young (2002), *Beyond Borders: The Global Biotechnology Report 2002*, Toronto: Ernst & Young.

Granberg, A. and R. Stankiewicz (2002), 'Biotechnology and the transformation of the pharmaceutical value chain and innovation system', paper presented at workshop The Economics and Business of Bio-Sciences and Bio-Technologies: What can be Learnt from the Nordic Countries and the UK?, September 25–27, Chalmers University of Technology, Gothenburg, Sweden.

Katz, J. and D. Sattelle (1991), *Biotechnology for All*, Cambridge: Hobsons Scientific.

McKelvey, M. (1996), *Evolutionary Innovations: The Business of Biotechnology*, Oxford: Oxford University Press.

OECD (Organisation for Economic Co-operation and Development) (2001), 'Biotechnology statistics in the OECD member countries: compendium of existing national statistics', STI working paper no. 2001/6, Directorate for Science, Technology and Industry, OECD.

OECD (2002), 'Statistical definition of biotechnology', website http://www.oecd. org/EN/document/0,,EN-document-617-1-no-21-31006-617,00.html, viewed May 2003.

OECD (2003), Report on *ad hoc* meeting in 2001 in Paris, website http//www. oecd.org/doc/M00003000/M00003698.doc, viewed May 2003.

WHO (World Health Organization) (2001), *WHO Health Report 2001*, website www.who.int/whr 2001, viewed 10 May 2003.

2. Conceptualizing and measuring modern biotechnology

Johan Brink, Maureen McKelvey and Keith Smith

1. INTRODUCTION

This chapter addresses issues related to conceptualizing and measuring modern biotechnology, thereby raising questions of definitions, methodology and data.[1] This chapter argues that behind many of the empirical and comparative issues lie fundamental problems of conceptualization of the empirical phenomena. Addressing these conceptual problems is important for analytical progress, but also for decision-makers in government policy, universities and firms.

Within this context, the chapter has three purposes. The first purpose is to highlight some key issues about the conceptual choices possible in operationalization and the implications thereof. There is a vast number of different terminologies regarding the usage of the concept 'modern biotechnology' as well as different methodologies, indicators and data. Such diversity is neither surprising nor strange – given that the dynamic nature of the underlying object under study by definition introduces inconsistencies. Still, inconsistencies prevent comparisons, and choices must be made.

The second purpose is to introduce a conceptual matrix to structure one way of thinking about how to conceptualize the empirical phenomena. The objective here is to structure concepts in a way that is useful for analytical work. The approach here is based on the idea of distinguishing clearly between two axes of 'product and sector', on the one hand, and of 'knowledge bases', on the other hand.[2] The concept of 'knowledge base' refers to areas of scientific and technological knowledge, including both the knowledge itself as well as its embodiment in techniques and instrumentation. This conceptual matrix follows on from the first purpose, but aims to take a step further in providing a novel and useful structure of interpretation.

The third purpose is to form a common starting-point in terms of the weight given to methodological considerations in this book, given that all chapters include significant empirical data. Each chapter makes choices about definitions, methodology and data and these choices together influence 'what is seen' and 'what is concluded'. While diversity exists, nevertheless, authors of this book are united by an awareness of the importance of methodological considerations as well as by a common desire to find more appropriate ways of understanding the empirical phenomena.

Although these issues may appear to be mundane on the surface, dissonance between different definitions and different measurements of the empirical phenomena form the second paradox introduced in Chapter 1. The general problem is that modern biotechnology is an emerging, generic technological area, containing many broad (and internally differentiated) knowledge bases. How then to capture it, without either capturing only irrelevant details or missing the main point?

This chapter begins by comparing and contrasting definitions in Section 2. A conceptual matrix for understanding modern biotechnology is then proposed in Section 3, based on the two axes of 'product (sector)' and 'knowledge base'. Section 4 raises some central issues related to the operationalization of this concept through data and statistics. This chapter does not purport to offer a universal solution – instead, the purpose is to raise awareness of the choices (and implications thereof) of definitions, methodology and data for analysing the empirical phenomena.

2. DEFINING MODERN BIOTECHNOLOGY – AND RELATED CONCEPTS

While Chapter 1 gave a quick insight in the context of this book, this section compares definitions of modern biotechnology and related concepts in more detail. This is done partly in order to provide a broad framework for the book and partly in order to be able to discuss operationalization of concepts in the subsequent sections.

Roughly speaking, biotechnology is defined as the application of knowledge of living organisms and their components, to industrial products and processes. It has already been claimed that a diversity of definitions exists – and even more so for operationalization. Some studies focus almost exclusively on a particular industry or type of firms, whereas others claim that modern biotechnology should by now be widely regarded as a diverse set of knowledge bases and an enabling technology, rather than as a distinct industry or sector *per se*. Clearly, the knowledge, techniques and tools of modern biotechnology have diffused and affected quite different application areas and industries. Examples of areas affected include human health care, drug development and pharmaceuticals, food production and processing, new materials and fine chemicals, energy, sensors and such environmental applications as bioremediation.

Many different definitions exist in the literature, ranging from reports published by internationally influential bodies such as the Organisation for Economic Co-operation and Development (OECD), through government agencies and consultancies to studies undertaken by academic researchers in the social sciences. A list of definitions follows, containing examples from government reports, industry associations and the social sciences. While the OECD definition is currently among the most frequently cited, the point here is that many other definitions exist as well, and these open up considerable space for diversity in interpretation, measurement and policy ideas.

- The OECD (OECD 2003):'Biotechnology is the application of scientific and engineering principles to the processing of materials by biological agents to provide goods and services.'
- The Office of Technology Assessment in the USA in 1984 (OTA 1984): 'Modern biotechnology incorporates a specific focus on industrial usage of rDNA, cell fusion and novel bioprocessing techniques. Industrial use of living organisms rDNA: New drugs, food, chemicals, degradation of toxics.'
- The European Federation of Biotechnology (EFB 1982 as referred in OTA 1984): 'Biotechnology is the integration of biochemistry, microbiology and engineering sciences in order to achieve technological (industrial) application of the capabilities of micro-organisms, cultured tissue cells and parts thereof.'
- The Office of Technology Assessment in the USA in 1991 (OTA 1991): OTA introduces the commercial activities around biotechnology as 'a new set of techniques that can be used in basic research, product development, and manufacturing in several different industries'.
- The Lord Sainsbury report in the UK (Sainsbury 1999):[3] 'Biotechnology companies are those whose primary business focus is the commercialization of these new technologies . . . The sectors for which biotechnology holds most promise includes pharmaceuticals, agriculture and food.'
- The Biotechnology Industry Organization in the USA in 2003 (BIO 2003): 'New biotechnology – the use of cellular and molecular processes to solve problems or make products . . . Biotechnologies capitalizing on the attributes of cells and biological molecules.'
- (Powell, et al. 1996): 'In many respects, biotechnology is not an industry *per se*, but a set of technologies with the potential to trans-

form various fields. Many researchers treat the wide array of biotechnology companies as comparable. In contrast, we intentionally
restrict our attention to only those for-profit firms engaged principally in human therapeutics and diagnostics, hereafter referred to as
dedicated biotechnology firms, DBFs.'

- Zucker and Darby (1997): 'The revolution in the biosciences
  has transformed technologies used in many other industries (including medical supply, chemical, agricultural, food-processing, and
  brewing), but none so rapidly and dramatically as in drug discovery'.
  They further argue that a biotechnology is generally defined more
  narrowly in terms of using breakthrough technologies such as
  genetic engineering.
- Feldman and Ronzio (2001): The definition includes combinatorial
  chemistry and liposomes to fields of chemistry and fermentation,
  large-scale cell culture and tissue culture.
- Carlsson (2002): Bio-industries transition is due to the new biotechnology and information technologies. Bio-industries is defined as
  such: 'Fundamentally dependent on the generation, processing or
  manipulation of biological systems and materials. This includes
  health care and medical services, agriculture and food technology,
  environmental technologies, biomaterials, large sectors of chemical
  technology, parts of the energy sector and several others.' The study
  includes biomedicine more generally, medical devices, instruments
  and supplies as well as applications outside the health care sector
  such as food, agriculture and forestry.

The OECD definition is widely used but is quite narrowly delineated in relation to a techno-economic context. One problem with the OECD definition
is that the convergence of modern biotech research with other fields implies
that it is becoming increasingly difficult – and perhaps meaningless – to distinguish traditional chemical and medical research from those dependent
on modern biotechnology. Therefore, the OECD is currently reworking
their biotech definition to improve the ability to characterize different
forms of biotechnology, based on the five areas of knowledge listed in
Chapter 1. Looking at other definitions, one can see a strong variety in
terms of concentration on the fields of application, types of companies and
economic sectors as well as in terms of the underlying knowledge and technologies.

Problems of definition feed directly into problems in the development of
international statistics to capture the phenomena of modern biotechnology. Some of these challenges are similar to the earlier challenges in understanding information technology (IT) and informtion and communication

technologies (ICT) industries within modern society. Here, one problem was an often sharp dichotomy between the small size of manufacturing ICT sectors, and the large claims made for the economic impacts of IT in terms of productivity impacts in other industries. Statistical agencies coordinated through the OECD responded to this challenge by expanding the scope of what was regarded as ICT industries. Similar problems arise with respect to modern biotechnology. Thus, the attempts by national statistics agencies and the OECD to harmonize definitions and statistics are quite important in a longer-term perspective of analysing the phenomena.

One way to help sort out the difficulties of giving the 'right' definition is to understand that modern biotechnology is part of a knowledge frontier, with a past, a present and a future (see Chapter 3). Modern biotechnology is more than the current scientific discoveries and current products in the market. It also includes a past trajectory of development of knowledge, techniques and tools – and where all the accumulated knowledge is included. It includes a current 'state of the art' and in the future, new discoveries and interpretations will change what we understand and what we can do. The historical processes matter, and thereby draw our attention to the span of relevant knowledge, techniques and instrumentation. The earlier generations of biotechnology have also continuously co-evolved with 'modern' ones, since the process of knowledge generation is both cumulative and self-reinforced. This also means, moreover, that the knowledge base and its evolution make borders between sub-fields blurred and intertwined over time.

The 'modern' in modern biotechnology is often used as a way to differentiate the present from the past. First generation, or traditional, biotechnology has been a part of human history ever since the early usage of yeast and bacteria for food processing, and selective animal breeding for desired traits. In the early twentieth century, the second generation of biotechnology became a tool in the hands of engineers when biologically based production processes became industrialized. Examples of first- and second-generation biotechnology include the usage of biological-based mechanisms in food processing; alcohol and diary production; and bioprocessing in order to make biopharmaceuticals and fine chemicals, such as penicillin and citric acid respectively. The age of modern, or third generation, biotechnology is dated back to the discovery of the DNA molecule by Watson and Crick in 1953. Their discovery – that is, their solving of the scientific puzzle in understanding DNA – became the starting point of a new era. The third generation is explicitly based on underlying scientific progress whereas the first and second generation were more technological applications, without a solid scientific understanding of the underlying biological processes.[4]

Not even 'modern biotechnology' can be taken for granted, however, as a definition set in stone. Modern biotechnology depends on advances in many fields of medical science, natural science and engineering. During the latest 30 years, the scientific and technological knowledge about biological mechanisms has grown tremendously. Modern biotechnology is more than knowledge – indeed its impacts come about through the combination of increasing knowledge with techniques and instrumentation. Two generic techniques have been particularly influential in terms of enabling the diffusion of biotechnology into new areas, namely the recombinant DNA technique by Boyer and Cohen in 1973 and the monoclonal antibody or hybridoma technology by Milstein and Kohler's discovery in 1975. These two techniques rapidly found industrial applications, for example, in diagnostics and in biopharmaceuticals like insulin and human growth hormone made from genetically modified bacteria (or yeast). More recently, the sequencing of the human genome and the genomes of several other organisms has been more or less fully completed. This systematic genetic information and related IT tools are claimed to represent a milestone, and in this post-genomic era (PGE), an increasing number of scientists around the world are engaged in the study of the function of genes and the proteins they are coding for. Thus, proteomics and metabolomics have become expanding research fields that contribute to enhance our understanding of, and ability to modify, life processes. The merging of modern biotechnology with other areas like IT has also led to new fields, such as DNA chips and bioinformatics.

Such developments matter for industrial applications in specific settings. They have fundamentally changed the drug development process in pharmacueticals (Nightingale 2000). In doing so, it has enhanced the efficiency and effectiveness of the research and development (R&D) process, especially related to finding and testing potential target molecules and candidate drugs. Hence, the R&D process is changed also for new drugs that are manufactured by more conventional means (e.g. organic chemical synthesis).

Multiple definitions and confusion about terminology is not surprising – given this highly dynamic process, where both knowledge and applications are continuously being developed. However, the lack of a clear definition is unfortunate, because not having one means it is difficult to conduct stringent analysis and to compare and contrast cases. This situation is likely compounded by the fact that analysts have different backgrounds – and the importance of diverse backgrounds may be obvious if we consider that analysts includes natural scientists, medical researchers, firm managers, government policy-makers, journalists, management consultants, social scientists (including economists) and so on.[5] This suggests that neither shared worldview nor generally accepted terminology exists, and hence definitions and statements must be critically examined with great care.[6]

Thus, when moving from these broad official definitions to actual studies and the interpretation of results, it is clear that the broad definition and understanding has to be turned into criteria, which can be used to categorize a process, event or outcome as modern biotechnology. This holds in general as well as for each chapter within this book. Such criteria matter because, at this point, the decisions about what to actually include – or exclude – will have a large impact on many follow-on issues. It can affect how many firms are counted; the relative specialization into sub-sectors; the type, quantity and quality of science assessed; which industries are understood as affected by modern biotechnology and so on.

3.   A CONCEPTUAL MATRIX FOR MODERN BIOTECHNOLOGY

This section presents a conceptual matrix, aimed at outlining the essential dimensions of modern biotechnology, along the two axes of 'product (sector)' and 'knowledge base'. It proceeds by discussing possible definitions, methodology and data for each axis in turn. These two axes then intersect to provide the conceptual matrix. The suggestion is that the conceptual matrix ought to be useful to clarify the lines of debate and to nuance the interpretation of research results, both of this book and in future research.

The first axis is that of the product-based (sector) one. The discussion first addresses how the definition of similar products or product groups is usually used to identify an industrial sector. From there, the discussion applies this axis to modern biotechnology, based on reasoning about the relationship between 'core' biotechnology and associated sectors.

The product-based (sector) axis requires the concept of an industry, with roots in economics and industrial organization. Such a definition starts from the firm's ability to supply specific products (used in certain ways) and from that, classifies each firm into a certain industry. This type of economic reasoning is traditionally linked to governmental data and statistics, and the Standard Industrial Classification system (SIC) from the US Bureau of the Census is the most commonly used way to classify firms and industries for statistical data.[7] Once the industrial sectors are set, categories and comparisons can be made across industries, based on characteristics of the production process.

A different way to proceed to define an industry could be based on characteristics of the 'market' on the demand side. For example (Payson 2000) argues that a function-based definition of classes of products is suitable as a complement to the traditional product and industry definitions. In other words, in his argument, the function of a product relative to the needs of the user would allow us to group sets of apparently different products (industries) based on 'use'. If the demand side were further emphasized, then 'markets' could further be defined from a customer perspective, using product characteristics and functions (Lancaster 1966), and could specify certain geographical considerations. The human health care sector as related to modern biotechnology would fit this perspective, that is the aggregated biomedical industry including medical technology, instruments and medical supplies fulfilling the various health care demands (Laage-Hellman 1998). However, this way of defining markets should be most suitable for tangible products. By focusing only on product characteristics and functions, the definition is also difficult to implement in practice.

A comprehensive definition of a product (sector) could be based on a combination of supply and demand sides, thus including the competing product, characteristics of production knowledge and the specific demands of users. A relevant example could be those biopharmaceuticals, which are supplied with the help of modern biotechnology and fulfil the need for therapeutics in human health care.

As discussed further in the next section, trade data refers to a group that is almost entirely based on biologics. This definition both includes many products that are not part of modern biotechnology and excludes other important products based on biotechnology.

Moreover, in national surveys as well as in specific studies, the concept of dedicated biotech firms (DBF) is often erroneously identified as the biotech industries *per se*. DBFs can be delimited in a number of ways, based either on the product, production knowledge and/or use. Delimitations can be set to products towards therapeutics and diagnostics for human needs (Powell et al. 1996) or more widely to modern biotechnology as used in all possible product groups (OECD 2001). A reason for capturing not only these small, start-up firms is that the DBFs are in fact often developing products in both cooperation and competition with firms in existing industries (OTA 1991). (See also Chapter 3.)

The proposal here is that the product-based (sector) axis should include both 'core' biotech as well as associated sectors, as illustrated in Figure 2.1.

Figure 2.1 provides a means of identifying and distinguishing the products (sector) of 'core' biotechnology from the use and intersection of biotechnology with various industrial sectors such as pharmaceuticals, food and medical technology. As argued in McKelvey (forthcoming) and in Chapter 10 of this book, linking products (sectors) based on the core knowledge to various existing sectors matters in terms of results.

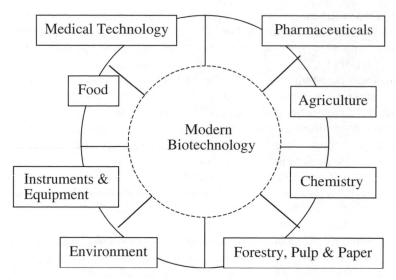

*Figure 2.1   Axis of product-based (sectors) in biotechnology: A
combination of 'core biotechnology' and sectors*

In this way, Figure 2.1 can capture elements found in the broad definition of modern biotechnology, such as the OECD definition. This way of
defining modern biotechnology from the product (sector) thereby forms
one of the two axes in the conceptual matrix.

The second axis is formed by the knowledge bases. This discussion briefly
presents the arguments, then mainly addresses the special case of modern
biotechnology.

The argument for this axis arises from the conceptualization that knowledge bases – including knowledge, techniques and tools organized by
various ways such as disciplines – affect the economic uses and impacts of
science and technology. For emerging technological areas, it is rather
common to begin with a list of relevant fields. The problems, however, often
arise when such a list is applied to a purpose. It may be difficult to compile
as well as to maintain, once the dynamic aspect is considered. A list may be
difficult to use, once industrial applications are considered since they
usually require a variety of fields and interdisciplinarity for problem-
solving. This often requires classification based on knowledge bases,
thereby leading to problems of classification of firms and/or products.

In general, one problem of only using knowledge bases to define an area
like modern biotechnology is that at any time, a long list could be made. It
should include all relevant scientific and technological fields of knowledge,
techniques and tools, as has been exemplified previously. For example,

Chapter 1 presented the five areas that the OECD *ad hoc* group is considering using to define the boundaries of modern biotechnology.

Hence, a related complication is the growth and increasing scope of relevant knowledge bases. Even a well-defined and unambiguous list of relevant knowledge will be outdated as further developments occur in the future.

Another problem is that any one product or firm will often use a variety of knowledge bases. While multiple knowledge bases are an inherent fact of economic life for firms, this presents some problems of categorization when used in order to categorize actual firms into product (sector) classifications.

A related observation is that biotech knowledge bases often have many general areas of applications, which cut across traditional industries. This leads to some measurement difficulties due to heterogeneity of activities.[8] Such a diversity of firms to be classified implies a methodological problem. Should only new biotech firms be considered or should firms diversifying into biotechnology be included? If diversified firms are included, is it the fraction specialized into biotechnology that is interesting or is it the whole firm (Swann and Prevezer 1996), due to its access to existing complementary resources?

The generic nature of many knowledge bases relevant to modern biotechnology implies that the companies involved can be found in different business sectors. More than that, it implies that these firms and sectors are to a varying degree specialized in biotech-related applications. Some may mostly be focused around biotech knowledge whereas others use it as input into research, production processes and/or products. As with the first axis of product (sector), this second axis of knowledge base is useful for certain purposes. It helps identify a changing list of relevant knowledge, techniques and tools at any given period and to identify their spread and impacts across a variety of other sectors over time.

The conceptual matrix proposed here is a combination of the product (industrial sector) axis and the knowledge axis. The combination should provide advantages as compared to using only one or the other approach, and thereby be useful for conceptualizing and measuring modern biotechnology. The conceptual matrix is based on the two axes, which provide classifications based both on product (industrial sector) as well as on knowledge.

Figure 2.2 illustrates the conceptual matrix, followed by examples relevant to modern biotechnology.

The conceptual framework in Figure 2.2 should be useful in order to identify and group together various definitions, methods and data, and thereby also to interpret results and design new studies.

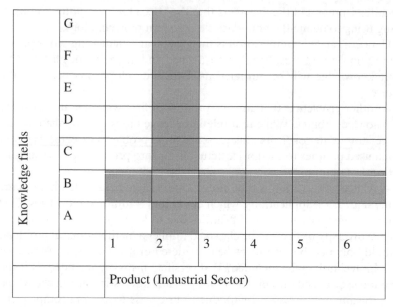

Figure 2.2   Conceptual matrix based on the two axes

Turning to the first axis of Figure 2.2, the classification into product
(industrial sector) implies a focus on products, production technologies and
uses. The example can be drawn from the two most obvious product groups
or industrial sectors under development today, namely biologics and
pharmaceuticals. Both include mainly products to meet human health care
needs but the companies also sell somewhat different product and rely on
somewhat different production technologies and usage. The axis of 'knowledge' can be used to define fields and/or more general concepts. Examples
of fields would include specific disciplines like bioinformatics or molecular
biology. Example of more general concepts could be life science (bioscience).

The conceptual matrix becomes interesting, when a combination of
product (sector) and of knowledge bases are used to further specify the
phenomena studied. Take the first example of biologics and pharmaceuticals as products (sectors). While both product classes involve the development and use of modern biotechnology in the research process, the firms
making them also rely on many other areas of knowledge. Most pharmaceuticals are mainly produced with traditional chemistry, not least as producing pharmaceuticals based on genetically modified organisms (GMOs)
proved costly and/or impossible for certain categories of drugs. Similar
arguments can be made for medical technology, biotech devices, agriculture

and environmental engineering. They face certain market conditions, but also rely on both generic and more specific biotech knowledge bases.

One way to allow for both precision and shifting boundaries is to include both the product (industry) axis and the knowledge axis within the same conceptual matrix. This is one way to allow for both rigidity and plasticity, given that a precise and complete definition of modern biotechnology will never occur since the field of modern biotechnology is highly dynamic. New products and applications will be introduced and intermediary products and biotechnology-based production processes will diffuse to quite different industries[9] – thereby affecting their products as well as their productivity and profitability.[10] New technologies and applications will continuously rupture agreement over classification into product and sector (industry).

4. METHODOLOGY AND DATA

At present, there are serious difficulties in gauging the overall dimensions and dynamics of modern biotechnology within the economic system, mainly because of limitations in available indicators and statistics. This section discusses some crucial conceptual and practical problems of moving from a concept (definition) to operationalization. Any attempt to grasp modern biotechnology faces these problems, and the discussion first takes it from the viewpoint of economic statistics, then looks at available data and what they tell us, and finally discusses economic indicators of biotechnology.

The basic difficulty with quantitative approaches lies precisely in the fact that this is not a sector but a technological area (see Chapter 3). Moreover, modern biotechnology has many product dimensions, different underlying knowledge bases and wide fields of application. These complex product and sector aspects of biotechnology preclude any straightforward economic measurement. This is important, since although biotechnology is an emerging technology, it is one for which significant economic claims are being made. So although this section discusses all areas of biotech data, the focus is on areas of actual or potential economic measurement, especially in fields where existing data limitations might be overcome.

The main data sources for biotechnology activities or products at the present include:

● trade data classified by product group;
● specialized surveys of firms engaged in some form of biotechnology production, in terms of output, employment, alliances and so on;

- surveys of 'technology use' at firm level;
- scientific publications' data;
- patent data, either United States Patent and Trademark Office (USPTO) or European Patent Office (EPO);
- R&D data covering expenditure and personnel, classified by socio-economic objectives and fields of research;
- databases on specific topics, such as alliances, venture capital, firms and so on.

Before turning to these specific types of data, some general comments. One immediate need is for agreed cross-country definitions to use for statistical purposes. This should encompass not only agreement about what constitutes a biotech process, but should also relate to biotechnology activities such that it is possible to distinguish between sectors on the basis of their dependence or incorporation of these activities. Such a definition could then be used in relation to more than one data source.

Such a common approach does not exist mainly because national statistics offices do not always collect biotechnology data, and there is little coodination between the organizations who do collect. These collecting organizations also have quite different motivations and objectives. Within the OECD countries, data are currently collected by a disparate group of agencies – statistical offices, consulting companies (particularly Ernst & Young), business associations, regulatory agencies, industry ministries, universities and research-funding organizations. Only four countries (Canada, Italy, The Netherlands and New Zealand) carry out industry surveys through their national statistical offices. Data for at least four countries comes from consulting companies, for three from industry associations and for another four from ministries or R&D funding agencies. As we might expect, definition and collection methodologies differ sharply, often reflecting the specialized interests of whoever is collecting the data. So some data collectors may be interested in new firm creation, and some may be interested in R&D performance. In Norway, for example, data collection is in the hands of the agency that regulates genetically modified organisms. These different interests of course shape sample selection and hence statistical coverage. In addition, of course, specific studies may rely on their own selection and sample of data, which they have collected themselves, but that data is only sometimes available publicly.

Turning to specific types of data, trade data classified by product groups, and this obviously requires a standardized definition before this type of data can be collected. One way out of this dilemma for trade statistics has been to focus only on a subset of biotechnology. As there is no real disagreement on one area, namely biotechnology based on recombinant

DNA, some sources confine themselves to this only. However, this tends in practice to lead to a focus on biologics, namely therapeutic products derived from living organisms, and this is used in trade statistics. This emphasis excludes, for example, 'traditional' biotech processes, particularly in food products. But this definition also excludes such emerging areas as plant or agricultural biotechnology, and environmental biotechnology. So, one result of focusing on biologics in trade data is that biotechnology trade is extremely small (about 1 per cent of US product trade) and very limited in geographical focus (most of it goes to a limited group of countries).

The second and third types of data require surveys of firms, and many countries have carried out surveys of firms engaged in some dimension of modern biotechnology.[11] Most of these surveys consist of counts of companies, of R&D within them, of sales and sales growth and of personnel. In some cases there are venture capital data related to seed stage funding committed to biotechnology companies. Similar studies may be carried out by individual researchers, but the discussion here mainly focuses on more public data, carried out by official sources and/or used for government purposes.

These surveys are far from comparable internationally. The best survey is OECD (2001). The basic problems start with firm definitions, that is the problem of defining a biotech company. Furthermore, some data gatherers report data on biotech companies without ever reporting their definitions or selection criteria, and this of course suggests a strong probability of biases in coverage. Many seem to rely on a definition of 'core biotech' firms that rests on whether firms are using some form of recombinant DNA technology. In Australia, the definition of a biotech firm is one that 'is entirely or substantially biotechnology related and that has a significant commitment to technological innovation' (OECD 2001). Canada uses a similar definition, but adds all firms that perform any biotechnology R&D. Germany uses a similar definition of core companies, but classifies them as 'entrepreneurial' or not, and adds 'extended' companies that are not purely in life sciences. The picture that emerges focuses heavily on firms involved in pharmaceuticals and health care applications.

The central problem with these definitions of 'core' biotech firms is that they sharply constrain what is counted as modern biotechnology. In those countries that look more specifically for the application of biotechnology methods across industries, a different picture emerges. Italy has no definition of biotech research but collects data by asking questions about biotechnology research through its R&D survey. This has the interesting result of showing that eight major sectors perform biotech R&D including motor vehicles and fabricated metal products (the latter is interesting since it is one of the largest industries in Europe). The largest expenditure on biotech

R&D in Italy occurred in the water supply industry. A similar result can be found with Japan, which focuses its survey on biotech activities. This leads to the biggest proportion of 'biotech' firms being located in the food and drink industry, with substantial presence also in metals and machinery industries. Presumably these results reflect the multi-technology nature of modern firms, but in any event the results are sharply different from those in countries that focus on 'core' biotech start-ups.

Similar results emerge from Canada, which has shifted away from gathering data based on the notion of 'core biotech firms' to one based on 'innovative biotechnology firms. The latter means a firm that 'uses biotechnology for the purpose of developing new products and is engaged in biotechnology related R&D' (McNiven et al. 2003). This definition enables Statistics Canada to extend into different forms of biotechnology, and to extend the sectoral scope of the investigation, covering health, agriculture, natural resources, environment, aquaculture, bioinformatics and food processing. Part of the reason for this redefinition is that previous surveys had shown considerably more firms using biotechnology outside the 'core' biotech definition than inside it – 358 core biotech firms, but 784 user firms in 1996 (McNiven 2002). Health is certainly the largest revenue-earning sector from biotech products (generating 71 per cent of all biotech revenues in Canada) but all sectors mentioned above are generating revenues from innovative products, particularly food processing, environment and agriculture.

So the point that emerges here is that restricted definitions of biotech firms not only limit the scope for international comparability, but they can also present a quite misleading picture of the extent of biotech research and production. Where countries such as Australia focus on a core biotech sector and exclude 'traditional' biotechnology or more importantly agricultural biotechnology, then a quite distorted pattern emerges.

A fourth type of data is scientific publications. A central feature of modern biotechnology is its close connection with the relevant science base, and its close organizational connection to universities. This means that bibliometric data, emerging from analyses of scientific publications patterns, is more than usually relevant for understanding innovation patterns. Once again, there are limitations, which are here discussed in terms of publishing strategies and problems concerning journal coverage.

The strategy issue is that publishing requires disclosure, while innovation usually requires limits to disclosure in the interests of appropriability. On the one hand, secrecy is a very basic strategy for innovating firms – often far more important than patenting as an appropriablity measure. On the other hand, firms may have strategic reasons to publish, such as to signal their involvement in a field and gain access to the international research community, attract partners and future employees and so on.

A related difficulty is journal coverage. Most biotech publication data in fact covers predominantly health-related areas – microbiology, oncology and so on. There seem to be persistent problems in defining appropriate biotechnology coverage in agricultural or environmental applications, and this exacerbates the partial picture referred to above. So while bibliometric data is extremely useful in mapping both search patterns and impacts (measured via citations) in areas where publication is an important strategy for firms, it is not a general indicator of the trajectories of change in biotechnology.

A fifth type of data is patents. Within innovation studies, counts of patent applications and granted patents are a major source of quantitative data. A well-known limitation is that there are more or less non-quantifiable variations in the propensity to patent, across firms, industries and countries.

In the case of modern biotechnology there are also classification differences across patent offices. The most important differences are between USPTO and EPO. There are differences between when a patent is acknowledged (in the year of application or year of grant) that affects the data, but more important are sometimes subtle differences in definition. The US definitions begin from a very wide view: 'technologies related to the analysis and application of the genomes of all creatures', while the European definitions relate more strictly to microorganisms and enzymes.[12] Once again, while patent data are immensely useful, they are far from standardized or definitive. Various categories of other patents may be included within a count of biotechnology patents, a choice which affects which types of technologies to include or exclude and thereby the results.

A sixth type of data is R&D surveys, a more recent development. Of course it must always be borne in mind that R&D is only one among many innovation inputs. Despite limitations, R&D surveys are the only data source for biotechnology that are coordinated and indeed standardized (via the OECD's *Frascati Manual*). The problems here lie not so much in data quality as in differences in country collection practices and data availability.

Most R&D data are reported along two dimensions, namely expenditure and personnel. These are normally reported in terms of sources of funding (i.e. business, government, foundations etc.) and sectors of performance (i.e. business, universities, government etc.). Among the advanced economies, the surveys often go far beyond this and have some more or less unique features that permit a detailed understanding of the structure of R&D performance. This is because, as well as collecting data on sources of funding and sectors of performance, the surveys for R&D data also provide four other types of breakdown of R&D expenditure. For the business sector the data breaks down performance by International Standard Industrial Classification (ISIC) category, that is by the industry which is performing research. For all sectors, there are three further ways of breaking down R&D expenditure and personnel resources. These are:

1. by socio-economic objective (such as economic development, defence, health, environment etc.);
2. by type of research (that is, pure basic research, strategic basic research, applied research or experimental development);
3. by field of research (meaning the specific area in which new knowledge is sought, such as molecular biology, applied mathematics, electronic engineering and so on).

The latter make it possible to gain a detailed picture of how R&D is prioritized towards specific areas of research.

As a concrete example, in Australia the research funding of the private non-profit sector (i.e. charitable foundations) is overwhelmingly directed at one socio-economic objective, namely health. Within this objective, we can look at fields of research – the two largest are clinical medicine and biological sciences. It is then possible to look in detail at the specific research fields that characterize biological sciences. These turn out to be genetic, cell biology and biotechnology. Turning to the disaggregated data we can then identify specific priority areas, as in Table 2.1.

Table 2.1 suggests that national statistics can be used to identify rather specific investments in different fields within – or related to – modern bio-

Table 2.1   Expenditure on Australian private non-profit R&D by research field biological sciences, rank of top 10 performers, 2000–01

Research field – biological sciences	$'000
Gene expression	12331
Cell development (incl. cell division and apoptosis)	8541
Genome structure	8352
Genetic engineering and enzyme technology	4915
Protein targeting and signal transduction	4116
Infectious agents	3293
Gene therapy	3210
Cellular interactions (incl. adhesion, matrix, cell wall)	3129
Genetics not elsewhere classified	3126
Neurogenetics	2940

Source: Australian Bureau of Statistics, R&D statistics.

technology. It is also possible to cross-classify such data by type of research – most of this turns out to be 'strategic basic' research – fundamental research but with application areas in mind.

R&D data exploration of this kind can be extended into business R&D, government-funded R&D, and so on, and thus offers a rather comprehensive picture of a country's overall research effort. It is one of the best ways to identify the complex overall pattern of biotech development in ways that are comparable internationally.

A seventh type of data is collected in databases, which address a specific topic like alliances, venture capital investments, firms and so on. These databases may be public, commercial and/or developed for specific research studies. Examples of alliance databases include ones held by MERIT/CATI, Recombinant Capital and PharmaDeal. An example of a venture capital database include statistics from the European Private Equity and Venture Capital Association (EVCA). An example of firm databases is the BioSweden database on Swedish biotech-pharmaceutical firms (McKelvey et al. 2003). The firm databases of total populations are often compiled, using a variety of public and private sources, including databases, firm surveys, public data on firms, reports to national accounting offices and so on. Other databases may be built as comprehensively as possible, based on a combination of public, commercial and private databases. Each has advantages and disadvantages in terms of coverage, scope and time, but offer at least some possibilities for quantitative studies. Still, important choices must be made, in terms of whether to use public or commercial databases – or invest in a new, privately held database – as well as in terms of the biases introduced by a particular sample included in a database.

In summary, crucial issues about methodology and data must be considered. There is a range of other biotechnology data available – on international trade in biotechnology products, on inter-firm alliances in biotechnology, on firm creation and so on. However all of this data, without exception, suffers from the limitations outlined above. There are serious problems of definition, sampling strategies and coverage that preclude any quantitative generalizations about modern biotechnology either as a technology or as an economic activity, at the present time. Hence, whether public data is used – or whether unique data is developed in an individual study – choices about operationalization need to be carefully considered.

In conclusion, this chapter has sought to discuss the choices possible in the operationalization of the biotechnology concept; to introduce a conceptual matrix; and to form a common starting-point about definitions, methodology and data for the book as a whole. A central concern of social science research is the validity of empirical studies and comparisons for both theoretical explanations and recommendations for government policy and firms. The suggestion here is that despite the current diversity, greater harmony for comparisons as well as progress in finding ways to analyse new phenomena, is possible.

NOTES

1. This chapter is the second introductory chapter within the book, yet it also initiates the transition to Part III 'Setting the scene'.
2. Sector and industry are used as interchangeable concepts.
3. Note. This report does not specify the technologies referred to.
4. A further distinction can be made between biotechnology as production technology and biotechnology as a searching technology (Henderson et al. 1999). In this taxonomy the first and second generation will be regarded somewhat nearer production technologies and the third generation as more of a search technology. This taxonomy is however not as straightforward as it first seems and relates to the distinction between product and process innovations.
5. Many fields such as neo-Schumpeterian economics, evolutionary economics, science and technology policy, innovation studies, corporate strategy research and regional development have used the example of modern biotechnology to develop new theory.
6. The variety of definitions used to explain any one concept can be exemplified by the different meanings of the term 'life science'. 'Life science' stands for, at the one extreme, the common underlying scientific base for every biologically related activity and at the other extreme, the short-lived business idea to gather both therapeutic and agricultural applications of modern biotechnology under one roof. During the 1990s several mergers between chemical, agricultural and pharmaceutical firms resulted in large diversified life-science firms, for example AstraZeneca, Aventis, DuPont, Monsanto and Novartis.
7. The SIC is from 1997 and was replaced by the North American Industry Classification System (NAICS) but the structure is not drastically different.
8. One solution to this problem would be to deliberately focus even more deeply into specific applications such as one part of diagnostics or for drug delivery and so on. While useful for some purposes, only using such a definition would mean that the more generic nature of the technology would be lost in the details.
9. New non-biotech-based but highly intertwined knowledge areas are exemplified by; bioinformatics, an information technology for biotech applications and biomaterials, through new biotech-based research methods and materials in the emergence of tissue engineering.
10. During the 1970s a strictly genetic view dominated and modern biotechnology was defined by genetic engineering and hybridoma technologies only. This eventually led to a problem since other aspects of biotechnology continuously were coevolving. A present example is the case of tissue engineering, which has increased the interest in different aspects of modern cell culturing, a high-tech application of the traditionally bioprocessing technology.
11. OECD (2001), 'Biotechnology statistics in OECD member countries: compendium of existing national statistics', STI working paper 2001/6 presents a comprehensive overview of these surveys.
12. See OECD (2003 op. cit., p. 10) for a detailed analysis.

REFERENCES

Biotechnology Industry Organization (BIO) (2003), website www.bio.org/er/, viewed 30 May 2003.

Carlsson, B. (2002), *Technological Systems in the Bio Industries: An International Study*, Boston and Dordrecht: Kluwer Academic Publishers.

Feldman, M.P. and C. Ronzio (2001), 'Closing the innovative loop: moving from the laboratory to the shop floor in biotechnology manufacturing', *Entrepreneurship & Regional Development* **13**, 1–16.

Henderson, R., L. Orsenigo and G. Pisano (1999), 'The pharmaceutical industry and the revolution in molecular biology: exploring the interactions between scientific, institutional and organizational change', in D.N. Mowery and R. Cambridge (eds), *The Sources of Industrial Advantages*, Cambridge: Cambridge University Press.

Laage-Hellman, J. (1998), *Den biomedicinska industrin i Sverige*, Stockholm: NUTEK.

Lancaster, K. (1966). 'A new approach to consumer theory', *Journal of Political Economy* **74**(2), 132–57.

McKelvey, M. (forthcoming), 'Biotechnology industries', in H. Horst and A. Pyka (eds), *Companion on Neo-Schumpeterian Economics*, Cheltenham UK and Northampton, USA: Edward Elgar.

McKelvey, M., H. Alm and M. Riccaboni (2003), 'Does co-location matter for formal knowledge collaboration in the Swedish biotechnology-pharmaceutical sector?', *Research Policy*, **32**, 483–501.

McNiven, C. (2002), 'Use of biotechnologies in the Canadian industrial sector: results from the biotechnology use and development survey – 1999', Statistics Canada working paper, March.

McNiven, C., L. Raoub and N. Traoré (2003), 'Features of Canadian biotechnology innovative firms: results from the biotechnology use and development survey – 2001', Statistics Canada working paper, March.

Nightingale, P. (2000), 'Economies of scale in experimentation: knowledge and technology in pharmaceutical R&D', *Industrial and Corporate Change*, **9**(2), 315–59.

OECD (Organisation for Economic Co-operation and Development) (2001), *Biotechnology Statistics in OECD Member Countries: Compendium of Existing National Statistics*, Paris: OECD.

OECD (2002), *Frascati Manual*, Paris: OECD.

OECD (2003), 'Statistical definition of biotechnology', website www.oecd.org/EN/document/0,,EN-document-617-1-no-21-31006-617,00.html, viewed 20 May 2003.

OTA (Office of Technology Assessment) (1984), *Commercial Biotechnology: An International Analysis*, Washington, DC: US Congress.

OTA (1991), *Biotechnology in a Global Economy*, Washington, DC: US Congress.

Payson, S. (2000), *Economics, Science and Technology*, Cheltenham, UK and Northampton, MA, USA, Edward Elgar.

Powell, W.W., K.W. Koput and L. Smith-Doerr (1996), 'Interorganizational collaboration and the locus of innovation: networks of learning in biotechnology, *Administrative Science Quarterly*, **41**(1), 116–45.

Sainsbury, Lord (1999), *Biotechnology clusters*, London: Office of Science and Technology.

Swann, P. and M. Prevezer (1996), 'A comparison of the dynamics of industrial clustering in computing and biotechnology', *Research Policy*, **25**, 1139–57.
Zucker, L. and M. Darby (1997), 'Present at the biotechnological revolution: transformation of technological identity for a large incumbent pharmaceutical firm', *Research Policy*, **26**, 429–46.

PART II

Setting the scene

3. Stylized facts about innovation processes in modern biotechnology

Maureen McKelvey, Annika Rickne and Jens Laage-Hellman

1. INTRODUCTION

This chapter argues that modern biotechnology has been – and continues to be – developing at a rapid pace and is connected to fundamental changes in the economy over time, with new products, firms and activities starting up and with existing ones being significantly modified or disappearing over time. The stylized facts of innovation processes proposed in Chapter 1 are used to structure an overview of the development of modern biotechnology. These are stylized facts in the sense of summarizing a set of propositions and facts that many researchers within a community take for granted.[1] This chapter draws on existing research about modern biotechnology as well as illustrates through a case of genomics companies and commercialization of human biobanks.

In doing so, this chapter marks the transition from the introduction of the book (Part I) to this section 'Setting the scene' (Part II). It thereby moves the book from a critical assessment of definitions, methodology and data to an understanding of the emergence and complexity of actors and processes involved in science, technology and innovation. As such, this chapter paints a broad picture of modern biotechnology, and is designed to be a contribution in its own right as well as relevant to the book as a whole. It can also be used to place in context subsequent chapters found in 'Challenging the existing' (Part III) and 'Forming the new' (Part IV).

This chapter is structured around the four stylized facts as presented in Chapter 1:

1. innovations emerge from uncertain, complex processes involving knowledge and markets;
2. new scientific and technological areas create economic value in many different ways. The impacts of the economic value accrue within global actors as well as for geographically situated agglomerations;

3. societal linkages exist among diverse actors. Such networks as well as societal institutions affect innovation processes;
4. the firm as an organizational form plays a particularly important role in knowledge exploration and exploitation.

These four stylized facts may be applicable in general to other technologies and industries, but this chapter considers them explicitly in relation to modern biotechnology.

In general, literature in neo-Schumpeterian economics, evolutionary economics, science and technology policy, and innovation studies has been influential in arguing that the complexity of knowledge and markets requires an approach to innovation processes that includes actors as well as broader innovation system aspects (Edquist and McKelvey 2000). Literature on innovation systems as well as on the firm's research and development (R&D) and innovations has affected government policy and firm strategy in recent years.[2] One line of reasoning often relies on analytical frameworks composed of a variety of elements, such as actors, knowledge bases and competences, industrial dynamics, government policy, innovation systems and demand. One way of seeing this is that diverse actors and processes are linked during knowledge creation, exploitation and utilization.[3] Another line of reasoning focuses on internal-to-the-firm decision-making and strategy.[4] The two perspectives are combined in this book. The firm plays a particularly important role in knowledge exploitation – and the firm makes decisions, acts, follows routines, and operates linked to broader knowledge and market processes through resources, incentives, feedback mechanisms and so on.

This combined view fits well with our knowledge about the empirical details of modern biotechnology as such. A key issue is the complexity of actors and their roles (functions) within the broader processes (innovation system) – but also the special role accorded to incumbent and start-up firms, intellectual property rights and commercialization of science. This empirical area provides many clear examples of the interrelationships between the knowledge and economic aspects of innovation processes. Scientists start firms – but keep working at universities. Large pharmaceutical firms pay large sums to universities and dedicated biotech firms (DBFs) to carry out explorative R&D and to license patents. Moreover, innovations as well as firms are inherently part of a broader socio-economic-political system – given that changes in one aspect of the system often have effects across it.[5] The Bayh-Doyle Act of 1980 allowed US universities to patent discoveries based on research that had been funded by public grants. As such, this change to the legal system also changed the institutional setting in such a way as to facilitate – even promote – the subsequent explosion of patents.[6] Many of the most profitable – as well as the most controversial ones – have been in areas related to modern biotechnology. This chapter does not have an overarching ambition to describe or explain all such aspects.

The aim of this chapter is to structure a myriad of empirical details around the four stylized facts. Section 2 applies this to provide a general overview of contemporary developments in modern biotechnology, while Section 3 applies it to a specific empirical probe. The case is of three firms in genomics and the commercialization of human biobanks. It is given as an illustration of events as they play out within one specific part of the broader phenomena of modern biotechnology.[7] Without giving precise answers to questions of 'how' and 'why', the overview does provide an analysis that shows that the development of knowledge and ideas interacts with market processes and with the formation and disappearance of industries and firms. Section 4 concludes this chapter, by reflecting on what the analysis has provided as well as implications for future research and policy/strategy agendas.

2. INNOVATION PROCESSES IN MODERN BIOTECHNOLOGY

This section structures a myriad of literature providing empirical details. It is not primarily a literature review of the major debates and topics, even though the reader may find the compilation useful.[8]

2.1 Innovations Emerge from Uncertain, Complex Processes Involving Knowledge and Markets

Modern biotechnology has been defined as a broad concept, including a range of rapidly changing products (sectors) and knowledge bases (see Chapter 2). It is widely claimed that the twenty-first century will be 'the century of biotechnology' – both with regard to scientific progress and industrial development. The assumptions behind this claim rest on the accelerating knowledge development and on the range of possible societal impacts. New knowledge, techniques and instrumentation in a range of related fields have allowed an increasingly detailed insight of the life processes both at a molecular level and as part of biological systems. This has, in turn, led to a better understanding of biological processes and mechanisms related to humans, plants, insects and animals. These processes demonstrate points of 'science and technology push' and 'market pull' as well as significant impacts on industrial structure (see Chapter 4). These processes involve uncertainty.

One aspect is interdisciplinarity, in that modern biotechnology includes separate disciplines, and yet since the relevant knowledge, techniques and instrumentation affect a range of related disciplines, much interdisciplinarity occurs. Kay (1993) and Bud (1993) provide excellent historical insight into the development of biotechnology and the new biology. There is a long list of potentially relevant disciplines, such as biochemistry, organic chemistry, microbiology, bioprocess engineering, genomics, proteomics, bioinformatics, nanotechnology and so on. The course of development is propelled by the internal logics of science and technology as well as by market factors and behaviour of various actors and networks.

One of the more unique features of the innovation process of modern biotechnology appears to be the density – and importance – of linkages between innovations with developments in science and technology (Kenney 1986). The industrial applications and societal impacts of modern biotechnology rely on basic science as well as technology. This holds if we consider basic science as discovering new knowledge and technology as problem-solving and modifications of the natural and manmade worlds.[9] Even a brief history indicates that discoveries made within the context of basic research have many times paved the way for economic impacts – albeit in a complex way – if one examines specific cases within genetic engineering and a range of related disciplines.[10] Indeed, modern biotechnology appears to be an excellent example of 'high-tech' or 'science-intensive' industries, whether measured in terms of R&D expenditures, academic qualifications of employees, citations in patents to scientific literature or network density between universities and firms.[11]

Much more is going on than simply the 'linear model' suggests, for example that investment in scientific research and firm R&D will result in products ready for the market. Section 3 presents a good illustration, using genomics firms and commercialization of human biobanks. Thus in contrast to the linear model, the modern view of innovation processes does not assume a straightforward progress from pouring in money to science, to transfer and imitation as technology, to products ready in the market (see Fagerberg et al. forthcoming). Instead, the innovation process involves many facets, with multiple explanatory variables. It must be understood as a dynamic, uncertain process linking actors and systemic effects. This view of innovation is important to keep in mind, particularly when we consider what it means to say that modern biotechnology is – or is not – heavily science-based. The chain-link model of Kline and Rosenberg (1986) stresses that innovation and developments in science and technology are separate but ongoing parallel processes with feedback loops. Moreover, as has already been argued, these types of knowledge bases are cumulative within a broader structure of knowing – and thereby science as a societal activity must be equated broader than only the latest scientific results.

Science policy matters – especially the negotiations over resources between governments, research councils, universities, firms and scientists to achieve long-term societal benefits (Bush 1945). Governments and other financiers in many countries have always expected that their increasing investments in this type of research will solve important societal problems. Potential societal impacts are claimed to exist in many areas, especially human health care, agriculture, food production and environmental protection. Many researchers have tended to argue that such societal impacts will occur in the very long run – while financiers and firms demand, at times, results with more immediate and direct effects. In recent decades, the scientists have been more likely to argue that effects are also more direct, as seen in spin-off firms, technology transfer offices, patent licensing and so on.

Another aspect is the opening up of new innovation opportunities over time, due to both 'science and technology push' as well as 'market pull'. This holds whether we consider cases from human health care, agriculture, or any other industry where modern biotechnology may be relevant – but the histories of development are different, implying the relevance of characteristics specific to sectors (see Chapter 5). As innovations and the development of science and technology progress over time, new technological opportunities open up, providing possibilities for new products and new ways to transform inputs into outputs demanded within society (Orsenigo 1989). The scientific results are not 'ready' to be directly implemented within products and services for sale on the market nor as public health and welfare measures. Many parallel and complementary developments and testing – in many related aspects from regulation to production to sales – are in practice always necessary before more widespread use is possible. Still, the experimental and uncertain character of science and technology imply that it is often difficult to foresee the kind of research findings and product ideas that will come out of such activities.

Without some form of functioning distribution mechanism – often carried out through markets, firms, property rights and so on – modern biotechnology will have no, or only a very limited, impact in improving health, food supply, environment and so on. Markets are one way of distributing resources in society, although alternatives such as the state also exist (Lindblom 1977; Senghaas 1985). Firms play a role in transforming inputs into outputs by, for example, linking scientific findings to innovation opportunities and societal problems, by combining ideas with complementary resources and by developing new solutions that are brought into use across society. The results of R&D may affect costs and access to resources in such a way as to link them to changing organizational forms (Nightingale 2000).

One current trend within developments in science and technology is that the total cost of R&D is increasing (OTA 1993; DiMasi et al. 2003). This type of research in firms and universities is characterized by increasing average R&D costs – even if certain marginal costs decrease rapidly. On the one hand, techniques and instrumentation often drastically reduce the cost per test – and also open up new possibilities to study life processes on the molecular and cellular levels. On the other hand, modern equipment is costly and often requires large-scale laboratory facilities as well as highly skilled/educated personnel. Similar things could be said of other R&D resources like human biobanks, large databases, information technology (IT) support and so on. On the whole, then, the average expenditures thereby increase. As it plays out, there is an increasing need to invest and organize large, complex and expensive research teams.[12] Researchers may also rely on large research facilities, which in scale may soon rival that of physics.

2.2 New Scientific and Technological Areas Create Economic Value in Many Different Ways. The Impacts of the Economic Value Accrue Within Global Actors as Well as for Geographically Situated Agglomerations

Modern biotechnology may create economic value in different ways, like other rapidly changing and emerging technological areas. Over time, many different ideas – including knowledge, techniques, instrumentation, biological materials and so on – have economic impacts (see Chapter 2). Some ideas are realized and turned into goods and service products for sale on markets. Others are used directly by the inventing firms and/or may be used to provide supplies and equipment to other firms, industries and final users. These may accrue in different ways to global actors as well as to geographically situated agglomerations.

One aspect related to the complexity of the innovation process is that there is not usually a one-to-one relationship between a specific scientific discovery and a certain industrial application. That is to say, any given biotech research finding, technique or instrumentation may be used in a variety of applications and industries. Genetic engineering, for example, has been used both to manufacture certain types of pharmaceuticals and as an input into research for pharmaceuticals as well as for seeds, improved livestock and so on. Depending on the use, the firm or other organization may need to access and organize complementary knowledge and complementary technology. The ability to organize such processes is likely to affect a firm's performance as well as national specialization and competitiveness.[13]

Another aspect – which is becoming more and more relevant as the technological area matures – is that modern biotechnology involves products (sectors) and knowledge bases that are well known as well as others that are in the process of being developed. As a consequence, innovation processes in some areas will be more focused on adaptation to specific industrial applications, while in other areas they are more based on extensive experimentation. The latter have more substantial scientific, technological and market-related uncertainties than the former.

A related observation is that the types of knowledge in specific applications and specific sectors may have effects on the industrial structure in specific sectors.[14] Orsenigo et al. (2001) argue that new developments in knowledge will have differential effects on the likelihood of whether small firms will continue to survive or whether they will be bought up by large firms. They categorize the new knowledge in terms of scope of application, and their arguments can also be related to debates about how interdisciplinarity and coordination issues affect the industrial structure in the biotechnology-pharmaceutical sector. The type of knowledge and the search processes are likely to affect market concentration as well as the relative advantages of incumbents versus DBFs to compete.

Finally, the benefits from creating economic value may accrue to global actors as well as to geographically situated agglomerations, which are known as clusters, regional innovation systems and so on.[15] In some instances, the large firms have an advantage in accessing knowledge and markets world-wide while in other instances, a high concentration of small actors with many linkages may emerge over time, appearing to give competitive advantage to specific regions (see Chapters 10 and 13). The largest pharmaceutical and chemical-agricultural companies keep merging thereby building knowledge and resource advantages (Henderson 2000) whereas regions like Boston and San Francisco seem to lead the world in spinning off companies and linking universities to such firms.

However, the picture which is emerging from social science research is more complex than this. Some small firms are fundamentally international from the beginning – which can be seen through international R&D contracts, personal contacts with researchers, co-authorship of academic papers and so on. US DBFs sell R&D contracts abroad – which has been a dominant view of US DBFs supplying the world with relevant knowledge. Moreover, there is also counter evidence of US large and medium firms buying R&D contracts from European DBFs.[16] Some large firms are highly dependent on specific regions within countries, in terms of R&D facilities, local linkages and recruitment. Especially during mergers and acquisitions, however, the large companies may suddenly decide to move out their research facility, as happened when Pharmacia pulled out of the Uppsala-Stockholm region in Sweden after an international merger with UpJohn.

2.3 Societal Linkages Exist Among Diverse Actors. Such Networks as Well as Societal Institutions Affect Innovation Processes

Much research about modern biotechnology has examined topics related to collaboration, strategic alliances, networks, innovation systems and the like.[17] Large firms as well as small firms that are involved in modern biotechnology are often engaged in the scientific work and often collaborate with academic research partners as well as with other firms. Much biotechnology research is carried out within networks of collaborative relationships among geographically distributed research groups in firms as well as universities and research institutes.

In other words, an intricate web of relations exists between actors generating, exploiting and using innovations. Indeed, the interdisciplinary character of modern biotechnology already mentioned implies that many types of organizations will need to develop strategies to access a variety of emerging fields. They may do so through internally mastering a multitude of competences and/or through developing collaboration. The nature of the knowledge bases and the further development of science and technology are likely to affect the pattern of collaboration over time (see Chapters 7 and 9). Importantly, these relations are unevenly distributed across the population of actors, where under some conditions the smaller firms have an advantage while under other conditions, it is the large firms. If one understands that societal linkages can be used to the advantage of a single actor, then it is clear that not all actors can find appropriate strategic alliances nor have access to an existing network.

Moreover, the empirical evidence indicates that diverse types of actors may somehow be involved in innovation processes. These range from universities and research institutes, individual scientists, financiers of research as well as firms, government agencies, other policy-making organizations, bridging organizations, incumbent firms, hospitals, end-users and so on. (See Chapters 4, 7, 9, 11, 12, 13). Public and private actors may play complementary roles in the innovation process through various networks of exchange relationships – but at times, they seem to compete more than cooperate. While some of these relationships are related to business decisions, others are more purely related to R&D.

One aspect of these societal linkages in modern biotechnology is that scientists – at universities, research institutes and firms – play an important role in all parts of the process and also significantly affect commercial exploitation of modern biotechnology. Universities and research institutes are the places where much of the scientific work is carried out, and thereby constitute one of the most important types of research organization. University research teams are influential by producing the knowledge that constitutes the starting-point for forming new firms and industries (see Chapters 11 and 13). There are various 'bridging organizations' set up by universities, such as technology transfer and licensing offices, or separately created by government policy-makers (Cetindamar and Laage-Hellman 2003, p. 298). Strengthening industry-academia collaboration in order to facilitate technology transfer is a popular trend in many countries. Irrespective of organizational origin, size or age, firms engaged in modern biotechnology – especially in the medical field – tend to have close relationships with the academic community.[18]

Finally, societal institutions related to regulation, ethics and public debate affect innovation and the development of science and technology in modern biotechnology. One reason is that many ethical issues as well as questions about the positive and negative societal impacts become evident, when one considers the future for humans, animals and nature.[19] The use of human stem cells for therapeutic cloning exemplifies a research area that created debate and a sense of the urgent need for regulation in order to clarify what the scientists are allowed – or not allowed – to do. Another example is the use of human biobanks as an input into genetics research and drug development (Rynning 2003). Issues raised include protecting the privacy and integrity of sample donors. Recently, some countries in northern Europe have enacted specific biobank laws whereas some other countries have no legislation and are likely to not have it soon. This has implications for actors' perception of the future. Actors involved in the early phase may face uncertainty about their future opportunity to act, given that the implementation of ethical guidelines, legislation and regulation tends to lag behind science and technology.

Government regulation and consumer perception of safety affect innovations and products in areas such as health care, agriculture and food production. Especially drugs have to go through extensive testing to reach market approval, where clinical trials are expensive and involve high risk of failure.[20] This has effects on the developing company's profitability, growth potential and survival, since the high attrition rate in drug development projects largely contributes to the high average cost for developing a new drug. In other sectors, the regulation-related costs are lower, but in many cases the basic problem of the trade-offs of incentives to innovate and costs with compliance is similar. With regard to medical equipment, for example, a global trend in developed countries is the increasing requirements from regulatory authorities, seen as a tool to improve patient safety.

Recent history shows that both debates and ethical regulations are unevenly distributed over all possible uses, with some receiving much attention and some largely neglected. Public debate and private opinions may affect the immediate rate and direction of ongoing innovation processes –

as well as potential future products in the market. In particular, the risks associated with genetic engineering for food and agriculture have attracted a great deal of attention, although much more so in Europe than in the USA (KSLAT 1997). The widespread largely negative public opinion towards the use of genetically modified organisms (GMOs) in Europe is a case in point. Without doubt, scepticism and in some cases direct resistance of consumers towards tests and potential food products has delayed the diffusion of GMOs in Europe – compared with the USA where the public so far has been more positive. Differences such as labelling requirements for GMOs has led to disputes over international trade, including US complaints to the World Trade Organization (WTO) about unfair protection of markets by European countries.

While the public in most countries seems more positive to developments within human health care, criticism and opposition have arisen here as well. One example is genetic testing for hereditary diseases, where there is a worry that sensitive information about individuals will be used to their disadvantage by insurance companies and employers. In the meantime, companies must respond to the effects of debate on businesses even in sectors such as insurance – and they may make choices which shift the burden of risk-taking between the public and private sectors (see Chapters 8 and 6 respectively).

2.4 The Firm as an Organizational Form Plays a Particularly Important Role in Knowledge Exploration and Exploitation

Chapter 2 has already discussed the diversity of firms involved in innovation processes in modern biotechnology – including large incumbents as well as start-up firms and including firms in a variety of sectors. The following discussion focuses on characteristics that are relevant to the role of the firm within the innovation processes in society.

One important feature is that the firms are actively involved not only in commercialization activities but also in research. Triggered *inter alia* by the high R&D costs mentioned previously, industry has increasingly taken on responsibility for carrying out applied research (Cockburn et al. 1999b). To a lesser extent, firms and private institutes may also perform basic research in their own laboratories. Even when firms carry out in-house research, they tend to interact with universities and individual scientists, for example, for the purposes of sourcing new technologies, inventions and product ideas and to recruit skilled personnel.[21] These collaborative ventures take a wide variety of forms, ranging from truly shared work to the purchasing of licences for work already completed and patented by the academic partners (Nilsson 2001). In modern biotechnology, firms are also significantly involved in the development of science and technology for business purposes. Continuing firm R&D is required in order to improve the technology, validate findings, carry the vision, develop applications, assist in the market introduction and so on.

Another aspect is that, even in this heavily science-based industry, early and continuously updated knowledge about actual and potential customer preferences is crucial for successful development projects. Users may also participate directly in research projects and testing, organized by firms and driven by market considerations. Examples include clinical testing of drugs for new indications as well as field tests of seeds. Hence, users and intermediary experts affect demand – and thereby also diffusion.

Furthermore, the diffusion of radically new products may be slow, particularly if the target is markets where current solutions are very different from existing ones, or where no solution currently exists for that problem. Final consumers and intermediary experts may need to learn and accept the new product before widespread use is possible – as is particularly evident in human health care. For example, Nobel Biocare, today's world leader in dental implants, had to struggle for more than 10 years to reach a significant share of the potential market – this in spite of access to favourable clinical results for its innovative titanium-based implants (Rickne 2000). Over the years, as an important part of its market introduction strategy, the company has spent large resources on training dentists all over the world, turning them into lead users. Thus, the first mover firm must have strategies to link with lead users, and this strategy is associated with extended efforts and costs. Or, if the product is part of a new method for diagnosing or treating patients, the health care personnel need to be retrained and convinced to change their daily working habits, in addition to purchasing goods and services.[22] For some observers, one side of the debate over agriculture and food could also be interpreted in these terms (KSLAT 1997), in the sense that if the public learns that these products are safe, cheaper and require less chemicals in the fields, they will more likely adopt them.

Since the 1980s, 'academic entrepreneurship' has produced a large number of 'dedicated biotech firms' both in the USA and Europe.[23] Without doubt, this phenomenon is partly due to the frequently short distance, intellectually, between biotech research findings and industrial applications in some disciplines, at some time. When there is a large overlap, this had made it relatively easy for university researchers and other scientifically trained individuals to identify product and business ideas based on scientific discoveries and start-up companies. Especially in the USA, the incentive structures have stimulated academics to engage in business activities, usually with support of venture capital or business angels. Similar evidence may be found within some European universities, although other research identifies significant differences between the USA and Europe.[24] These start-up firms do not only start from roots in academia, but may instead be spun off from other organizations like large corporate research, hospitals or other user organizations.

Moreover, close collaboration with the scientific community and with other organizations is not enough to build a successful and growing business (Niosi 2003). Relationships also need to be established, for example, with potential users and customers, suppliers of goods and services, distributors and, not least, with financiers. As to the latter, good relationships with venture capitalists that can provide 'competent capital' is particularly valuable.[25] That is because the start-up firms as a rule are in great need not only of money but also of active management support and advice on business matters.

Incumbent companies in certain sectors – such as pharmaceuticals, diagnostics, medical equipment, food, agriculture, chemicals and scientific instruments – use the knowledge bases of modern biotechnology to renew and expand their product and business lines. Firm learning and performance attributes may differ, based on choices related to trade-offs between performing their own R&D and collaborating with others.[26] Thus, the small firms and the large incumbent companies may collaborate for strategic reasons. [27]

Finally, the waves of investments, booms and bust in the history of modern biotechnology show us several waves of start-ups. Although not yet as strongly as in the IT and telecom fields in the 1990s, the biotech industry is subject to the same type of up and down cycles as other businesses. The venture capital market for biotech companies and the confidence of investors have yet not collapsed to the same extent – but historical data shows similar (if less dramatic) bubbles and bursts. Following the upswing biotech boom in the 1990s, many small biotech companies are now experiencing great difficulties both in realizing commercial and economic success and in attracting new capital. Access to capital matters greatly in sustaining the small firm's innovation process – and hence its long-term survival.[28] The long-term nature of the commercialization process makes the access to persevering finance crucial for firm survival – given the few numbers of products on the markets and the fickle markets for explorative R&D, patents and licensing. Thus, as access to external resources fluctuate, so too do the fortunes of many companies. Many of the newer and younger firms within the field of modern biotechnology have yet to prove their sustainability and growth.

In summary, there is no doubt that developments in modern biotechnology are to a large extent driven by rapid scientific progress and related technical development providing efficient research methods. This discussion also indicates that science, technology, markets, industry structure, networks and knowledge flows coevolve in an interactive, iterative fashion over time.

3. ILLUSTRATION: GENOMICS FIRMS AND COMMERCIALIZATION OF HUMAN BIOBANKS

This section presents an empirical example with the purpose of illustrating a specific empirical probe of the development of modern biotechnology. The example is based on a study of biobanks' commercialization, which compares three clinical genomics companies and their interactions with society (Laage-Hellman 2003). The illustration of genomics companies and human biobanks provides details of how emergence is played out within one specific part of modern biotechnology, and how it is loosely structured around the four stylized facts.

Biobanks are structured collections of human biological material, such as tissue specimens, blood samples and extracted DNA. Most samples currently stored in biobanks have been taken for health care purposes (e.g. diagnosis), but sometimes samples are collected and stored specifically for scientific purposes. During the last decade, the interest in using biobanks in medical research has increased dramatically, mainly as a result of the great progress in genetics and molecular biology. In particular, as a complement to genomic studies carried out on various model organisms, biobanks can be used in the search for 'disease genes' and, more broadly, in research on the relationship between inheritance, environment and disease susceptibility. Besides providing new fundamental knowledge about biological processes and the molecular mechanisms behind various human disorders, knowledge gained through biobank-based research can be taken as a starting-point for the development of new industrial products, such as diagnostic/prognostic tests and drugs. That is why in recent years large pharmaceutical companies, as part of their drug discovery activities, have engaged in research projects where biobanks are used as a tool to find disease genes and related drug targets. This interest in developing 'genomics-based' drugs and markers, in combination with the outsourcing trend in the pharmaceutical industry, has created new business opportunities for biotechnology companies. Over the last 10 years, a number of dedicated 'clinical genomics companies' specializing in human genetic studies have been founded both in North America and Europe. For these companies, biobanks constitute a key resource which they gain access to either by collecting proprietary patient materials or by collaborating with hospitals or universities.

The core of the above-mentioned biobank study consists of three cases, namely the commercialization ventures centred on deCode Genetics in Iceland, Oxagen in the UK and UmanGenomics in Sweden, respectively. These firms specialize in clinical genomics research and act as an intermediary between the pharmaceutical and diagnostics industry on the one hand and public health care and academic institutions on the other. The three ventures exhibit similarities as well as differences. A commonality is, for example, that all three companies were founded for the purpose of commercializing new knowledge and other valuable resources originating in the public sector. Although not all of them are direct university spin-offs, this makes them typical representatives of the emergent biotech sector. Important differences exist, for example, with regard to the organization of the relationship between the company and the public institutions.

The ongoing and complex and uncertain character of the biotech innovation process (stylized factor 1) is visible here. One explanation for this result is partly due to the multitude of competences and technologies that need to be combined all along the innovation process. In the case of biobanks, collections of biosamples (and related clinical information) cannot be effectively used unless they are fruitfully combined with other complementary resources of different kinds. In the first instance, clinical genomics research requires a range of scientific competences. Thus, molecular biologists cannot work in isolation. Besides various specialists mastering a range of gene technological methods, the research teams must include or have access to complementary competencies in areas such as genetics, bioinformatics, epidemiology, bioethics, and not least the clinical fields covered by the projects. Usually, a genomics company cannot afford to have all these resources in-house, so it needs to link up with other organizations providing these competences. It can be, for example, hospitals, universities, independent research institutes, biotech supply companies, drug discovery firms and pharmaceutical companies.

Furthermore, the identification of the genetic basis of disease is just the initial stage of the pharmaceutical (or diagnostics) innovation process. First, the knowledge gained through genetic studies must be integrated with downstream research activities leading to the discovery and validation of 'druggable' targets and, as a next step, the development of new therapeutic compounds (or diagnostic markers). Thereby, a whole range of other R&D competences come into play, including cell biology, proteomics, chemistry and pharmacology. A current trend is that many biotech companies originally focusing on genomics are integrating forward into drug discovery and development. This strategic shift, triggered by environmental changes, requires acquisition of additional resources and competences as well as the establishment of collaborative relationships with new types of external resource providers. Second, needless to say, the discovery of a candidate drug must be followed by considerable investments in clinical trials, registration, production and marketing. These related activities are invariably carried out by industrial firms. They have the technologies and competences necessary to link the scientific advances to market opportunities and bring about innovations in a real sense.[29]

As clearly illustrated by the biobank study, high R&D costs are another typical feature of modern biotechnology. It is expensive both to build and operate biobanks and to use them for genomics research. It has been increasingly recognized within the scientific community that in order to validate research findings there is a need not only for very large numbers of biosamples but also for detailed and reliable information about the donors. Since it is often difficult for university researchers to raise enough money to finance collection and storage, industrial involvement in biobanking has become common. However, this is not an uncontroversial development, since this may be perceived to threaten 'the academic freedom' of the university researchers.[30] Besides increasing collaboration between industry and public institutions, the escalating costs also stimulates cooperation among scientists, nationally and internationally. For example, by pooling several local biobanks the cost-efficiency can be improved with increasing scientific productivity as a welcomed result.

Proceeding to the second stylized fact, according to which new scientific and technological areas may create economic value in many different ways, it can be concluded that the outcome of a certain biobank-based gene discovery project may be useful for developing both therapeutic products and diagnostic products. In addition, the results of such a project may be applicable to several different diseases that are affected by polymorphisms in the same gene. Both of these examples illustrate that there is no one-to-one relationship between a certain scientific discovery and a specific industrial application. Furthermore, it can be concluded that modern biotech research tools used in human genetic studies, such as micro arrays and DNA-sequencing machines, can be used for different purposes not only in medical research but also in other fields, for example food production and environmental control.

As with modern biotechnology more broadly, human biobanks relate both to well-developed knowledge fields and industries, as well as other ones that are still science-based and in emerging phases. As illustrated, biobanks have become a key resource for clinical genomics firms representing 'a new class of economic agents' (see Chapter 9). At the same time, the end-users of the results obtained by using biobanks are primarily mature firms in the pharmaceutical and diagnostics industries, some of which may build up their own biobanks for internal projects. It can be noted that one of the early sponsors of the research resulting in a big population-based biobank in Sweden was a bread-maker, a truly well-established company indeed (due to the early focus on nutrition aspects).

That societal linkages exist among diverse actors and affect both knowledge and economic processes (stylized factor 3) is well illustrated by the biobank case. There are many different types of firms, organizations and individuals that are actively involved in the commercial use of biobanks. To a large extent the interaction among these actors takes place within the frame of exchange relationships that may be more or less long term and close. Figure 3.1 shows how the clinical genomics company can be seen as a node in a network of relationships comprising a multitude of public as well as private actors. The study showed, *inter alia*, that the organization of the relationship between the genomics company and the biobank-

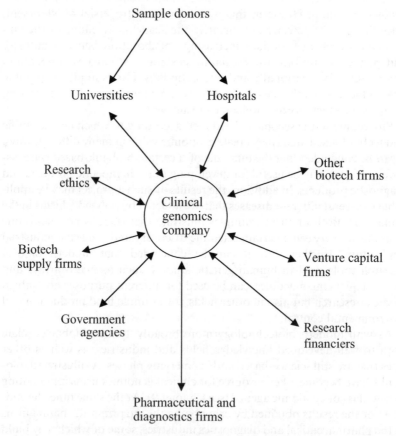

Figure 3.1   Actors involved in commercialization of biobanks.

administering units is crucial in order to achieve a successful commercialization. For example, there are important legal issues related to various intellectual property rights that need to be sorted out by the actors in advance. Failure to do so may lead to serious conflicts hampering the innovation activities.

Another type of societal linkage has to do with regulation and public debates. Definitely, the use of biobanks for research raises a number of ethical and legal issues of relevance to authorities as well as to the general public. Given the sensitive nature of the information that can be extracted from biosamples, the protection of the donors' privacy and integrity, and how to weigh these interests against those of the society, has become a key issue in biobank research. The question is for example what kind of consent should be required from the donors, or what kind of security systems should be used to handle samples and information. For some biotech companies, the regulatory uncertainty has been perceived as a barrier to commencing biobank-based research. Now, several countries have enacted specific Biobanks Acts, but it is not always clear how to apply the new legislation in practice.

Given the sensitivity of these issues and the rapidly increasing use of biobanks for scientific purposes, at universities and in industry, it can be maintained that, with some exceptions, surprisingly little public debate has occurred, at least so far – compared for example with the attention devoted to GMOs. However, in the late 1990s the Icelandic deCode Genetics became the target of a heated debate, both locally and internationally, when the Icelandic parliament decided to grant the company an exclusive licence to build and operate a national health sector database. As it now seems, the security arrangements required by the government, influenced by the heavy criticism from the scientific community as well as the general public, are so strict that the whole project has been jeopardized. More recently, the Swedish UmanGenomics has attracted a great deal of media attention, due to a severe conflict regarding an existing biobank, for which the company has been granted exclusive commercialization rights. As a consequence of this conflict, concerned for example with the control over the biosamples, and involving many interested parties (including university researchers and sample donors), the very survival of the company is threatened. It can be noted that the Swedish Biobanks Act that came into force in January 2003 plays an important role in this context (e.g. it explicitly prohibits the transfer of biosamples with a view to financial gain). It seems that the British national project to build a large population-based biobank is learning from these cases by taking a cautious approach to commercial involvement.

In line with our fourth stylized fact, regarding the firm as a particularly important organizational entity, the possibilities of using biobanks as a resource for improved health are strongly dependent on the genomics and pharmaceutical companies and their actions. For example, as in other parts of modern biotechnology a large and increasing share of the applied research is carried out by industrial firms and commercial laboratories. Given the perceived economic opportunities of clinical genomics research, similar types of studies are carried out both by academic institutions and private companies. However, as already pointed out, collaboration with hospitals and academic institutions is a common way both for large and small firms to gain access to clinical materials.

Generally, it is the role of firms, involved in the commercialization processes, to know the market and the potential customers. This is one of the reasons why it is advantageous for the genomics companies to establish early-stage collaboration with industrial partners. This enhances the probability that the research projects pursued by these firms will lead to products (drugs or diagnostics) with a commercial potential. Now when many genomics companies have chosen to run their projects longer in-house, before out-licensing, the business risk increases. To compensate for the lack of direct customer cooperation they will need to spend a considerable amount of time and money on contacts with prospective licensees. Otherwise, they may end up with research results that nobody is interested in paying for.

The slow introduction of new products into the market is a general characteristic of the innovation process in modern biotechnology. That this is true for clinical genomics companies goes without saying, given that they are working in the early phase of drug discovery. But also in relation to their direct customers in the pharmaceutical industry, the marketing of genomic R&D services has proved to be more difficult and time-consuming than many of these firms had expected. Despite the hype around genomics it has in reality been difficult to convince 'big pharmaceuticals' to invest resources in early-stage genomics projects. Managers of the genomics firms often complain that the pharmaceutical companies have not correctly understood the great opportunities offered by genomics; but their counterparts, on the other hand, want to see proof that the genomic approach to drug discovery really works and therefore often pursue a wait and see policy. This illustrates a typical dilemma for young science-based firms focusing on the commercialization of revolutionary research findings.

Another characteristic of modern biotechnology related to the key role of firms, especially the small start-up companies, is the cyclical fluctuation of the capital market, which affects the innovative activities and their outcome. Most of today's clinical genomics companies, including the three investigated firms, were established during the second half of the 1990s, when there was an abundant supply of venture capital looking for investment opportunities in biotechnology. Now, as we all know, the conditions in the financial markets are very tough, and this has put many of the genomics companies in a difficult situation. They need more money to finance their forward integrating strategies and to compensate for the lack of income from customers.[31] But at the same time, the tight equity market has forced several companies to modify their strategic plans and launch cost-cutting programmes. For example, in October 2002 deCode, one of the world's largest genomics companies, decided to reduce with immediate effect its workforce from 650 to 450. Most of the 200 people laid off worked in the company's gene research laboratories. In other cases, such as the Swedish UmanGenomics, the failure to access new venture capital may lead to closure of the company.

In summary, the analysis of innovation processes provides a structure to understand the specific empirical details about genomics companies and commercialization of human biobanks. These processes of science, technology and innovation have led to tensions and problems of coordination and interaction among diverse actors – but have also led to the opening up of opportunities for cooperation, commercialization and firm formation. In understanding dynamic innovation processes, this empirical probe clearly indicates that much experimentation occurs, resulting in successes as well as failures in various dimensions. No 'ideal' outcome or organizational form can be identified. Instead, a combination of factors underlies both the rate and direction of experimentation as well as the outcome.

4. CONCLUSIONS

This chapter has used the four stylized facts about innovation processes in modern biotechnology to provide an overview and illustrate a case of genomics companies and human biobanks. This analysis helps to set the scene for specific issues about challenging the existing and forming the new.

The following list summarizes a number of specific characteristics, which were identified through the empirical material presented in Sections 2 and 3.

1. Innovations emerge from uncertain, complex processes involving knowledge and markets.
   - Interdisciplinarity exists, and hence there is a need to combine a multitude of competences and technologies.
   - Science policy matters. Both 'science and technology push' as well as 'market pull' impact developments.

- High costs of R&D can be observed, and are likely to increase over time. This is related to the increasing need to organize complex and expensive research processes, often with expensive research facilities.

2. New scientific and technological areas create economic value in many different ways. The impacts of the economic value – in terms of property rights, firm profitability, employment, productivity increases, economic growth and so on – accrue within global actors as well as for geographically situated agglomerations.
   - No one-to-one relationship exists between discovery and application.
   - Both mature and emerging products (sectors) and knowledge bases can create economic value.
   - Developments in knowledge also affect the industrial structure within a specific sector.
   - The resulting economic value from innovation may be captured by global actors as well as agglomerations.

3. Societal linkages exist among diverse actors. Such networks as well as societal institutions affect innovation processes.
   - A multitude of diverse actors are involved, and they tend to interact with each other.
   - Scientists, universities and firm R&D laboratories have played a particularly important role.
   - Regulation and public debate influence innovation and developments in science and technology.

4. The firm as an organizational form plays a particularly important role in knowledge exploration and exploitation.
   - Firms and commercial laboratories also carry out significant research activities.
   - The firm needs to know potential markets and customers.
   - Radically new markets may have very slow market introduction and diffusion.
   - Cyclical fluctuations affect in particular the population of small- and medium-sized firms.

These characteristics were embedded within the empirical material for each of the stylized facts. Taken together, they provide a more detailed understanding of innovation processes in modern biotechnology.

This perspective raises many relevant and broad questions. How do policies influence the kinds of competences that are accessible to actors and/or present in the system? How does knowledge flow within and across spatial and actor boundaries? What factors affect the extent to which scientific researchers and/or students are 'entrepreneurial'? What are the sets of factors that influence what kinds of research are undertaken, where it is located, who has access to it and on what terms? What influences the extent and type of collaborative linkages and competition? These types of questions differ from standard analyses of the economy and policy options based on 'market failure' arguments. The questions posed and answers given matter because they affect our thinking. They affect our analyses and explanations of innovation processes, and can be used to influence the decisions of practitioners.

This perspective also has more immediate implications, in relation to the objectives of social science research, the concerns of government policy and the strategies of universities and firms.

First, this broad perspective about the context within which modern biotechnology arises allows us to develop theory. Concepts are related to theories, which link together concepts and data into explanations and recommendations for practical action. This perspective highlights the need to further develop theory to explain, for example, developments in science and technology, the impacts on firms and regions, industrial dynamics, networks, market development, development of dynamic capabilities, as well as the firm as a particularly crucial actor. Subsequent chapters in this book address different issues, whereby different theories are integrated with empirical insights and testing. In each case, one or several of the stylized facts and characteristics form an implicit context to accept or to call into question, related to their specific research question.[32]

Second, in terms of policy concerns, a number of societal institutions are clearly important to innovation in modern biotechnology. In most developed countries today, decision-makers in government, universities and in related businesses see biotechnology as an important driver of industrial development and as a way to create employment and welfare (OECD 1989).[33] The future opportunities afforded by basic research, development activities and commercialization all help to explain the increased public spending on science and the increased private spending on R&D. This trend is visible in many countries. The USA has taken the lead, and the amount of federal money spent on the life sciences, notably biomedical research, has effectively doubled over the last five years (NSF 2002; NIH 2003). The arguments for increasing the budget are that generally, these investments will not only produce good science but also support industry and enhance its competitive position in the world market. Modern biotechnology has also become a prioritized target of industrial policy in many other regions or countries like the European Union (EU),[34] Singapore, Australia and a long list of others. In other words, many countries and regions see the strengthening of biotechnology-based innovation systems and clusters as an industrial political tool to increase international competitiveness.

Government policy may also be understood in a broader sense, if a new policy for modern biotechnology is to be developed. The government policies to stimulate modern biotechnology may include various direct measures aimed at stimulating and facilitating commercialization of research findings. Examples include financial support of biotech start-up firms, organizational changes in the university system to facilitate property rights and entrepreneurship, and regional support organizations like science parks. It may also include more indirect measures. The educational system needs to train individuals, to supply timely human capital at relevant volume and profile. Regional differences of the labour market as regards mobility, availability of skilled scientists and expert competences influence entrepreneurship as well as firm growth. Moreover, the financial market creates specific conditions in the various national innovation systems, as does the regulatory system.

Third, in terms of strategy for universities and firms, important implications relate to the effects of university–industry relationships on developments in science and technology; the effects of interdisciplinarity, complexity and large research facilities for 'being in the game'; and the strategies of collaboration between large firms and start-up firms.

In terms of impacts on science, there have been numerous debates about the effects of societal linkages and institutions on innovations within firms as well as basic science within universities and research institutes. Patents are often seen as positive for innovation – and hence the tightening of the patent regime and the high number of patents related to modern biotechnology are taken as positive signs. In contrast, one argument has been that the need to obtain licences for large numbers of patents in up-stream industries may raise the cost of coordination and of investment to such a level that it reduces the overall incentives to participate in innovation (Eisenberg 1996). The effects on science have also been questioned. Some critics argue that the increasing commercial interests in academic research constitute a threat to 'free research' and the independence of universities (Kenney 1986). The risk, the argument goes, is that objectivity and credibility of scientists are jeopardized if they have too strong ties with individual companies – and current debates among international scientists present reasons for some concern.

In a related debate, the increased density of network relationships between university and industry has been the cause of concern. One side sees that it may affect the very nature of science itself, by orienting scientists to trivial tasks seen from the internal logic of basic science – but which happen to be profitable. The other side replies that much of the development of engineering science has been precisely this, for example that companies had specific problems which university scientists could make more abstract, general and later created the possibility to train students in novel areas.

Moreover, there are implications for decision-makers of the decreasing marginal cost, interdisiciplinarity and the need to organize expensive and complex research teams – especially in a time when these innovation processes appear to span societal networks and institutions. The implications for universities and research institutes are that in order to be internationally competitive, they need to achieve a critical mass. This can be done by acquiring large internal capabilities and/or by teaming up with external R&D partners who can supply complementary resources. The Human Genome Project offers many enlightening stories about alternative possible choices. Hence, basic scientists face the same 'make or buy' decision as firms – with collaborative ventures within science and between firms and scientists as one option. The implications for both universities and firms may be that the best and largest leaders will keep their advantage in some areas, due to their complementary knowledge and other resources and their ability to make larger investments. Moreover, actors may find it increasingly impossible to be in the game at all, if they do not have the financial resources, highly qualified personnel and physical access to facilities.

Finally, due to the previously mentioned uncertainties, costs and rapid developments in science and technology, not even the largest among the established firms, can afford to perform R&D in all disciplines of possible relevance to their business. Instead, they can choose to pursue an 'outsourcing strategy' for some knowledge areas and keep open 'windows of opportunity'. In other words, the large firms may use biotech firms as subcontractors of certain R&D services, technologies and new products under development, as documented in extensive literature. The start-up biotech firms can provide the established firms with valuable resources and ideas, and may thereby help them to overcome technological discontinuities.

Obviously, this kind of strategic behaviour on the part of established companies has contributed to create favourable conditions for the creation and growth of small, dedicated biotech firms. Hence, from the perspective of large incumbent firms, the DBFs may play a complementary role rather than that of a direct competitor. At times, though, the DBFs have directly competed in product markets. Thus, in the long run, the DBFs may constitute threats to incumbent firms, especially when they develop and commercialize innovative products substituting for established products in the market.

NOTES

1. In that sense, stylized facts play a similar role as paradigms Kuhn (1970) but stylized facts are not as complete in structuring our view of the world. Dosi (1988), for example, has used them to organize a view on innovation processes, and they are perhaps most used within parts of the somewhat overlapping communities of neo-Schumpeterian, evolutionary economics, science and technology policy and innovation studies. Stylized facts seem to arise particularly in this field, when researchers keep confronting theories and empirical data with each other. As such, it can be related to the discussion on 'appreciative theorizing' as proposed by Nelson and Winter (1982) and the role of empirical work in theorizing (see Downward et al. 2003; Horst and Pyka forthcoming 2004).
2. See OECD (1997); Kornberg (1995); Oliver (1999); Wheale et al. (1998).
3. See McKelvey (1997a); Carlsson (1997); Carlsson (2002).
4. See Pisano (1997; Tidd et al. (1997); Grant (2002).
5. See Orsenigo et al. (1998); McKelvey and Orsenigo (2002); Mangematin et al. (2003); McKelvey et al. (forthcoming).
6. See Eisenberg (1996); Eisenberg (2001); Mowery et al. (2001); Cohen et al. (2002); Mowery and Sampat (forthcoming).
7. As such, one could place them at a particular location on a developed version of the conceptual matrix introduced in Chapter 2.
8. For literature reviews and a reference collection, see McKelvey (2002); McKelvey(forthcoming); and McKelvey and Orsenigo (forthcoming). A variety of sources have been used to compile the arguments found in this chapter. Some sources are of a more general nature, and they may reflect popular scientific and/or public debate sources such as magazines, popular books, newspapers and so on. Given the large number and nature of these sources, they have not been directly referenced, given that they together are used to provide a broad view of debates. Other sources involve specific arguments of relevance here, and these are explicitly referenced in the text. Likely, omissions of some relevant sources have been made – yet some 'accepted wisdom' is difficult to assign to one (or a first) reference. The reader is also referred to reference collections within neo-Schumpeterian economics such as Horst and Pyka (forthcoming) and within innovation systems approaches (Edquist and McKelvey 2000). Yet other sources provide accounts of the historical processes. While they certainly provide evidence on points of controversy, these works have mainly been used here to support a generally accepted view of the main events. In particular, the reader is referred to Bud (1993) and Kay (1993). Moreover, reference is made to other papers presented at the workshop *The Economics and Business of Bio-Sciences & Bio-Technologies: What can be learnt from the Nordic Countries and the UK?*, Chalmers University of Technology, Hällsnäs and Gothenburg, Sweden, 25–27 September 2002.
9. An alternative is to view both science and technology as a continuum in the knowledge dimension. This view implies that there are no clear borders between science and technology and it is therefore difficult to have clear terminology.
10. See cases detailed in Collins and Wyatt (1988); Rabinow (1996); McKelvey (1996a); Kay (1993); Harvey et al. (2003). See McKelvey (1997c) for the argument that differentiating individuals from the organizations in which they work is necessary to understand what is going on within university–industry–government relationships.
11. See Kenney (1986); Audretsch and Stephan (1996); Liebeskind et al. (1996); Powell et al. (1996); Cockburn et al. (1999); Zucker et al. (1999); Henrekson and Rosenberg (2000); Owen-Smith et al. (2002); Zucker et al. (2002).
12. Special thanks to V. Mangematin on this point. Otherwise, most of these types of arguments can be found in the literature in the relevant fields of natural science and engineering.
13. See OTA (1984); Arora and Gambardella (1990); OTA (1991); Senker and Sharp (1997).
14. See Saviotti (1999); Stankiewicz (2002); Malerba and Orsenigo (2002)
15. There is an extensive literature on various issues related to this topic. See Audretsch and

Stephan (1996); Audretsch and Feldman (1996); Prevezer (1997); Braunerhjelm et al. (2002); Fuchs (2003); Cortright and Mayer (2002). For research on this topic in relation to Europe, see Walsh et al. (1995); Bartholomew (1997); Crowther et al. (1999); Swann et al. (1999); Giesecke (2000); Lemarié et al. (2001); Fransman (2001); Cooke (2001); Zeller (2001); Casper and Kettler (2001); McKelvey et al. (2003); Holmén and McKelvey (2003). For more general arguments about this topic, see also Feldman (1994); Breschi and Lissoni (2001).

16. For evidence on both sides, see Sharp (1991); Senker and Sharp (1997); McKelvey et al. 2003).

17. There is an extensive literature on various issues related to this topic. See Pisano et al. (1988); Shaw (1990); Arora and Gambardella (1990); Pisano (1991); Whittaker and Bower (1994); Liebeskind et al. (1996); Powell et al. (1996); McKelvey (1997b); Orsenigo et al. (1998); Powell (1998); Galambos and Sturchio (1998); Henderson et al. (1999); de la Mothe and Niosi (2000); Baum et al. (2000); Orsenigo et al. (2001); Carlsson et al. (2002); McKelvey et al. (2003); Mangematin et al. (2003); McKelvey et al. (forthcoming); Gault and de la Mothe (2002). Workshops in Luukkonen (2002) and Alm (2002) addressed related issues. For more general arguments and review of this topic, see also Nooteboom (1999); Narula and Zanfei (forthcoming).

18. Indeed, universities are often involved in the innovation process at an organizational level through, for example, being holders of intellectual property, holding equity in start-up companies and engaging in contract research for companies.

19. See OTA (1984): Ch. 21; Rifkin (1998); *Red Herring* (2000); Persson and Welin (2002); Hansson and Levin (2003).

20. See Grabowski and Vernon (1994). DiMasi et al. (2003), by studying the US pharmaceutical industry, have estimated that 21.5 per cent of new drugs that begin phase I clinical trials eventually reach the market. Furthermore, they found the average cost of bringing a new chemical entity to the market to be $802 million, when accounting for the time between investment and marketing.

21. A trend resulting from these collaborations is that universities are becoming increasingly dependent on research grants and contracts from industry – with some national variations.

22. Especially in health care, the well-motivated 'conservatism' of doctors and the increasing demand for evidence-based medicine makes market penetration a cumbersome process, requiring long-term and tenacious efforts.

23. An alternative option for the inventor would be to link up with an established company that is willing to commercialize the new idea. When the idea has a direct overlap with existing product lines, it is often easy to find partners, including licensing patents or carrying out explorative R&D under contract. But when this overlap does not exist, established firms may not be interested in collaboration. In that case, the founding of a spin-off company may be the only way to commercialize the idea.

24. See Faulkner and Senker (1995) and Lawton-Smith (2002) on evidence of university–industry and science–technology relationships in Europe, whereas other research identifies differences, see especially Henrekson and Rosenberg (2000); Genua (1999); Owen-Smith et al. (2002).

25. See Braunerhjelm et al. (2002); Wedin (2002).

26. See Shaw (1985); Pisano (1991); Dodgson (1991a); Dodgson (1991b); Powell and Brantley (1992); Wiendt and Amin (1994); McKelvey (1996c); Baum et al. (2000).

27. See Gambardella (1992); Wald (1996); McKelvey (1996b); Tapon and Cadsby (1996); Galambos and Sturchio (1998); Rothaermel (2000); Malo (2003); Hara (2003); Nesta and Saviotti (2003).

28. There are implications for firm strategy. Companies need owners and creditors with a good understanding of the conditions for successful innovation and who are prepared to take the risk of supporting the company over a long and uncertain period of time.

29. Despite the apparent linearity of the pharmaceutical innovation process, partly a consequence of regulatory demands, the entire process is in reality interactive. The four

aspects of a drug – the compound, the application, the organizational authorization and the market – are interdependent and shaped interactively and simultaneously (Hara 2003).

30. The problems and conflicts linked to the commercial use of biosamples collected through public sector institutions is one example of the complexities involved in the private-public interaction in science.
31. For deCode Genetics, for example, this may be related to their acquisition of MediChem Life, an agreement signed in January 2002.
32. Most chapters also delve into ways to further methodology and data to capture the moving phenomena of modern biotechnology.
33. See also Acharya (1999).
34. Especially under the EU 6th Framework Programme (Commission of the European Communities, 2002).

REFERENCES

Acharya, R. (1999), *The Emergence and Growth of Biotechnology: Experiences in Industrialised and Developing Countries*, Cheltenham, UK and Northampton, MA, USA: Edward Elgar.

Alm, H. (2002), 'The Swedish bio-pharma industry: how and why small- and medium-sized firms use collaboration in relation to innovation', paper presented at workshop *The Economics and Business of Bio-Sciences & Bio-Technologies: What can be learnt from the Nordic Countries and the UK?*, Chalmers University of Technology, Gothenburg, Sweden, 25–27 September.

Arora, A. and A. Gambardella (1990), 'Complementarity and external linkages: the strategies of the large firms in biotechnology', *The Journal of Industrial Economics*, **XXXVIII**(4), 361–79.

Audretsch, D. and M. Feldman (1996), 'R&D spillovers and the geography of innovation and production', *American Economic Review*, **86**, 630–40.

Audretsch, D. and P. Stephan (1996), 'Company-scientist locational links: the case of biotechnology', *American Economic Review*, **86**(3) 641–52.

Bartholomew, S. (1997), 'National systems of biotechnology innovation: complex interdependence in the global system', *Journal of International Business Studies*, **28**(2), 241–66.

Baum, J.A.C., T. Calabrese and B.S. Silverman (2000), 'Don't go it alone: alliance network composition and startups' performance in Canadian biotechnology', *Strategic Management Journal*, **21**, 267–94.

Braunerhjelm, P., D. Cetindamar and D. Johansson (2002), 'The support structure of the biomedical clusters in Ohio and Sweden', in B. Carlsson (ed.), *Technological Systems in the Bio Industries: An International Study*, Boston: Kluwer Academic Publishers.

Breschi, S. and F. Lissoni (2001), 'Knowledge spillovers and local innovation systems: a critical survey', *Industrial and Corporate Change*, **10**(4), 975–1005.

Bud, R. (1993), *The Uses of Life: A History of Biotechnology*, Cambridge, MA: Cambridge University Press.

Bush, V. (1945), 'Science: the endless frontier', a report to the President on postwar scientific research, Washington, DC.

Carlsson B. (ed.) (1997), *Technological Systems and Industrial Dynamics*, Boston, MA: Kluwer Academic Publishers.

Carlsson, B. (ed.) (2002), *Technological Systems in the Bio Industries: An International Study*, Boston, MA: Kluwer Academic Publishers.

Casper, Steven and H. Kettler (2001), 'National institutional frameworks and the hybridization of entrepreneurial business models: the German and UK biotechnology sectors', *Industry & Innovation*, 8.

Cetindamar, D. and J. Laage-Hellman (2003), 'Growth dynamics in the biomedial/biotechnology system', *Small Business Economics*, **20**, 287–303.

Cockburn, I.M., R. Henderson and S. Stern (1999), 'The diffusion of science-driven drug discovery: organizational change in pharmaceutical research', NBER working paper, no. 7359, September.

Cockburn, I.M., R. Henderson and S. Stern (1999b), 'Balancing incentives: the tension between basic and applied research, NBER working paper, no. 6882.

Cohen, W., R. Nelson and J. Walsh (2002), 'Links and impacts: the influence of public research on industrial R&D', *Management Science*, **48**(1), 1–23.

Collins, P. and S. Wyatt (1988), 'Citations in patents to the basic research literature', *Research Policy*, **17**, 65–74.

Commission of the European Communities (2002), 'Life science and biotechnology – a strategy for Europe', communication from the commission to the council, the European parliament, the economic and social committee and the committee of the regions, 23 January.

Cooke, P. (2001), 'New economy innovation systems: biotechnology in Europe and the USA', *Industry & Innovation*, **8**, 267–89

Cortright, J. and H. Mayer (2002), *Signs of Life: The Growth of Biotechnology Centers in the U.S.*, Washington, DC: The Brookings Institution.

Crowther, S., M. Hopkins, P. Martin, E. Millstone, M. Sharp and P. van Zwanenberg (1999), 'Benchmarking innovation in modern biotechnology', report for the European Commission, DG III – Industry, 1999, Brighton: SPRU, University of Sussex.

De la Mothe J. and J. Niosi (eds) (2000), *The Economic and Social Dynamics of Biotechnology*, Boston, MA: Kluwer Academic Publishers.

DiMasi, J.A., R.W. Hansen and H.G. Grabowski (2003), 'The price of innovation: new estimates of drug development costs', *Journal of Health Economics*, **22**, 151–85.

Dosi, G. (1988), 'Sources, procedures and microeconomic effects of innovation', *Journal of Economic Literature*, **XXVI**(3) 1120–71.

Dodgson, M. (1991a), 'Strategic alignment and organizational options in biotechnology firms', *Technology Analysis and Strategic Management*, **3**(2), 115–26.

Dodgson, M. (1991b), *The Management of Technological Learning: Lessons from a Biotechnology Company*, Berlin and New York: Walter de Gruyter.

Downward, P., J. Finch and J. Ramsay (2003), 'Seeking a role for empirical analysis in critical realist explanation' in P. Downward (ed.), *Applied Economics and the Critical Realist Critique*, London: Routledge.

Edquist, C. and M. McKelvey (eds) (2000), *Systems of Innovation: A Reference Collection*, Cheltenham, UK and Northampton, MA, USA: Edward Elgar.

Eisenberg, R. (1996), 'Public research and private development: patents and technology transfer in government-sponsored research', *Virginia Law Review*, 1663–1727.

Eisenberg, R. (2001), 'Bargaining over the transfer of proprietary research tools: is this market emerging or failing?', in D. Dreyfuss and H. First (eds), *Expanding the Bounds of Intellectual Property: Innovation Policy for the Knowledge Society*, Oxford: Oxford University Press.

Fagerberg, J., D. Mowery and R. Nelson (eds) (forthcoming), *Handbook of Innovation*, Oxford: Oxford University Press.

Faulkner, W. and J. Senker (1995), *Knowledge Frontiers: Public Sector Research and Industrial Innovation in Biotechnology, Engineering Ceramics and Parallel Computing*, Oxford: Clarendon Press.

Feldman, M. (1994). *The Geography of Innovation*, Boston, MA: Kluwer Academic Publishers.

Fransman, M. (2001), 'Designing Dolly: interactions between economics, technology and science and the evolution of hybrid institutions', *Research Policy*, **30**(2), 263–73.

Fuchs, G. (ed.) (2003), *Biotechnology in Comparative Perspective*, London: Routledge.

Gambardella, A. (1992), 'Competitive advantage from in-house scientific research: the US pharmaceutical industry in the 1980s', *Research Policy*, **21**, 391–407.

Galambos L. and J. Sturchio (1998), 'Pharmaceutical firms and the transition to biotechnology: a study in strategic innovation', *Business History Review*, **72**(2), 250–78.

Gault, F. and J. de la Mothe (eds) (2002), *Alliances, Networks and Partnerships in the Innovation Process*, Boston, MA: Kluwer Academic Publishers.

Genua, A. (1999), *The Economics of Knowledge Production: Funding and the Structure of University Research*. Cheltenham, UK and Northampton, MA, USA: Edward Elgar.

Giesecke, S. (2000), 'The contrasting role of government in the development of biotechnology industry in the US and Germany', *Research Policy*, **29**(2), 205–23.

Grabowski H. and J. Vernon (1994), 'Innovation and structural change in pharmaceuticals and biotechnology', *Industrial and Corporate Change*, **3**(3), 435–50.

Grant, R. (2002), *Contemporary Strategy Analysis: Concepts, Techniques and Applications*, Malden, MA: Blackwell Business.

Hansson, M.G. and M. Levin (eds) (2003), *Biobanks as Resources for Health*, Uppsala: Uppsala University.

Hara, T. (2003), *Innovation in the Pharmaceutical Industry: The Process of Drug Discovery and Development*, Cheltenham, UK and Northampton, MA, USA, Edward Elgar.

Harvey, M., S. Quilley and H. Beynon (2003), *Exploring the Tomato: Transformations of Nature, Society and Economy*, Cheltenham, UK and Northampton, MA, USA: Edward Elgar.

Henderson, R. (2000), 'Drug industry mergers won't necessarily benefit R&D', *Research Technology Management*, July/August, 10–11.

Henderson, R., L. Orsenigo and G. Pisano (1999), 'The pharmaceutical industry and the revolution in molecular biology: interactions among scientific, institutional and organizational change', in D. Mowery and R. Nelson (eds), *Sources of Industrial Leadership: Studies of Seven Industries,* Cambridge: Cambridge University Press, pp. 267–311.

Henrekson, M. and N. Rosenberg, (2000), *Akademiskt entreprenörskap: universitet och näringsliv i samverkan* (Academic Entrepreneurship: University and Industry in Collaboration), Stockholm: SNS Förlag.

Holmén, M. and M. McKelvey (2003), 'How to systematically study regional restructuration? IT and biotech case studies in Sweden', special issue of *European Urban and Regional Studies Journal*.

Horst, H. and A. Pyka (eds) (forthcoming), *Companion on Neo-Schumpeterian Economics*, Cheltenham, UK and Northampton, MA, USA: Edward Elgar.

Kay, L. (1993), *The Molecular Vision of Life: Caltech, the Rockefeller Foundation, and the Rise of the New Biology*, Oxford: Oxford University Press.

Kenney, M. (1986), *Biotechnology: The University-Industry Complex*, New Haven, CT: Yale University Press.

Kline, S. and N. Rosenberg (1986), 'An overview of innovation', in R. Lanau and N. Rosenberg, *The Positive Sum Strategy: Harnessing Technology for Economic Growth*, Washinton, DC: National Academy Press.

Kornberg, A. (1995), *The Golden Helix: Inside Biotech Ventures*, Sausalito, CA: University Science Books.

KSLAT (1997), *Transgenic Animals and Food Production*, Kungl: Skogs- och Lantbruksakademiens, 136 (20).

Kuhn, T. (1970), *The Structure of Scientific Revolutions*, 2nd edn, Chicago, IL: University of Chicago Press.

Laage-Hellman, J. (2003), 'Clinical genomics companies and biobanks: the use of biosamples in commercial research on the genetics of common diseases', in M.G. Hansson and M. Levin (eds), *Biobanks as Resources for Health*, Uppsala: Uppsala University, pp. 51–89.

Lawton-Smith, H. (2002) 'The biotechnology industry in Oxfordshire: dynamics of change', paper presented at workshop *The Economics and Business of Bio-Sciences & Bio-Technologies: What can be learnt from the Nordic Countries and the UK?*, Chalmers University of Technology, Gothenburg, 25–27 September.

Lemarié, S., V. Mangematin and A. Torré (2001), 'Is the creation and development of biotech SMEs localised?' *Small Business Economics*, **17**, 61–76.

Liebeskind, J.P., A.L. Oliver, L. Zucker, and M. Brewer (1996), 'Social networks, learning, and flexibility: sourcing scientific knowledge in new biotechnology firms', *Organization Science*, 7(4), 428–43.

Lindblom, C. (1977), *Politics and Markets: The World's Political-economic Systems*, New York: Basic Books.

Luukkonen, T. (2002), 'Building competencies in biotechnology firms: the influence of firm age and origin on knowledge networks and business skills', paper presented at workshop *The Economics and Business of Bio-Sciences & Bio-Technologies: What can be Learnt from the Nordic Countries and the UK?*, Chalmers University of Technology, Gothenburg, Sweden, 25–27 September.

McKelvey, M. (1996a), *Evolutionary Innovations: The Business of Biotechnology*, Oxford: Oxford University Press.

McKelvey, M. (1996b), 'Discontinuities in genetic engineering for pharmaceuticals? Firm jumps and lock-in in systems of innovation', *Technology Analysis & Strategic Management*, **8**(2).

McKelvey, M. (1996c), 'Redefining transfer in biotechnology and software: multiple creation of knowledge and issues of ownership', in A. Inzelt and R. Coenen (eds), *Knowledge, Technology Transfer and Forecasting*, Boston, MA: Kluwer Academic Publishers, pp. 33–47.

McKelvey, M. (1997a), 'Using evolutionary theory to define systems of innovation', in Charles Edquist (ed.), *Systems of Innovation: Technology, Institutions and Organizations*, Pinter/Cassell Publishers, pp. 200–22.

McKelvey, M. (1997b), 'Coevolution in commercial genetic engineering', *Industrial and Corporate Change*, **6**(3), 503–32.

McKelvey, M. (1997c), 'Emerging environments in biotechnology', in H. Etzkowitz and L. Leydesdorff (eds), *Universities and the Global Knowledge Economy: A Triple Helix of University–Industry–Government Relations*, London: Pinter Publishers.

McKelvey, M. (2002), 'The economics of biotechnology', entry into *The International Encyclopedia of Business and Management* and entry into *The Handbook of Economics*, International Thompson Business Press.

McKelvey, M. (forthcoming), 'Biotechnology industries', in H. Horst and A. Pyka (eds), *Companion on Neo-Schumpeterian Economics*, Cheltenham, UK and Northampton, MA, USA: Edward Elgar.

McKelvey, M. and L. Orsenigo (2002), 'Biotech-Pharmaceuticals analysed as sectoral systems of innovation', extended report to the EU-funded ESSY Project, website www.cespri.it.

McKelvey, M., H. Alm and M. Riccaboni (2003), 'Does co-location matter for formal knowledge collaboration in the Swedish biotechnology-pharmaceutical sector?' *Research Policy*, **32**, 483–501.

McKelvey, M. and L. Orsenigo (eds) (forthcoming), *The Economics of Biotechnology: A Reference Collection*, Cheltenham, UK and Northampton, MA, USA: Edward Elgar.

McKelvey, M, L. Orsenigo and F. Pammoli (forthcoming), 'Pharmaceuticals as a sectoral system of innovation', in F. Malerba (ed.) (2004), *Sectoral Systems of Innovation*, Cambridge: Cambridge University Press.

Malerba, F. and L. Orsenigo (2002), 'Innovation and market structure in the dynamics of the pharmaceutical industry and biotechnology: towards a history-friendly model', *Industrial and Corporate Change*, **11**(4), 667–703.

Malo, S. (2003), 'Coping with a turbulent environment through technological learning: the case of combinatorial chemistry', paper presented at the annual EAEME conference, April, arranged through the International Schumpeter Society Augsburg, Germany.

Mangematin, V., S. Lemarié, J. Boisson, D. Catherine, F. Corolleur, R. Coronini and M. Trommettter (2003), 'Sectoral systems of innovation: SME development and heterogenity of trajectories', *Research Policy*.

Mowery, D., R. Nelson, B. Sampat and A. Ziedonis (2001), 'The growth of patenting and licensing by U.S. Universities: an assessment of the effect of the Bayh-Dole Act of 1980', *Research Policy*, **30**(1), 99–119.

Mowery, D. and B. Sampat (forthcoming), 'Universities in the national innovation system' in J. Fagerberg, D. Mowery and R. Nelson (eds), *Handbook of Innovation*, Oxford: Oxford University Press.

Narula, R. and A. Zanfei (forthcoming), 'The international dimension of innovation', in J. Fagerberg, D. Mowery and R. Nelson (eds), *Handbook of Innovation*, Oxford: Oxford University Press.

National Science Foundation (NSF) (2002), 'Changing composition of federal funding for research and development and R&D plant since 1990', InfoBrief NSF 02-315 April 2002, website www.nsf.gov/sbe/srs/infbrief/nsf02315/start.htm

National Institutes of Health (NIH) (2003), 'Summary of the FY 2004 President's budget', 3 February.

Nelson, R. and S. Winter (1982), *An Evolutionary Theory of Economic Change*, Cambridge, MA: Belknap Press of Harvard University Press.

Nesta, L. and P. Saviotti. (2003), 'Intangible assets and market value: evidence from biotechnology firms', paper presented at the Chalmers workshop Innovation and Entrepreneurship in IT/Telecommunication and in Biotech/ Pharmaceuticals, May.

Nightingale, P. (2000), 'Economies of scale in experimentation: knowledge and technology in pharmaceutical R&D', *Industrial and Corporate Change*, **9**(2), 315–59.

Nilsson, A.S. (2001), 'Interaction between researchers, firms managers & venture capitalists: the essence of biotechnology business', PhD thesis, Stockholm, Karolinska Institutet, Sweden.

Niosi, J. (2003), 'Alliances are not enough: explaining rapid growth in biotechnology', *Research Policy*, **23**(5), 737–50.

Nooteboom, B. (1999), *Inter-Firm Alliances: Analysis and Design*, London: Routledge.

OECD (Organisation for Economic Co-operation and Development) (1997), *National Systems of Innovation*, Paris: OECD.

OECD (1989), *Biotechnology: Economic and Wider Impacts*, Paris: OECD.

OTA (Office of Technology Assessment) (1984), *Commercial Biotechnology: An International Analysis*, OTA-BA-218, Washington, DC: US Congress.

OTA (1991), *Biotechnology in a Global Economy*, OTA-H-494, Washington, DC: US Congress.

OTA (1993), *Pharmaceutical R&D: Costs, Risks and Rewards*, OTA-H-522, Washington DC: US Congress.

Oliver, R. (1999), *The Coming Biotech Age: The Business of Bio-Materials*, New York: McGraw-Hill.

Orsenigo, L. (1989), *The Emergence of Biotechnology: Institutions and Markets in the Industrial Innovation*, London: Pinter Publishers.

Orsenigo, L., F. Pammolli and M. Riccaboni (2001), 'Technological change and network dynamics: Lessons from the pharmaceutical industry', *Research Policy*, 30, 485–508.

Orsenigo, L., F. Pammolli, M. Riccaboni, A. Bonaccorsi and G. Turchetti (1998), 'The evolution of knowledge and the dynamics of an industry network', *Journal of Management and Governance*, 1, 147–75.

Owen-Smith, J., M. Riccaboni, F. Pammolli and W.W. Powell. (2002), 'A comparison of US and European University–Industry relations in the life sciences', *Management Science*.

Persson, A. and S. Welin (2002), 'Swedish human embryonic stem cell research: issues in regulation and commercialization', paper presented at workshop *The Economics and Business of Bio-Sciences & Bio-Technologies: What can be learnt from the Nordic Countries and the UK?*, Chalmers University of Technology, Gothenburg, Sweden, 25–27 September.

Pisano, G. (1991), 'The governance of innovation: vertical integration and collaborative arrangements in the biotechnology industry', *Research Policy*, **20**, 237–49.

Pisano, G. (1997), *The Development Factory: Unlocking the Potential of Process Innovation*, Boston, MA: Harvard Business School Press.

Pisano, G.P., W. Shan and D.J. Teece (1988), 'Joint ventures and collaboration in the biotechnology industry', in D.C. Mowery (ed.), *International Collaboration Ventures in US Manufacturing*, Cambridge: Ballinger.

Powell, W.W. (1998), 'Learning from collaboration: knowledge and networks in the biotechnology and pharmaceutical industry', *California Management Review*, **40**(3), 228–40.

Powell, W.W. and P. Brantley (1992), 'Competitive cooperation in biotechnology: learning through networks', in N. Nohria and R. Eccles (eds), *Networks and Organizations*, Boston, MA: Harvard Business School Press, pp. 366–94.

Powell, W.W., K.W. Koput and L. Smith-Doerr (1996), 'Interorganizational collaboration and the locus of innovation: networks of learning in biotechnology', *Administrative Science Quarterly*, **41**(1), 116–45.

Powell, W.W. and J. Owen-Smith (1998), 'Universities and the market for intellectual property in the life sciences', *Journal of Policy Analysis and Management*, **17**(2), 253–77.

Prevezer, M. (1997), 'The dynamics of industrial clustering in biotechnology', *Small Business Economics*, **9**, 255–71.

Rabinow, P. (1996), *Making PCR: A Story of Biotechnology*, Chicago, IL: University of Chicago Press.

Red Herring (2000), 'Bioethically speaking: biotech companies navigate murky waters in the 21st century', April, pp. 294–95.

Rickne, A. (2000), 'New technology-based firms and industrial dynamics – evidence from the technological system of biomaterials in Sweden, Ohio and Massachusetts', doctoral thesis, Department of Industrial Dynamics, Chalmers University of Technology, Gothenburg, Sweden.

Rifkin, J. (1998), *The Biotech Century: How Genetic Commerce Will Change the World*, London: Phoenix.

Rothaermel, F.T. (2000), 'Technological discontinuities and the nature of competition', *Technology Analysis & Strategic Management*, **12**(2),

Rynning, E. (2003), 'Public law aspects on the use of biobank samples: privacy versus the interests of research', in M.G. Hansson and M. Levin (eds), *Biobanks as Resources for Health*, Uppsala: Uppsala University, pp. 91–128.

Saviotti, P. (1999), 'Industrial structure and the dynamics of knowledge generation in biotechnology', in J. Senker (ed.), *Biotechnology and Competitive Advantage*, Cheltenham, UK and Northampton, MA, USA: Edward Elgar.

Senghaas, D. (1985), *The European Experience: A Historical Critique of Development Theory*, Dover, UK: Berg Publishers.

Senker, J. (ed.) (1999), *Biotechnology and Competitive Advantage: Europe's Firms and the US Challenge*, Cheltenham, UK and Northampton, MA, USA: Edward Elgar.

Senker, J. and M. Sharp (1997), 'Organizational learning in cooperative alliances: some case studies in biotechnology', *Technology Analysis & Strategic Management*, **9**(1), 35–51.

Sharp, M. (1991), 'Pharmaceuticals and biotechnology: perspectives for the European industry', in C. Freeman, M. Sharp and W. Walker (eds), *Technology and the Future of Europe: Global Competition and the Environment in the 1990s*, London: Pinter Publishers.

Shaw, B. (1985), 'The role of the interaction between the user and the manufacturer in medical equipment innovation', *R&D Management*, **15**(4), 283–92.

Shaw, B. (1990), 'Developing technological innovations within networks', *Entrepreneurship & Regional Development*, **3**, 111–28.

Sorensen, O. and T. Stuart (2001), 'Syndication networks and the spatial distribution of venture capital investments', *American Journal of Sociology*.

Stankiewicz, Rikard (2002), 'The cognitive dynamics of biotechnology and the evolution of its technological systems', in Bo Carlsson (ed.), *Technological Systems in the Bio Industries*, Boston, MA: Kluwer Academic Publishers, pp. 35–52.

Swann, P., M. Prevezer and D. Stout (eds) (1999), *The Dynamics of Industrial Clustering: International Companies in Computing and Biotechnology*, Oxford: Oxford University Press.

Tapon, F. and C.B. Cadsby (1996), 'The optimal organization of research: evidence from eight case studies of pharmaceutical firms', *Journal of Economic Behavior and Organization*, **31**, 381–99.

Tidd, J., J. Bessant and K. Pavitt (1997), *Managing Innovation: Integrating Technological, Market and Organizational Change*, Chichester: Wiley.

Wald, S. (1996), 'Introduction: on the pervasiveness of biotechnology', *STI Review*, **19**, special issue on biotechnology.

Walsh, V., J. Niosi and P. Mustar (1995), 'Small firm formation in biotechnology: a comparison of France, Britain and Canada', *Technovation*, **15**, 303–27.

Wedin, T. (2002), 'Combinatory resources and weak ties: the development of an investment network around the Stockholm-Uppsala biomedical sector', paper presented at workshop *The Economics and Business of Bio-Sciences & Bio-Technologies: What can be learnt from the Nordic Countries and the UK?*, Chalmers University of Technology, Gothenburg, Sweden, 25–27 September.

Wheale, P., R. von Schomberg and P. Glasner (eds) (1998), *The Social Management of Genetic Engineering*, Aldershot: Ashgate Publishing Company.

Wiendt, A. and N. Amin (1994), 'Biotechnology: the emerging battlefield for US and Japanese pharmaceutical companies', *Technology Analysis & Strategic Management*.

Whittaker, E. and J. Bower (1994), 'A shift to external alliances for product development in the pharmaceutical industry', *R&D Management*, **24**(3).

Zucker, L., M. Darby and B. Brewer (1999), 'Intellectual human capital and the birth of the US Biotechnology Enterprise', *American Economic Review*, **88**(1), 290–306.

Zucker, L., M. Darby and J. Armstrong (2002), 'Commercializing knowledge: university science, knowledge capture, and firm performance in biotechnology', *Management Science*, **48**(1), 138–53.

Zeller, C. (2001), 'Clustering biotech: a recipe for success? Spatial patterns of growth of biotechnology in Munich, Rhineland and Hamburg, *Small Business Economics*, (17), 123–41.

4. The post-genome era: rupture in the organization of the life science industry ?

Michel Quéré, CNRS-IDEFI; (*)

1. INTRODUCTION

This chapter intends to discuss the organizational conditions chosen by firms to implement innovative behaviours in a specific context: the life science industry. More specifically, this chapter is about how the post-genome era (PGE), defined as the scientific, technological and economic opportunities stemming from the ongoing identification of genomes (human and non-human), has to be considered as a matter of rupture in the organization of the life science industry. Section 2 characterizes the major features of that industry and a specific focus is put on the characteristics of its economic working, contrasting Europe and the USA. From this description, Section 3 elaborates on the importance of the human genome project (HGP) as a mark of transition in the organization of the life science industry. As such, Section 3 makes more explicit the major characteristics of the post-genome era (PGE) and develops the reasons why the latter can be thought of as a structural shock for knowledge dynamics and innovation patterns of firms acting in that industry. Section 4 develops analytical implications from these recent trends in the evolution of the life science industry. The importance given to 'innovation networks' will be stressed by discussing the 'metamorphosis' (Penrose 1959/1995, p. xix) in industrial organizations which consists of considering alliances and cooperation as a necessary and essential feature to ensure firms' innovation process. Finally, the concluding Section 5 suggests some speculative remarks related to innovative behaviours of firms in the life science industry.

2. THE LIFE SCIENCE INDUSTRY: STYLIZED FACTS AND DEFINITIONAL ISSUES

The term 'biotechnology' is somewhat vague and has been evolving through the course of time. The extensive development of applications from the use of rDNA techniques characterizes the so-called third generation (see Bud 1993 for a historical overview). Previous biotech generations were respectively based on fermentation and organic chemistry. Note the peculiarity of the third generation of biotechnology which is to be directly concerned with biological engineering. In what follows, I refer to the life science industry as a generic term, similar to that of biosciences, in order to qualify the overall set of commercial applications stemming from recent progress in biology research. Both terms ('life science industry' and 'biosciences') apply to a large range of industrial sectors among which agriculture and pharmaceuticals are certainly central, but sectors like agro-foods, environment, computer sciences, instrumentation, mechanics, materials, chemistry and so on, are also partially concerned. This is why I prefer to guard a broad definition, incorporating the overall set of scientific disciplines helpful in the understanding of living organisms through the extensive use of rDNA techniques and related derivatives (see IDEFI/ULB 2002).

The life science industry encapsulates industrial sectors that were for a long time characterized by oligopolistic markets where large diversified firms had a prominent role. However, these sectors are also among those which have extremely benefited from interactions with the scientific community and, consequently, they question the role of science in modern economies (see McKelvey 1996). This has resulted in a set of consequences that have progressively made the life science industry very unique.

First, it largely displaces the frontiers and the characteristics of public versus private goods (see Gibbons et al. 1994; Thackray 1998). Biology research exhibits various and complex arrangements among public and private resources and this observation has had important consequences on the organization of the industry. In particular, the so-called dedicated biotech firms (DBFs) appear as a specific means of combining public and private resources and play a very important role in innovation dynamics (see Saviotti 1998). Figure 4.1 shows the growing importance of DBFs. There is a reasonable presumption that this increase is correlated to the development of the third biotech generation as it follows the 'Genentech effect'. Genentech goes public in October 1980 and this results in a change in the business vision for entrepreneurs and financial investors in the USA. Moreover, DBFs are now playing a higher strategic role in the organization of the life science industry from the 1980s, at least in the USA.

Second, the life science industry is very unstable but global in aim. Due

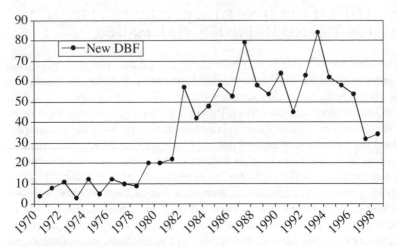

Source: Biotechnology Guide USA (1999).

Figure 4.1   Creation of DBFs in the USA.

to a rapid obsolescence of scientific truth and the related volatility of scientific progress in biology research, the characteristics of the industry are
changing very fast. In this turbulent environment, firms are facing a dramatic search for flexibility and a rapid evolution of productive opportunities and commercial applications (Grabowski and Vernon 1994).

Third, two stages in the use of rDNA techniques are clearly identified
and structure the characteristics of the life science industry (Quéré 2003).
Within a first phase, small and mainly academic start-ups have tried to
compete with large pharmaceutical corporations as well as large agro-food
corporations on well-established markets, be it the production of human
proteins in the pharmaceuticals sector or resistant-seeds in the agro-food
industry. As noted by McKelvey and Orsenigo (2001), that first stage also
includes research efforts dedicated to improve screening methods and
enhance the productivity of discovering new molecules. The generation of
firms derived from that first stage wanted to be 'all things to all markets';
they were research boutiques with great ambition as they were expecting to
compete with large established corporations. Genentech and Amgen are in
the USA the emblematic but contrasted examples of that period: whereas
the former engaged in an active business licensing strategy to face development costs of new molecules, the latter keeps these costs internally and
develop new molecules on its own budget.

However, a second stage begins with the successful implementation of the
so-called HGP. The HGP is not essential by itself, largely because the whole
set of targets under decoding (be it human or non-human objects) in

general, and the sequencing of human genome in particular, offer limited potential for direct learning. However, even if knowledge directly derived from that project is not currently so significant, the HGP induced a significant change in the organization of the life science industry that will be further discussed (Section 3). Therefore, the HGP has to be thought of as a symbolic event to trace a shift in the organization of the life science industry and the progressive emergence of the PGE. One can reasonably consider the latter being started in the 1990s but the wording (PGE) has been widely used since the mid-1990s. Those two stages differ importantly because they induce distinctive research methods and economic consequences. On the one hand, the first stage was based on discovering and optimizing new processes for firms to compete in well-established markets; on the other hand, the second stage allows firms to explore analytical understanding of life mechanisms. This difference induced quite an important change in the organization of the life science industry and more especially in the relationships between small and large firms.

Fourth, there are some difficulties to identify markets' characteristics for many DBFs as they have to adapt to continuous change due to the improvement of scientific progress. Many of those DBFs become service providers for other firms but the exact characteristics of the services they are providing are identified with some difficulty. This is especially true when considering the generation of academic start-ups that emerged in the 1990s to explore commercial opportunities from the PGE. For instance, Robbins-Roth (2000) identifies more than 30 US firms competing for the production of micro-arrays in the 1990s. But, even on what can be thought of as a well-defined market, those DBF firms are exploring various types of improvements (density of probes, quality of probes, treatment speed, customization of arrays, accuracy concerns and so on) that render that market fast evolving and very unstable in its characteristics. Thus, even for tangible market segments, firms are exploring competing techniques and developing alternative strategies that make their market environment very unstable and fast changing. This observation seems very accurate in all the areas (bioinformatics for instance) where DBFs act as service providers for other firms (large pharmaceutical corporations especially).

Fifth, large multinational corporations have been very passive for a long time and very slow to respond to the set of opportunities stemming from the extensive use of rDNA techniques (see Cockburn et al. 1999). To some extent, passive behaviours of large corporations during the first phase have largely contributed to keeping windows of opportunities open for small biotech firms and academic start-ups especially. Consequently, large agro and pharmaceuticals corporations have accumulated transition costs and lost revenues that some authors estimated at $3–4 billion by 1998 for the pharmaceutical sector alone (see Galambos and Sturchio 1998). By the early 1990s, large pharmaceutical corporations reacted and became very active in mergers and acquisitions of small biotech firms. Obviously, there was a first mover advantage, as firms like Eli Lilly, Roche and/or Schering-Plough that moved first performed quite better in economic terms than those that moved later (see Robbins-Roth 2000).

Sixth, the life science industry exhibits contrasted patterns, as regards the USA versus European contexts. For a long time, Europe was considered to be lagging behind, as far as statistics were concerned. It was not only the number of bio-tech firms, but variables like sales, employees, research and development (R&D) expenditures; they show explicitly a continuous increase in the US market whereas the European context was much less developed (see IDEFI/ULB 2002). However, until a few years ago, Europe highly increased the number of DBFs, which was mainly the result of a blast of academic start-ups exploring economic opportunities derived from new applications and services characterizing the PGE. Those European DBFs are essentially young firms, exploring the potentialities of the PGE. This is why the number of European biotech firms is currently comparable to the USA but their structural patterns are highly different. Tables 4.1 and 4.2 express basic characteristics for the top five biotech firms in Europe and the USA. It obviously appears that not only biotech firms are generating revenue more impressively in the USA, but they are also much more mature in the importance with regard to R&D expenditures and employment levels.

The maturity of the US biotech firms can be thought of as a validation of the usual hypothesis of a relative decline of European competitiveness in the life science industry, with regard to the US context (Orsenigo et al. 2001; McKelvey and Orsenigo 2001; Kopp and Laurent 2001). It is certainly true that US firms will continue to exhibit a huge amount of revenue in the next few years, as expected from the evolution of new drugs' approvals and of patents' portfolio (see IDEFI/ULB 2002). Disparities will not reduce significantly in the next few years. However, the rapid renewal in the knowledge base of biosciences can also be thought of as an opportunity to enhance European competitiveness with regard to research avenues characterizing the PGE. In that respect, DBFs are becoming crucial as they explore economic opportunities derived from the progress of science (see Pyka and Saviotti 2001).

Seventh, structural evolution of the life science industry also largely differ in the USA and Europe. The US life science industry is currently facing a consolidation phase with regard to the evolution of mergers and acquisitions (see Robbins-Roth 2000). That consolidation is characterized by a search for vertical integration among the smaller firms, a search for

Table 4.1 Largest US biotech DBFs, 2001

	Amgen	Genentech	Genzyme	Chiron	Biogen	Immunex
Sales (1)	3510	2212	1220	1141	1043	959
R&D expenditures (2)	865	526.2	244	344	314	204.6
Employees	7000	5000	5200	3736	2000	1550

Notes: (1) Direct subsidies from large pharmaceutical corporations are not taken into account.
(2) million US$.

Source: IDEFI/ULB (2002).

Table 4.2 Largest European biotech firms, 2001

	Qiagen	Shire Pharmaceut.	Innogenetics	Powder Ject	Genset	Celltech
Sales (1)	294.39	979.43	59	176.85	29	216.40
R&D expenditures (2)	29.88	18.97	23.8	49.68	41.7	67.04
Employees	1557	NA	580	1000	487	NA

Notes: (1) Direct subsidies from large pharmaceutical corporations are not taken into account.
(2) million euros.

Source: IDEFI/ULB (2002).

increasing the financial capitalization of medium-sized firms (see Ostro and Esposito 1999) and an effort from large corporations to integrate new capa- bilities (see Malloy 1999). However, alliance characteristics are also distin- guishing the USA from Europe. In the USA, alliances are mainly related to marketing and distribution phases of the production process whereas alli- ances in R&D capabilities through the merger of academic start-ups searching for a critical mass effect seem to prevail in Europe. If the evolu- tion of the life science industry is slightly different in the USA and Europe, a common feature lies in the importance progressively taken by 'innovation networks'. The complex mix of interests between various categories of actors (including private and public as well) is a basic feature of innovation dynamics in the life science industry (see Section 4).

   Those sketchy remarks aim at characterizing the main features of a very typical science-based industry and establishing how the two stages iden- tified within the contemporary working of that industry differ with regard to the organization of the industry. These two stages particularly differ with regard to knowledge accumulation regime and related firms' commercial opportunities. The economic consequences of the HGP have resulted in a rupture in knowledge regimes, and I want to discuss the consequences of the HGP in order to emphasize the extent to which the life science indus- try is facing a structural shock in its current working.

3.   THE HGP AND ITS CONSEQUENCES

Here, the HGP is a generic term encompassing all the research effort devoted to the sequencing of genomes, whether they are plants, animals or human beings. Thus, the HGP refers to the definition in use by the US DOE (Department of Energy) and NIH (National Institutes of Health), that is, a wide range of scientific activities including studies of human diseases, experimental organisms, development of new technologies for biological research, computational methods and ethical, legal and social issues. Planned in 1988 and started two years later, that US research programme benefited from a huge amount of money from these two institutions (an estimate of the funds delivered by the DOE and the NIH to the project is around $3.3 billion over the 13-year period).

   In fact, the amount of information derived from the sequencing of differ- ent genomes cannot be easily translated into corresponding knowledge providing keys to understanding the mechanisms which are driving the pro- duction and regulation of proteins. However, the HGP is much more essen- tial with regard to indirect learning. It mainly results in interdisciplinary learning for the discovery of new industrial activities, new methods and techniques, new instruments, new equipment, new computer and software applications and the like. The HGP paves the way for a new knowledge regime in the life science industry based on an intertwined combination among different but complementary disciplines (biology and chemistry of course, but also materials, mechanics, robotics, equipment suppliers, software and computer industries and so on). Therefore, the HGP allows a huge amount of decentralized knowledge opportunities to occur. These opportunities are a further source of difficulty for innovative behaviours of large pharmaceuticals and agro-food incumbents. They have to deal with these opportunities without losing control of their overall production processes. This is considered as a risky business as large corporations exhibit a weak absorptive capacity to secure that process.

The HGP results in various economic opportunities defined as the PGE. A major characteristic of the PGE is that knowledge accumulation is no longer driven by large international corporations or small academic start-ups, but by complex interactive and changing sets mixing together public research institutions, large international corporations, academic start-ups and even 'consumer associations'. This complex mix intersects in various ways and requires specific organizational design, often labelled as innovation networks, alliances and cooperations, quasi-integration and so on. But, the PGE is basically challenging the current organization of that life science industry.

Initially, the industry focused research investigation on the mapping of DNA which structures the cells of organisms. That effort has improved very quickly (mixing public and private science is a basic explanation for that – implying cooperation and competition efforts as well) and the HGP, planned from the joint initiative of NIH and DOE in the USA, was completed in April 2003. Surprisingly, the mapping results in a significant decrease in the expected number of genes. Predictions of around 100000 genes were established from the work of Nobel laureate W. Gilbert in the mid-1980s and that number was printed in the collective vision of the scientific community as an acceptable benchmark. Now, the human genome is estimated to contain 30000–40000 genes, which is between 1 per cent and 2 per cent of its genetic material (3164.7 million chemical nucleotide bases – adenine, cytosine, thymine and guanine). The rest may be involved in managing the chromosomal structure and protein regulation (where, when and in what quantity). However, humans have on average 10 times more proteins than genes, due to mRNA transcript alternative splicing and chemical modifications to the proteins: obviously, a single gene contributes to encode different proteins.

This last comment stresses the importance of the HGP that appears as a kink of impressive structural change in the industry in terms of learning

regimes, knowledge accumulation and economic opportunities. Consequently, the most important learning from this result is the complexity of
human genes' actions and the related increase in puzzling issues. Let us pin-
point a few challenging problems resulting from the current stage of the
PGE:

1. It seems that human genes are activated in a more complex manner
    than those of other organisms. If their coding regions are about the
    same size as that of other organisms, they are combining and interact-
    ing in much more complex ways. On average, each exon (an active part
    of a gene) can be implied in the production of different proteins as well
    as each gene can express several proteins by using different combina-
    tions of the coding regions located in its boundaries.
2. It seems that protein manufacturing is far more complex for human
    genomes than for any other living organism. Humans use domains of
    proteins that have a particular shape or function in a complex and even
    creative way. This displaces the sequencing issue towards the transfor-
    mation of sequencing information into useful knowledge about the
    working of living organisms (see the 'Genomes to Life' project cur-
    rently supported by DOE and NIH).
3. The distribution of single nucleotide polymorphism (SNPs) within the
    genome is highly heterogeneous. SNPs are DNA sequence variations
    that occur when a single nucleotide (A, C, G or T) in the genome
    sequence is altered. SNPs seem relatively stable from one generation to
    another and influence how individuals respond to environmental
    aggressions as well as to drugs and other therapies. If two-thirds of
    SNPs involve the replacement of cytosine (C) with thymine (T), the
    density of SNPs varies considerably from one region to another.
    Currently, scientists have identified 1.4 million locations where SNPs
    occur in the human genome (estimates are 10 million SNPs). Better
    identification and understanding of SNPs is required, and this should
    result in a better prediction in the effectiveness of drugs (see the inter-
    national 'HapMap' project supported by NIH and DOE). SNPs' iden-
    tification is a typical target for which research efforts need to be
    coordinated in order to avoid waste of resources. Beyond its interna-
    tional character, that research activity also implies specific organiza-
    tional arrangements (research consortia) mixing public and private
    resources.
4. The HGP success underlies significant problems with intellectual prop-
    erty rights (IPRs). Without entering into a conflicting discussion about
    the definition of novelty (for instance, are gene fragments' expressed
    sequence tags (ESTs), whose utility is identified by very vague defini-

tions, a relevant material to be patented?), the major problem lies in the fact that the US Patent and Trademark Office (USPTO) applications are confidential. As there are currently around 3 million genome-related patents, it is difficult to know which sequencing information is part of those patent applications. Then, people (and firms) that are using information stemming from public databases are very uncertain on the expected profitability of their research efforts, as that sequencing information could have been included in previous patent applications.

The above set of characteristics of the HGP and its consequences (the transition to the PGE) represent the basic features for inducing a rupture in the economic working of the life science industry. All those characteristics open various windows of opportunities to be explored and transformed into profitable applications. A few basic areas or domains of research that are now structuring to improve our understanding of the working of living organisms are listed below. They not only indicate the variety of research avenues defining the PGE but they also parallel attempts to perform the discovery of new applications in diagnosis and treatments of diseases:

- functional genomics (looking at changes in animal models when a gene with an unknown function is turned on and off);
- structural genomics (looking at the 3D structure and evolution of proteins and detecting the influence of 3D variations onto their biological functions);
- positional cloning genomics (looking at regions of chromosomes containing genes that have shown up, according to a specific disease. Starting with the disease, work is performed in a backward direction in order to identify gene sequences that are activated in patients and not activated in healthy people);
- transcriptomics (large-scale analysis of mRNAs in order to determine where, when and under what conditions genes are expressed);
- proteomics (identifying the constellation of all proteins in a cell and studying their structures and activities; differences and changes in gene expression should allow for identification of proteins which might be involved in the shift from healthy to cancerous cells, for instance);
- Pharmacogenomics (looking at differences in gene sequences between individuals in order to identify SNPs and gene mutations that can be used to predict susceptibility to diseases and identify appropriate individual treatments).

All these research avenues can be thought of as promising opportunities for genetically based treatments of diseases. However, nobody currently knows which avenue will be the more promising and the most accurate for performing in commercial applications. What is certain is that modern biology and the life science industry are facing a huge increase in the necessary scientific and technological capabilities to deal with the functioning of living organisms. What is also certain and very significant in this new stage of the life science industry is the need to face new research methods, protocols and even scientific disciplines, that is, to shift the knowledge base of the life science industry.

Moreover, these research avenues are also resulting from huge progress in equipment and instrumentation techniques that complement research investigations and reveal a logistical innovative environment. Take here the polymerase chain reaction (PCR) technique as an example: invented by K. Mullis in 1983, it was basically an improvement of sequencing techniques previously developed in F. Sanger's laboratory (see Fields 2000). More generally, if sequencing techniques that allow for the deciphering of 3 billion nucleotides of human DNA were not fundamentally different in aim, their improvement has been quite impressive (namely the automation process and the conversion to a fluorescent-based technology to read off the sequence of bases). As noted by Fields (2000):

> The machines had to be improved for faster separations, smaller volumes, increased numbers of reactions, automated reloading, and the like. Programs were necessary to assign a quality score to every determined base, to assemble the data from the phenomenal ramp-up in output, and to co-ordinate the millions of clones and reactions and sequence reads.

The improvement in sequencing techniques is interesting as it confirms the hypothesis of a rupture in the organization of the life science industry due to the HGP. The latter cannot be thought of anymore as a project managed by biologists alone. The improvement required significant expertise from other disciplines, namely, physics, chemistry, computer science, engineering and even management. In addition, each discipline has to be aware of the characteristics and/or requirements of the others. In other words, the improvement reveals the importance of interdisciplinary projects as well as the importance of developing cross-combinatorial capabilities for which innovation networks are a corresponding organizational design.

Instrumentation techniques is another interesting example. The development of scanning tunnelling microscopes and force microscopes (or atomic force microscopes) allow scientists to describe and reproduce images of biological molecules or systems in their physiological conditions. As such, it is an impressive support to functional studies of biological systems.

Those instrumentation techniques are also improving quickly and, as they are non-contact force microscopy, they are very promising in the analysis of molecules or the working of biological systems: they allow descriptive analysis of the underlying phenomena without damaging (or at least minimizing the damage) due to observation. But, instrumentation techniques also require the combination of different disciplines and expertise to optimize these technological improvements.

Finally, beyond these research avenues and their instrumentation techniques, the management of all the information created by the HGP is resulting in the emergence of a new domain of industrial activities: bioinformatics. Currently, bioinformatics is becoming a cornerstone of the PGE. Bioinformatics is now becoming a generic term that expresses the need for tackling the huge amount of scientific calculations required to understand the working of the genome, that is, the protein fabrication and interactions. Different orientations can be listed here. The 3D modelling of proteins, their evolution and variations in time and space (and milieu) and the management of gene data banks are all sources of huge amounts of numeric expertise (related to calculation, simulation and/or inventory of data). The introduction and the industrial utilization of DNA chips in order to express mRNA and understand protein synthesis (proteomics) require enormous computing capabilities. If the number of human genes is now thought to be 30000 to 40000, the number of proteins is thought to be four to six times that number. Understanding the gap between the number of human genes and that of human proteins explains the importance of proteomics as well as the strategic importance of bioinformatics in order to match that challenge.

Even more than bioinformatics, the PGE is a source of higher diversification processes within and out of the life science industry, due to the introduction of combinatory techniques and multiple technologies' hybridization: 'the interface between biological and non biological substances is important for producing novel sensors and devices' (Hoch et al. 1996, p. xiv). Among the technical devices that are central to the life science industry, the role of electronics appear particularly critical: 'When bound to the gate regions of transistors, biological molecules can influence and/or alter the electronic characteristics, thereby providing switching and sensing capabilities' (Hoch et al. 1996, p. xiv). Nano-fabrication and sensor research appear among the most interesting areas for that pervasiveness of electronics in biology. On the one hand, controlling structures and suprastructures of biological molecules, surface interactions including the fabrication of sub-micron pores, the orientation of cell behaviours and/or neurons are promising areas of interfacing technologies coming from different disciplines (physics, mathematics, informatics and automation and so on). On the other hand, sensibility techniques are crucial for understanding biological mechanisms. For instance, when antibodies meet their antigens, mass changes occur and sensor research is extremely important in order to identify and understand the mechanisms at stake. Then, coming back to the combination of micro-fabrication and biology, it is really new research areas and disciplines that are now structuring:

> Such research requires scientists who are experts in their own areas of research but, at the same time, are able to traverse the gap between widely disparate fields. It requires an understanding of each other's language and capabilities. And, it requires a genuine cooperation between biologists and engineers and materials scientists (Hoch et al. 1996, p. xv).

To sum up, what currently occurs in the PGE largely challenges the process of knowledge accumulation that prevailed in the previous stages of the life science industry. Those multiple and combinatorial aspects of scientific and technical requirements importantly impact knowledge characteristics as well as knowledge sources, emitters, receptors and channels of diffusion. In fact, the purpose of the PGE is understanding life mechanisms. This has to do with the science of complexity, the manipulation of high-density data, probabilistic calculations and simulations. This requires a combination of different scientific disciplines and complex intertwining combinations of related resources and capabilities. It is a source of a huge increase in knowledge needs that favours the existence of several research avenues as well as commercial potentialities.

Take genomics as an emblematic example. Two main lines of specialization are clearly emerging: on the one hand, some DBFs are specializing in offering genomic services, dedicated to further improvement for basic research (including additional technologies, equipment, databases, methods and so on); on the other hand, other DBFs are focusing on the exploitation of the sequencing information, exploring some of the research avenues previously listed in order to develop applications (identification of protein profiles, identification of genotypes corresponding to particular phenotypes, identification of diseases' susceptibility, identification of drug responsiveness and so on). The range of new opportunities is impressive and this situation induces a real shock on the working of the life science industry. This implies quite a revolution in its organization, the mobilization of knowledge and resources, the interactive character among scientific disciplines and firms belonging to various industrial sectors, which will lead to an economic context contrasting with previous behaviours of DBFs that were competing on well-established markets.

The PGE will induce a higher specialization built on combinatorial experiences among complementary knowledge that will create more market niches for economic applications. This is why firms are creating innovation networks that are complex arrangements of cooperation and affiliation which seem to appear both as an appropriate mode of complementing capabilities to generate accurate knowledge, and as an appropriate insurance mechanism to survive in such a highly uncertain environment. Probably, the situation for large agro and pharmaceuticals corporations will be much less comfortable in the future because they will have to face an economic context where 'localized' knowledge, applications and markets will not only matter, but cause coordination problems in keeping under control global activity and related strategy. Indeed, large corporations will be importantly challenged by the PGE with regard to their managerial capabilities and the improvement of their scientific and technological capabilities.

4.   THE ECONOMIC IMPLICATIONS OF THE PGE

Two central economic issues arise from Sections 2 and 3. The first is the change in knowledge regime and knowledge accumulation that characterizes modern biology (the PGE). The second is the related change in organizational design chosen by firms to explore commercial opportunities in the PGE.

The first economic issue leads to consideration that the HGP has actually been a shock to the life science industry because the research effort to express the genomes of living organisms has provoked structural change in learning regimes, knowledge accumulation and related economic opportunities. Even if the amount of information expressing the genome is not associated to a clear-cut corresponding knowledge and a better understanding of life mechanisms, the HGP has been the source of discovering new industrial activities, new methods and techniques, new equipment and new complementary aspects among activities that were completely separated before that research effort. As a consequence, the PGE is characterized by a huge amount of technological opportunities that challenge firms' innovative behaviours. Complex webs of interactions among different types of firms (with regard to size or sector) characterize the PGE. Thus, the life science industry seems to be facing a situation where 'firms opt for sustaining the ability to learn, via interdependence, over independence by means of vertical integration' (Powell et al. 1996).

This leads to the second economic issue which is a rupture in that industrial context as that knowledge accumulation and evolution is now occurring through inter-organizational exchanges, R&D alliances and networks of learning. Currently, small and very small firms act as an exploratory infrastructure for well-established firms (large multinational corporations). A great number of DBFs are challenging the PGE research avenues and exploring technical opportunities in order to develop promising economic applications. This creates an unstable productive environment in which large firms try to keep under control these new opportunities by developing cooperation and affiliations with small firms (and academic start-ups more especially). This strategy is difficult to implement because of the uncertain outcome of the innovation process: 'knowledge is garnered from collaboration on a specific project, but this participation has unanticipated results not apparent at the outset of the relationship' (Powell et al. 1996).

Indeed, this open context introduces the following question: what organizational design is suited to the above conditions for innovation dynamics in the life science industry? To some extent, empirical observation tends to show the importance taken by 'technological platforms'. These platforms assemble firms and scientific institutions towards a common target. They aggregate complementary competences needed to improve knowledge in the domain under investigation. Technological platforms are essentially thematic-oriented and related to topics like combinatorial chemistry, bioinformatics, instrumentation techniques, genomics, proteomics and so on. Technological platforms seem to be identified as a collective acceptable device (and a suited organizational design) for matching technological specialization, global purpose and innovative results. Therefore, the collective set of resources embodied in 'technological platforms' illustrates the growing importance given to 'innovation networks', as attested by Powell et al. (1996) and Orsenigo et al. (2001).

From empirical observation, some analytical remarks and implications related to the theory of the firm can be addressed. The importance taken by innovation networks questions the nature of the firm itself in that industry. Innovation networks reveal an actual renewal in models of organizational learning due to the pressure of a continuous progress of science. Therefore, the characteristics of partnerships are difficult to depict through usual transaction cost analyses (Williamson 1985) or related managerial literature (see, for example, Teece 1986 and 1988): reliance on partners, assets' specificity, confidence and trust are not necessarily proper ingredients to be taken into account. In other words, the problem to be faced by a particular firm is not just a 'make-or-buy' arbitrage (even in a sophisticated manner) but the management of a complex intertwining network of capabilities shared to design specific and innovative activities. The situation faced by the life science industry is one of sharing knowledge to create new productive opportunities. This is really an innovative context where organizational design among partners is the essential factor for making collective organizational learning successful. Therefore, innovation networks seem to become a stable structural device in the current organization of the life science industry: they become the locus for innovation in a context of rapid technological change and everybody is concerned with them as 'neither growth nor age reduced the propensity [of firms] to engage in external relationships' (Powell et al. 1996).

The importance of innovation networks shows how an increase and diversification in knowledge required to face innovative behaviours address a traditional dilemma: to specialize on individual and specific knowledge or to improve capabilities by collective learning. Then, the importance taken by innovative networks seems to prove the dominance of the second option and reveals itself to be very proxy from the analysis suggested by Penrose (1959/1995) when she deals with understanding the growth of the firm.

Within the PGE, the life science industry is facing a context where 'the rate of internal expansion of firms is limited under any circumstances that encourages specialization of firms in periods of rapidly growing demand' (Penrose, 1959/1995, p. 72). But, with regard to the PGE, a demand growth explanation has to be complemented by a 'knowledge availability' explanation. If demand growth matters, the latter has nevertheless to be related to changes in knowledge structure and/or availability. The PGE is characterized by important changes in scientific progress and/or technological capabilities. The latter have induced the exploration of new productive opportunities and have been the source for distinctive advantages among firms, when they succeeded in creating associated demand. Therefore, the availability of new knowledge, and the underlying transformation of productive organizations in order to adapt and explore all kinds of productive opportunities from it, are the essential source of growth opportunities in the PGE. This obviously favours the existence of various interstices, referring to the term used by Penrose to express this kind of economic opportunity: 'the nature of the interstices is determined by the kind of activity in which the larger firms find their most profitable opportunities and in which they specialize, leaving other opportunities open' (Penrose 1959/1995, p. 223).

Moreover, the PGE seems to become an economic context that very much favours the sustainable entry of new firms because market opportunities grow faster than large corporations: to some extent, the life science industry in the 1990s has been characterized by a situation where large firms cannot explore all the perspectives opened by scientific progress. They have been unable to manage directly the availability of technological knowledge to their own benefit and have been obliged to rely on other partners (and on DBFs especially) through innovation networks. Moreover, the development of interstices has allowed a few DBFs to succeed in becoming 'large'.

There is another reason why Penrose's analysis is helpful in dealing with the evolution of the life science industry. In the preface of the third edition of *The Theory of The Growth of The Firm* Penrose insists on recent trends based on empirical observation, especially the importance of 'business networks':

> The rapid and intricate evolution of modern technology often makes it necessary for firms in related areas around the world to be closely in touch with developments in the research and innovation of firms in many centres. Formal relations among such firms may advance the competitive power of each of them; to make alliances may be not only a rational response, but even at times a necessary one' (Penrose 1959/1995, p. xix).

In Penrose's terms, business networks are very similar to innovation networks that characterize the current working of the life science industry. The evidence that 'business networks' are becoming an essential feature of the current working of industrial systems seems so obvious that Penrose uses the term 'metamorphosis' to qualify the change in industrial organization.

However, the evidence also requires a discussion about the boundaries of a firm. If firms are currently embedded in complex webs of cooperation and affiliations, understanding industrial organization implies not to consider the firm and its related growth as Penrose's initial attempt (1959/1995) was concerned with, but to consider the aggregates of firms interconnected with the aim of exploring collective learning and innovation processes (Antonelli and Quéré 2002). This is why Penrose's analysis of the growth of the firm stands to be complemented by Richardson's analysis of the organization of industry (Richardson, 1972) when dealing with the current economic working in the life science industry.

The interest of Richardson's analysis is to highlight the characterization of the evolution of industry by means of understanding the coordination of underlying activities. The shift from product to activity is very helpful with regard to innovation dynamics. Indeed, the central aim of the firm is not to combine inputs in order to deliver stable outputs incorporated in products or services; firms aim at coordinating peculiar types of activities. By addressing the importance of activities, the Richardson framework provides an original framework to deal with coordination mechanisms among firms within an industry, because firms no more have to be thought of as competing actors on the output markets, but have to be considered as agents interacting with knowledge in such a way that particular capabilities emerge from those interactions. Incidentally, the most obvious '*trait d'union*' between Penrose's and Richardson's analyses is the importance of (imperfect) knowledge. The former refers centrally to knowledge for understanding the growth of the firm whereas the latter is concerned with how imperfect knowledge affects the working of an economy in a more Hayekian perspective.

This is also why Richardson's analysis remarkably addresses the analytical existence of 'the complex networks of cooperation and affiliation that proliferate in market economies', that is the 'metamorphosis' noted by Penrose and the progressive importance taken by 'business networks'. In the Richardson framework, cooperation is a coordination device differentiated from the firm and the market because it exhibits specific characteristics which are to coordinate closely complementary but dissimilar activities:

> Here then, we have the prime reason for the existence of the complex networks of co-operation and association the existence of which we noted earlier. They exist of the need to co-ordinate closely complementary but dissimilar activities. This co-ordination cannot be left entirely to direction within firms because the activities are dissimilar, and cannot be left to market forces in that it requires not the balancing of the aggregate supply of something with the aggregate demand of it but rather the matching, both qualitative and quantitative, of individual enterprise plans (Richardson 1972, p. 892).

That last need, matching individual enterprise plans, quantitatively and qualitatively, is particularly important, with regard to the understanding of the organization of industry. In particular, cooperation requires industrial activities to be coordinated in such a way that knowledge produced by the set of firms involved in the cooperation process will match existing and new (or expected) demand.

Among the previous characteristics of the life science industry, the shift in the knowledge accumulation regime is quite consistent with the hypothesis that understanding the organization of industry requires one to deal with how competences and activities are coordinated among firms much more than with the analysis of the transformation of inputs in related outputs within established firms. In a context of rapid structural change (as in the PGE), the life science industry is no longer competing in well-established markets and related products as was the case in the 1980s (Quéré and Saviotti 2002; Quéré 2003). Competition is much more about knowledge advances and firms' capabilities. This is why it is helpful to deal with the coordination of economic actors in those areas by referring to an activity framework (Richardson 1972).

It comes closer to the understanding of knowledge dynamics for at least two reasons. First, by considering the shift in competition from product to capability, there is no reason why innovation should require a specific type of industry structure. Indeed, industrial dynamics (and related increasing returns) '*need* not produce vertical disintegration any more than they *need* produce straightforward market concentration' (Richardson 1998, p. 357, emphasis in original). Consequently, the organization of industry exhibits a diversity of patterns that must be understood through the type of under-lying activities and from the characteristics required to ensure their coordination. No optimal organizational model exists: all depends on knowledge requirements and firms' related interactive learning.

Second, in Richardson's framework, firms and markets have not to be considered as pure substitutes. Cooperation is not a hybrid mode; it is a full and distinctive coordination device. Consequently, the diversity of patterns in the organization of industry is also due to the complementary character among the three types of coordination devices: firm, market and cooperation. The organization of industry requires the simultaneous confrontation of all three kinds of coordination. In a context of market uncertainty and important volatility of capabilities embodied in industrial activities, the organization of industry can be particularly complex in that no stable design can be identified. The simultaneous confrontation of firm, market and cooperation modes of coordinating activities can only be the actual source of innovation processes. Consequently, those modes of coordinating activities are not substituting each other as the firm and the market do in traditional economic analysis. The richness of the Richardson explanation of the organization of industry lies precisely in that complex complementary effects exist among those coordinating devices, *at the same time* and, with regard to a single firm, the management of these complementary effects is central to deal with innovative behaviours in a suited way. The PGE offers an illustrative context through that growing importance of innovation networks.

With respect to the organization of industry, the life science industry in the 1980s contrasts with the life science industry in the 1990s. rDNA techniques have progressively offered powerful opportunities for a few entrants to compete with large incumbents; but during the 1990s, knowledge dynamics based on a stronger innovative regime develops and leads to an industrial structure where knowledge accumulation and evolution occur through inter-organizational exchanges, R&D alliances and networks of learning, that is, the development of innovation networks. The Richardsonian interpretation of cooperation largely applies to understanding the working of innovation networks and, consequently, to identifying the rate and direction of industrial dynamics. It provides appropriate guidelines to reflect on the diversity of current organizational patterns in the life science industry. Richardson's analysis based on the coordination of activities reveals itself to be powerful in understanding what is currently at stake within the life science industry.

5. CONCLUSIONS

A central aim of this contribution is to discuss the extent to which the PGE is a cause of a rupture in the organization of the life science industry. A first provocative conclusion is that the exploding opportunities induced by the PGE seem to essentially benefit the DBFs, even if the DBFs have not yet supplanted the incumbents (large corporations) because of the complexity of knowledge and activities to be coordinated in order to perform innovation processes in those areas (from research to manufacturing a new drug). Nevertheless, the progressive importance taken by DBFs since the 1990s proves that they are becoming essential drivers for knowledge dynamics. To some extent, they express a shift from an economics of research where large multinational corporations were quite effective towards an economics of knowledge where it is not only a matter of R&D budgets but is also, in a large sense, a matter of exploring, combining and adapting science opportunities to the understanding of life mechanisms. Certainly, DBFs have not yet all the capabilities in hand to compete seriously with large multinational corporations. However, for some cases, it can be questioned if what has appeared in other industries (like the 'Bill Gates success' in the computer industry) would not be reproducible in the life science industry.

Second, the main arguments developed in this contribution are centred on the R&D implications of the life science industry. As such, for instance, I believe that a common research effort for pharmaceuticals and agriculture would certainly make sense with regard to economies of scale in the availability of new techniques, equipment and methods: the extensive use of rDNA techniques requires similar knowledge expertise, at least partially in both areas. However, empirical observation shows explicit splitting between these two major sets of applications, due to important pressure from the demand side of those activities (De la Mothe and Niosi 2000). This is an obvious limitation to the argument of this chapter: the overall set of factors driving innovation dynamics in those sectors are clearly not fully taken into account. The effort of McKelvey and Orsenigo (2001) to approach the life science industry as a sectoral system of innovation is then a complementary (more than alternative) perspective, as they rightly insist on the importance of the demand and consumer side of such an industry. Nevertheless, concentrating on R&D constraints and related knowledge dynamics is also helpful to reveal specific characteristics in the working of that industry. More precisely, focusing on the R&D phase makes more explicit the importance of 'innovation networks' or 'business networks' as a common device in the renewed organization of the life science industry.

Third, depicting 'innovation networks' by using an activity framework in the line of the Richardson analysis of the organization of industry allows for

stressing the importance of cooperation for knowledge dynamics and innovation processes. Whatever the distinctive characteristics of firms (size and
sectors, especially), implementing innovation networks (here in the empirical
form of technological platforms) appears to be a common feature in the
organization of that industry. This does not result in a process of specialization but does result in a purpose of discovering new knowledge by complementing resources and activities. Indeed, exploring the specific characteristics
of activities developed in the life science industry allows for a better understanding of the rate and direction of the innovation process in those areas.

Fourth, one can wonder about how long the relative competitive advantage of the USA with regard to statistics will be sustainable, because of the
structural characteristics of the PGE, and notably, the search for flexibility
in technical and managerial capabilities due to the rapid knowledge renewal
in scientific progress. In that respect, the dominance of small and very small
firms that characterizes the European context can also be thought of as an
opportunity to face the emergence and the development of new combinatorial capabilities as well as an organization of the industry suited to higher
flexibility and mobility. In other words, there could be a relative advantage
of the European industry to match the current combinatory constraints
required by knowledge dynamics in the PGE, which lies in the importance
of small science-based firms, because of their higher flexibility and adapt-
ability to scientific progress. Obviously, US biotech firms will continue to
collect larger amounts of revenue in the next few years than European
biotech firms will do. However, with regard to the working of the PGE and
to the huge numbers of alternative research avenues, there could also be
opportunities for enhancing the competitiveness of the European context
in the life science industry.

NOTES

*   This contribution largely benefits from the support of the European Union Directorate
    General Research Key Action 'Improving the socio-economic knowledge-base' as it is a
    joint product of the project 'Technological knowledge and localized learning: What per-
    spectives for a European policy?', conducted under the contract HPSE-CT2001-00051.

REFERENCES

Antonelli, C. and M. Quéré (2002), 'The governance of interactive learning within
    the innovation systems', *Urban Studies*, **39**(5–6), 1051–63.
Bud, R. (1993), *The Uses of Life: A History of Biotechnology*, Cambridge:
    Cambridge University Press.

Cockburn, I.M., R. Henderson and S. Stern (1999), 'The diffusion of science-driven drug discovery: organizational change in pharmaceutical research', NBER working paper, no. 7359, September.

De la Mothe, J. and J. Niosi (2000), *The Economic and Social Dynamics of Biotechnology*, Boston: Kluwer Academic Publishers.

Dibner, M. (1999), *Biotechnology Guide USA: Companies, Data and Analysis*, Basingstoke: Macmillan.

Fields, S. (2000), 'The interplay of biology and technology', *Publications of the National Academic of Science*, **98**(18).

Galambos, L. and J. Sturchio (1998), 'Pharmaceutical firms and the transition to biotechnology: a study in strategic innovation', *Business History Review*, **72**(2), 250–78.

Gibbons, M., C. Limoges, H. Nowotny, S. Schwarztman, P. Scott and M. Trow (1994), *The New Production of Knowledge: The Dynamics of Science and Research in Contemporary Societies*, London: Sage Publications.

Grabowski, H. and J. Vernon (1994), 'Innovation and structural change in pharmaceuticals and biotechnology', *Industrial and Corporate Change*, **3**(3), 435–50.

Hoch, H.C., L.W. Jelinsky and H.G. Caighead, (1996), *Nanofabrication and Biosystems: Integrating Materials Science, Engineering, and Biology*, Cambridge: Cambridge University Press.

IDEFI/ULB (2002), 'Life sciences and biotechnology', Research Report, EU-TELL project, mimeo.

Kopp, P. and T. Laurent (2001), 'Biotechnologies et hautes techniques: le retard français', France Biotech Objectif 2010, mimeo.

Malloy, M. (1999), 'Mergers and acquisition in biotechnology', *Nature-Biotechnology*, **17**, May, 11–12.

McKelvey, M. (1996), *Evolutionary Innovations: The Business of Biotechnology*, Oxford: Oxford University Press.

McKelvey, M. and L. Orsenigo (2001), 'Pharmaceuticals as a sectoral system of innovation', ESSY working paper, mimeo.

Orsenigo, L., F. Pamolli and M. Riccaboni (2001), 'Technological change and network dynamics: lessons from the pharmaceutical industry', *Research Policy* (30), 485–508.

Ostro, M. J. and R.S. Esposito (1999), 'A rationale for consolidation among biotechnology micro-caps', *Nature-Biotechnology*, **17**, May, 16–17.

Penrose, E.T. (1959/1995), *The Theory of the Growth of the Firm*, 3rd edn, Oxford: Oxford University Press.

Powell, W.W., K.W. Koput and L. Smith-Doerr (1996), 'Interrorganizational collaboration and the locus of innovation: networks of learning in biotechnology', *Administrative Science Quarterly*, **41**(1), 116–45.

Pyka, A. and P. Saviotti (2001), 'Networking in biotechnology industries – from translators to explorers', SEIN project working paper, mimeo.

Quéré, M. and P.P. Saviotti (2002), 'Knowledge dynamics and the organisation of the life science industries', paper presented at DRUID Summer Conference on Industrial Dynamics of the New and Old Economy – 'Who is embracing whom?' Copenhagen, 6–8 June.

Quéré, M. (2003), 'Knowledge dynamics: biotechnology's incursion into the pharmaceutical industry', *Industry and Innovation*, **10**(3) 255–73.

Richardson, G.B. (1972), 'The organization of industry', *Economic Journal*, **82**, 883–96.

Richardson, G.B. (1998), *The Economics of Imperfect Knowledge*, Cheltenham, UK and Lyme, USA: Edward Edgar.

Robbins-Roth, C. (2000), *From Alchemy to IPO: The Business of Biotechnology*, Cambridge, MA: Perseus Publishing.

Saviotti, P. (1998), 'Industrial structure and the dynamics of knowledge generation in biotechnology', in J. Senker and R. van Vliet (eds), *Biotechnology and Competitive Advantage: Europe's Firms and the US Challenge*, Cheltenham, UK and Lyme, USA: Edward Edgar.

Thackray, A. (1998), *Private Science*, Philadelphia: University of Pennsylvania Press.

Teece, D. (1986) 'Profiting from technological innovation', *Research Policy*, **15**, 285–305.

Teece, D. (1988), 'Technological change and the nature of the firm', in G. Dosi, C. Freeman, R. Nelson, L. Soete and G. Silverberg (eds), *Technical Change and Economic Theory*, London: Pinter Publishers.

Williamson, O.E. (1985), *The Economic Institutions of Capitalism: Firms, Markets, Relational Contracting*. New York: The Free Press.

5. An overview of biotechnology innovation in Europe: firms, demand, government policy and research

Jacqueline Senker

During the 1980s and 1990s governments throughout Europe began to invest in biotech research in the hope of stimulating commercial exploitation (Enzing et al. 1999). Countries, however, have differed dramatically in the extent to which this research has stimulated biotech innovation and in the industrial sectors in which innovation has occurred.

Some studies attempting to explain the development of the biotech industry in specific countries focus on characteristics of the national system of innovation (for instance, see Kivinen and Varelius 2003; Giesecke 2000; Bartholomew 1997), but fail to pay attention to the sectoral distribution of biotech innovation. Other studies focus on specific sectors applying biotechnology, such as pharmaceuticals (e.g. see Henderson et al. 1999; Malerba and Orsenigo 2002) and yet others give more attention to the demand and supply factors affecting industrial development in several sectors (e.g. see Walsh 2002). None of these studies explain how and why national characteristics affect the sectoral application of biotechnology within a given country.

A project[1] designed to fill this gap reviewed biotech developments in eight countries: Austria, France, Germany, Greece, Ireland, The Netherlands, Spain and the UK. It focused on three sectors to which biotechnology is applied: biopharmaceuticals, agro-food biotech[2] and research equipment and supplies. For each country and sector, a comparison was made over time of trends in the science base, industrial development including the availability of finance capital and factors affecting demand. These comparisons highlight the diversity across sectors and between European countries and provide insights into rich and varied developments within Europe, based on a conceptualization that modern biotechnology is driven by firms, demand, government policy and research. This chapter helps to 'set the scene' by identifying the distinctiveness of European countries within wider global trends. It also identifies the factors that explain the great diversity of innovation patterns in the eight countries and three sectors studied. To set these findings in context, they are preceded by a brief introduction to the theories that contributed to a common framework for the study and the methodology employed.

1.   THEORETICAL CONSIDERATIONS AND METHODOLOGY

The common framework developed for the national case studies integrated complementary elements and relationships from several overlapping bodies of literature.[3] It also took into account a number of idiosyncrasies that characterize biotech innovation, such as the many research-based companies that have no products, international research collaborations between incumbent large firms and newly created dedicated biotech firms (Saviotti 1998) and a public hostile to certain applications (Gaskell et al. 1998; Gaskell et al. 2000). This latter aspect identified the importance of including demand- as well as supply-side factors in the framework.

The national system of innovation (NSI) literature (for main contributors see Lundvall 1992; Nelson 1993) provides the main elements of the conceptual framework for analysing country-specific factors influencing the innovative capabilities of companies. The key and interdependent elements of a NSI are the research and development (R&D) system, the role of the public sector including public policy, inter-firm relationships, the set-up of the financial system, the national education and training system and the internal organization of firms. National geopolitical boundaries and patterns of social and cultural values shape the institutions involved in the system of innovation but globalization and regionalization may affect the relevance of national boundaries for national systems of innovation.

The innovation systems approach, including technological systems (for main sources see Carlsson 1997; Edquist 1995; Edquist 1997) remind us to focus on specific techno-industrial areas, as well as on elements in the national system. It also points out that the boundaries of the system are defined by knowledge and/or technology, not by geography. Technological systems consist of dynamic knowledge and competence networks embedded in an institutional infrastructure. These networks are linked by flows of information. The presence of entrepreneurship and sufficient critical mass may transform the network into an innovative 'development block', but this depends on economic competence, defined as the use of knowledge and information to identify, exploit and expand business opportunities. More attention is placed on the application of knowledge than on its generation and diffusion.

Theories of socio-technical networks derive from heretogeneous attempts to integrate approaches from both economics and sociology to explain the organization of socio-economic relationships including Carlsson (1997), Lundgren (1995) and Håkansson (1987). The main contribution of this literature is its emphasis on public or private social actors and intermediaries in innovation networks. The controversies literature (Callon et al. 1986; Latour, 1989; Knorr Cetina 1982) provided an approach for mapping the 'social or public acceptability' of new technologies, and the public actors involved in decisions about them.

The literature review led to the design of an integrated framework for preparing national case studies. Figure 5.1 provides a simplified overview of the framework designed. It includes the networks within which relevant institutions and organizations are embedded and their interrelationships. As shown, the four main components of the framework are:

- networks of knowledge and skills;
- networks of industry and supply;
- factors connected with demand and social acceptability;
- and factors connected with finance and industrial development.

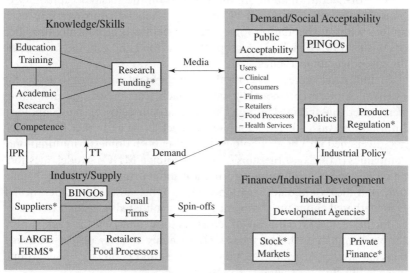

Notes: *International influence; TT = technology transfer; IPR = intellectual property rights; PINGOs = public interest non-government organizations; BINGOs = business interest non-government organizations

Notes: *International influences; TT = technology transfer; IPR = intellectual property rights; PINGOs = public interest non-government organizations; BINGOs = business interest non-government organizations

Figure 5.1   Networks of key factors influencing innovation

The national case studies aimed to provide information about these components for each sector being studied and about the elements that link them together.

To ensure that the eight national case studies were as comparable as possible, we employed a common framework and definitions for the phenomena being studied, used the same indicators and similar data sources to represent the factors under study and reported the findings in reports which conformed to a common structure. The common reporting structure for the national reports included a general introduction to the characteristics of each country which might affect the development of biotechnology such as strong national economic sectors, critical historical events, elements of the culture or political style, industrial policy since 1990 and any recent changes in direction in terms of the four components being considered. Chapters on each sector also followed a common framework and reviewed the science base, the industrial structure, the nature of the market including consumer attitudes and the prospects for the sector. The national case studies concluded by comparing the three sectors.

Secondary sources were used for collecting background national and sectoral information. In addition, information on the social acceptability of biotechnology was identified by studying a related controversy in each sector: xenotransplantation for biopharmaceuticals, genetically modified organisms (GMO) in foods or crops for agro-food and genome sequencing for equipment and supplies.

An industrial survey was carried out in the eight countries to assess the extent of biotech commercialization in each sector.[4] Sectoral definitions were based on specified product areas and sub-fields of biotechnology. Information was gathered about new biotech firms and other firms which had diversified into biotechnology, including both domestic multinational companies (MNCs) or the subsidiaries of foreign MNCs. Model questionnaires designed for collecting information about firms were for optional use only, and the required information could also be gathered by other means. Data gathered about the companies was consolidated in a single database[5] to facilitate the preparation of cross-sectoral analyses for each of the sectors being studied. The cross-sectoral analyses drew on the database, national studies, relevant secondary sources and contextual background provided by an assessment of European policy.

Section 2 presents the main results of the industry survey, focusing on differences between countries and sectors.

2. RESULTS OF THE INDUSTRY SURVEY

The industry survey, carried out in 1999, collected information on 724 small firms involved in biotechnology. An analysis of the main results of the survey, presented below, found that the commercial exploitation of biotechnology differs markedly between the three sectors and eight countries.

The greatest number of firms (337) were involved in pharmaceutical applications of biotechnology, followed by equipment and supplies firms (224). Only 162 firms involved in agro-food biotech applications were identified, less than half the number active in the pharmaceutical area. Figure 5.2 shows the distribution of firms by sector and country. This shows that France, Germany and the UK are the major biotech players, with The

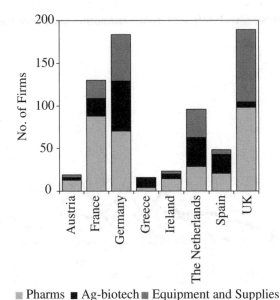

■ Pharms ■ Ag-biotech ■ Equipment and Supplies

Figure 5.2   Number of biotech firms by sector and country

Netherlands not far behind. Every country has firms involved in innovation in biopharmaceuticals and in agro-food biotechnology, but the focus on the latter activity is very limited in some countries. Equipment and supplies firms tend to concentrate their activities in the countries that are the major players, and scarcely exist in the other countries.

The industrial structure of firms in each country was also analysed by their age. Table 5.1 gives the median date for the creation of biotech firms in each country. In Austria, Greece and Spain, the number of biotech firms is small

Table 5.1   Median foundation date for biotech firms, by country

	A	DE	FR	GR	IRL	NL	SP	UK
Median foundation date	1974	1994	1992	1973	1993	1989	1974	1990

and a median creation date in the early 1970s suggests that the majority have diversified into biotechnology. The median foundation year for firms in Germany, France, UK, The Netherlands and Ireland is after 1989, suggesting that the majority have been created specifically to exploit biotechnology. The median date of creation of firms also differs by sector, with agro-food firms having the oldest median age (1987), equipment and supplies firms slightly more recent (1989) and biopharmaceuticals firms the youngest

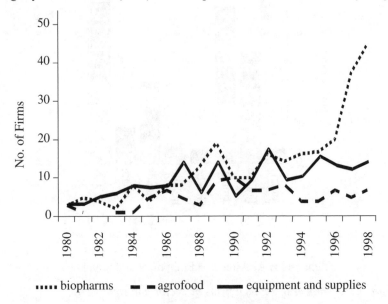

····· biopharms   ▬ ▬ agrofood   ▬▬▬ equipment and supplies

Figure 5.3   Number of firms created by sector and year

(1993). Moreover, it seems that some of the older firms (those founded before 1980) mainly diversified into biotechnology, rather than being established to exploit the technology. One-third of agro-food and 22 per cent of equipment and supplies firms are long-established, but only a minority of biopharmaceuticals firms (13 per cent) were founded before 1980.

Figure 5.3 shows the number of firms created each year since 1980 for each sector. The late 1980s were a period of rapid growth in the creation of biotech firms in all sectors. Since then growth in new agro-food firms has first levelled off and then slowed down. Biopharmaceuticals firms entered a second period of dynamic growth after 1995. Creation of firms in the equipment and supplies sectors has shown steady growth overall, with fluctuations from year to year.

Figure 5.4, the distribution of firms by the number of employees and sector, shows that agro-food biotech firms have the highest proportion of

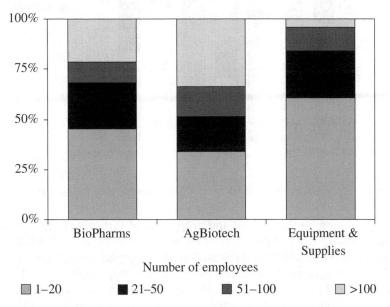

Figure 5.4   Distribution of firms by number of employees and sector

large firms (more than 100 employees). Equipment and supplies firms have the highest proportion of firms with less than 20 employees and over 80 per cent have less than 50 employees. Over two-thirds of biopharmaceuticals firms have less than 50 employees, but some of the new firms now appear to have grown; over 20 per cent have more than 100 employees.

Figure 5.5 shows how firms of different sizes are distributed between countries. The countries in which the number of employees concentrate in firms with over 100 employees are also those whose firms are the oldest – Austria, Greece and Spain, suggesting that exploitation of biotechnology is mainly by diversification, not the creation of new firms. By contrast, it appears that in countries with younger firms on average, and a high proportion of small and very small firms (UK, Germany, France and Ireland), commercial activity is led by dedicated biotech firms.

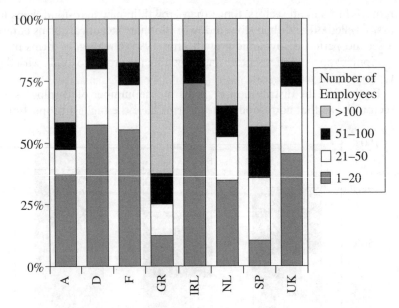

Figure 5.5   Proportion of firms by number of employees and country

Information on the amount and proportion of turnover attributable to biotechnology was not provided by all firms (see footnote 3). The highest total turnover (€3.989 million) came from the biopharmaceuticals sector, followed by agro-food biotech (€1.478 million) with equipment and supplies firms slightly behind (€1.281 million). Median turnover for each sector provides a different ranking. The highest median turnover is €1.87 million (equipment and supplies), followed by agro-food biotechnology (€1.199 million) with biopharmaceuticals firms in third place (€1.134 million).[6]

Only one-third of the biopharmaceuticals companies are involved in selling biotech products but over 50 per cent (175 firms) are offering a wide variety of specialized services including contract research, contract manufacturing, custom synthesis, development, sequencing, testing, software design and so on. There is patchy data only on companies' sales activities in the agro-food biotech sector. We have information on the main product markets of 61 firms only:[7] almost 60 per cent of these regard the domestic market as most important. Just under 20 per cent treat Europe as their main product market and the remainder focus on markets outside Europe and the USA. The majority of firms in the equipment and supplies sector sell biotech products (86 per cent) and the domestic market is particularly important for German and British firms. Dutch firms focused more on

Europe and other countries (outside Europe and the USA). In general, the domestic market was most important for 36 per cent of the firms and Europe was the main market for 32 per cent; only 16 per cent of the firms see their main product market in the USA. Since most of the patents (55 per cent) held by firms have national claims only, it is reasonable to assume that the majority of the equipment and supply firms focus mainly on the domestic market.

The sample of biotech firms included 139 subsidiaries (19 per cent of total) who provided information about the location of their parent company. As shown in Figure 5.6, a small number were subsidiaries of a parent company in the same country (16), but the majority were subsidiaries of firms in other European companies (67). There were 53 subsidiaries

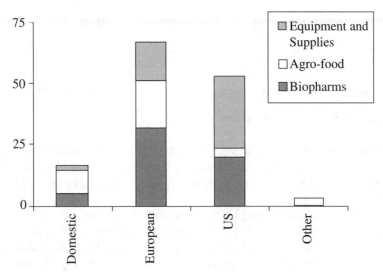

Figure 5.6   Number of subsidiaries by location of parent and sector

of US companies; 56 per cent of these were in the equipment and supplies sector, but very few (4) in agro-food. The majority of subsidiaries were located in France (29 per cent) and Germany (28 per cent); in terms of their share of all national companies, subsidiaries were most significant in Austria (32 per cent) and Greece (31 per cent).

The highest proportion of subsidiaries were in the agro-food (23 per cent) and equipment and supplies (21 per cent) sectors. They only represented 16 per cent of firms in biopharmaceuticals. Seventy per cent of firms in the agro-food sector with more than €150 million turnover were subsidiaries. It is suggested that the high turnover may be explained by the parent company acting as a market for its subsidiaries.

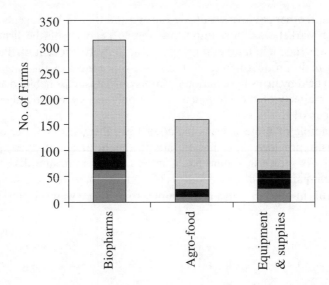

□ university spin-off ■ company spin-off □ independently established

Figure 5.7   Origin of companies by sector

In terms of their origins, the majority of firms were independently established, as shown in Figure 5.7. Industry spin-offs are negligible, and the highest proportion of university spin-offs are in the biopharmaceuticals sector.

Figure 5.8 shows the origins of companies by country. The countries with the greatest proportion of university spin-offs (just over 20 per cent) are Germany, Ireland and the UK. In Spain there are only independently established companies and, with a small number of exceptions, the same is true of France.

The majority of agro-food (96 per cent) and biopharmaceuticals (83 per cent) companies are involved in R&D collaborations; the proportion of equipment and supplies firms involved in R&D collaborations is much lower (44 per cent). In biopharmaceuticals and agro-food, collaborative partners in the public sector (PS) are more frequent than those with other firms. PS partners are mainly domestic and European. In biopharmaceuticals, firms in the USA formed the highest number of collaborative partners in industry, with domestic firms lagging slightly behind. In agro-food, R&D collaboration with the private sector focused on domestic and European partners. In equipment and supplies, we have data on R&D collaborations for Germany, The Netherlands and the UK (the countries where the sector is most developed). Firms in this sector tend to collaborate with domestic

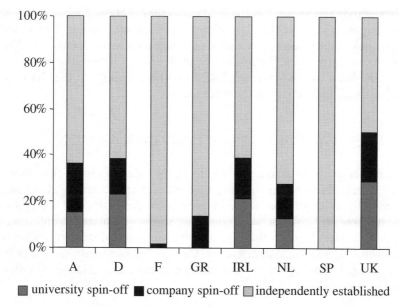

university spin-off ■ company spin-off □ independently established

Figure 5.8   Origin of companies by country

PS. On average, British and Dutch firms tend to team up more with other European firms than the German firms, which have a greater propensity to partner domestic firms.

To summarize the main points of the industry analysis, countries with the highest number of biotech firms are France, Germany and the UK, all countries with large populations. The Netherlands, one of the smaller countries, also has a significant number of firms. Austria, Greece and Spain have few firms and most were created 30–40 years ago, indicating that the majority have diversified into biotechnology. A more recent median year of foundation for Irish, French, German and UK firms suggests that the majority were created specifically to exploit biotechnology. These countries also have a high proportion of firms with less than 50 employees. France and Germany have the largest number of subsidiary companies, but they constitute the highest proportion of all biotech firms in Austria and Greece. University spin-off firms are rare, but they are most frequent in Germany, Ireland and the UK where they account for approximately 20 per cent of all biotechnology firms.

There are biopharmaceuticals companies in every country in our study. They are the most numerous (337) and they are younger than firms in the other two sectors, suggesting that they were set up specifically to exploit the potential of biotechnology. Every country also has some firms in the agro-food sector, but the number is small except in France, Germany, The

Netherlands and Spain. It is the smallest sector (162 firms), with the oldest and largest firms, and the highest proportion of subsidiaries. The equipment and supplies sector has a relatively large number of companies (224) which concentrate in a few countries only: Germany, The Netherlands and the UK. Most of the firms have less than 50 employees, but they have the highest median turnover of the three sectors.

3.   KEY FACTORS AFFECTING INNOVATION

In an attempt to understand the results of the industry survey, this section will review the key factors affecting innovation in terms of knowledge/skills, finance/industrial development and demand/social acceptability and how they differ between countries. Some of the national differences in the interfaces between these components (intellectual property, technology transfer etc.) are also discussed. Certain key factors do not differ by sector (e.g. technology transfer mechanisms), but where evidence of differences exists, this is mentioned.

3.1   The Knowledge/Skills Network

The knowledge/skills network is affected by government policies for the science base, including investment in public sector research (PSR) and postgraduate training, instruments to promote technology transfer and intellectual property regimes for public sector research.

The current skills and knowledge base in each country depends in part on the date at which national policy first focused on its development. The countries in our study can be roughly divided into two groups: the first group commissioned reports about biotechnology during the 1970s and began to introduce policies to enhance or provide a biotech research infrastructure around 1980 (France, Greece, The Netherlands and the UK). France and the UK also took steps to raise industrial awareness about the potential of biotechnology, to promote academic links with industry and to encourage the commercialization of academic research; this also occurred in The Netherlands from the mid-1990s. France, the UK and Greece also intervened in the formation of dedicated biotech firms. Two of the French start-ups (founded in 1979 and 1982) and one of the British start-ups (founded in 1980) focused on medical applications. The Greek start-up (which no longer exists) was intended to act as an intermediary between academic research and industrial application. The second group of countries (Austria, Germany, Ireland and Spain) did not introduce policies for biotechnology until the second half of the 1980s.

Collecting truly comparable data about investment in public sector research, national research manpower and postgraduate training in biotechnology proved almost impossible. Each country has a different system for organizing public sector research, and uses a different basis for assembling data on expenditure, research manpower[8] and postgraduate degrees. Even where figures are available for public research expenditure, these are underestimates in countries such as France, Germany and Spain because they do not include funds allocated by regional authorities and the block grant provided for university faculty to conduct research.

The available data indicates that the large countries[9] have numerous public sector institutes and university departments involved in research as well as numbering their public sector researchers and PhD awards in the thousands. This contrasts with lower activity in the smaller countries, which is compensated for by students who undertake postgraduate training abroad.

There is limited data on publicly funded biotech research for each sector.[10] Available data suggests that the biopharmaceuticals field commands a high proportion of the public biotechnology research budget. Non-profit organizations also make a significant contribution to this area in France, The Netherlands and the UK. A rough comparison of per capita public expenditure on biopharmaceuticals research shows Spain spends around half of the amount in the leading countries: France, Germany and the UK. Austrian and Greek expenditure is negligible.

Expenditure on agro-biotech research is only a small fraction of that allocated to the biopharmaceuticals area. The countries spending the most are France, Germany, The Netherlands and the UK. Greece, Austria and Ireland spend the least. Spain has an intermediate position with a fast-growing science base, especially in terms of the number of trained researchers.

Almost all universities in Europe with science or engineering faculties undertake research that has relevance for the equipment and supplies sector. Underpinning knowledge for this sector may arise during the course of almost any research that employs biotech techniques or concentrates on specific aspects of biotechnology. Thus it may be appropriate to suggest that almost the whole of the biotech science base in any country supports the equipment and supply sector. It is difficult, however, to estimate the number of skilled researchers or students in this area with any accuracy.

There are national differences in the approach to technology transfer and the arrangements for academic intellectual property. Technology transfer mechanisms for public sector research are very strong in the UK, quite good, but improving in Germany, and strong in Ireland. Technology transfer in France and The Netherlands used to favour large firms. In recent years the focus has shifted to encouraging the creation of new firms to exploit research results. Austria, Greece and Spain are characterized by poor links between academic research and industry. Academic entrepreneurship is inhibited by job security in every country except the UK. Some countries, notably France, have now changed rules that inhibit small firm creation by academics.

A variety of policies exist for dealing with the intellectual property (IP) that arises from academic research. The two main approaches vary both between and within countries. The main difference lies in whether IP belongs to the inventor or to the institution within which the inventor works. Academic patents are perceived to have a relationship to technology transfer but there is a lack of knowledge about the best method for promoting the application of results by industry. Where IP rests with the inventor, policies focus on overcoming academic reluctance to apply for patents. Where IP belongs to the institution, policy focuses on establishing agencies to handle IP.

3.2   Finance and Industrial Development

Most countries have a range of policies to support industrial innovation in general; sometimes these policies focus on biotechnology. Industrial policy in Greece and Austria, however, lacks any special focus on biotechnology and this was also true of Dutch industrial policy during some periods in the last 20 years.

Dutch industrial policy in the early 1980s focused on promoting companies' in-house research. In the 1990s it has focused on grants for collaborative research programmes between industry and public sector research and on stimulating the commercialization of public sector research. This mainly focused on large existing firms. From 2000 initiatives were put in place to promote the creation of start-up firms. Greek industrial policy is largely connected to technology transfer initiatives. It also funds industrial research projects undertaken by a consortium of firms or a single firm. Since 1998 Austrian industrial policy has also focused on initiatives including those to commercialize the results of public sector research, to promote knowledge transfer between public and private sector research and the creation of start-up companies.

Industrial policy in the other countries gives some emphasis to biotechnology. BioResearch Ireland (BRI) funds research in universities, supports the transfer of research results to existing industry, or to new university-based start-up companies. German industrial policy includes programmes to encourage firms to adopt new technologies and to raise industrial awareness about biotechnology. It also promotes applications-oriented research by distributing funds for collaborative research. France provides tax credits for companies conducting in-house research and the UK has a wide range of schemes to promote and fund university–industry research links with firms of all sizes, as well as encouraging the creation of start-ups. Spain funds joint research by firms and public research groups, development projects to take new products to the market and projects to help companies adopt new technology. Firms' costs are funded by credits at low (or zero) rates of interest. Biotechnology is identified as an action area, but in the period 1988–1999 it received only a tiny proportion of available funds.

The views of multinational pharmaceuticals companies contribute to shaping research programmes in France, Germany and the UK. In these countries some aspects of industrial policy focus on the biopharmaceuticals area and several programmes focus on relevant topics. France, for instance, has created a 'genopole' for genomics and gene therapy which brings together public research institutes, university teaching and small firms. The 'genopole' is intended to be a 'pole' of excellence which serves as a technological platform and promotes the creation of start-up companies.

For many years Europe had very poor availability of venture capital[11] compared with the USA (Rothwell and Zegveld 1982). The situation has changed rapidly in recent years, but in some countries it is still difficult for new biotech firms to find initial financing. For instance, although private equity investment in Spain is increasing, the majority goes to large Spanish companies. Start-ups received only 5 per cent, there was even less for seed capital (1 per cent) and under 0.02 per cent was directed towards biotechnology. The first Greek investment capital fund was set up in 1991, but it mainly invested in mature companies, or those able to produce high profits quickly. A new law was passed in 1995 to encourage private equity to increase the amount of investment in seed and start-up companies. As the venture capital firms lack expertise in biotechnology and are risk averse, they have not invested in biotech start-ups. There is also low availability of venture capital in Austria. Two venture capital firms are active in biotechnology but only a tiny proportion of available funds have been allocated to this sector. Irish venture capital firms are also reluctant to invest in the biotech sector because most companies are 'early stage' and cannot meet venture capital firms' investment criteria.

The situation is better in the UK, Germany, France and The Netherlands. The UK has one of the most favourable legal and fiscal environments for private equity in Europe with the Alternative Investment Market (AIM) and other stock exchanges providing the equity firms with a number of exit opportunities. Biotechnology receives a small proportion of available venture capital, but the share has risen over time. Tax relief is also available for private investors (business angels) who buy shares in some unquoted

trading companies. However, agro-food biotech companies have not generated enthusiasm within the investment community. Compared to other
biotech sectors, agro-food biotechnology is perceived as high risk and with
less potential premiums for its products (Crowther et al. 1999). There is an
abundance of venture capital seeking investment opportunities in Germany,
and a high interest in biopharmaceuticals firms because of the potential for
a high return, especially when a company is launched on the stock exchange.
The *Neuer Markt* in Frankfurt encourages biopharmaceutical firms to list
their stocks. Dutch venture capital is well developed and one firm specializes
in biotech investment. There are also many business angels and an agency
for introducing Dutch entrepreneurs to these informal investors. Most start-
up firms use their own resources or rely on business angels. France has
created the *Nouveau Marché,* a stock exchange for high-tech and rapidly
growing firms. There has also been a rapid growth of venture capital firms
who only invest in companies in the health sector. At the regional level
venture capital firms work together with local government agencies.

3.3   Demand and Social Acceptability

This section discusses the demand factors that affect the innovation process.
Each sector is discussed separately because these factors vary by sector.
Public procurement, for instance, is important in pharmaceuticals and standards have significance for equipment and supplies. The analysis of controversies undertaken in the three sectors to identify 'public acceptance' of the
new technology, and the public actors involved in decisions about them
found that there was no public debate in the equipment and supplies sector.
Although there was public debate in both the agro-food and biopharmaceuticals sectors, the influence of public interest non-government organizations
(PINGOs), and public acceptance of the application of biotechnology was
very different in these two sectors. Regulation is relevant to all three sectors,
but its impact also varies by sector.

3.3.1   Biopharmaceuticals
The potential market for pharmaceuticals products is affected by the size
of the country, per capita expenditure on pharmaceuticals products and the
public procurement regime. The largest markets for pharmaceuticals products are in France and Germany. Per capita expenditure[12] is highest in
France, with Austria and Germany not far behind. The Netherlands, Spain
and the UK are in an intermediate position, but per capita expenditure in
Greece and Ireland trails far behind. The market for some small countries,
notably Ireland and Austria, goes beyond the domestic market. Global
pharmaceuticals companies from the USA locate in Ireland, mainly for

manufacturing purposes, so as to access the European market. Similarly, one of the advantages of locating in Austria is the closeness to the expanding markets of Central and Eastern Europe.

Every European country uses public procurement policies to control pharmaceuticals prices, because the state is largely responsible for the costs of health care. Hard bargaining between companies and health ministries leads to considerable price variation between countries, because each country bargains in isolation and has its own regime for regulating prices. Thus, the market for pharmaceuticals products in the European Community (EC) is highly fragmented.

Germany and the UK promote innovation by pharmaceuticals companies, with Germany allowing companies freedom in setting prices for new products and the UK regulating profits, but rewarding R&D activity. Austria used to provide companies with a certain margin on their production costs for products produced inside the country, whereas international reference prices were used to regulate the price of imports. New procedures were introduced in 1999. Prices are highly regulated in the other countries; France requires approval of the manufacturer's price before a product is authorized for reimbursement. Price increases are rarely allowed and decreases may be introduced (Danzon and Chao 2000). The Netherlands, Greece, Ireland and Spain use international comparisons for setting pharmaceuticals prices. Spain pays much lower prices for pharmaceuticals products than other European countries and pricing policies in Greece disadvantage domestic companies and favour imports.

Regulation in every country mainly follows EC Directives and, since 1995, the approval of new biopharmaceuticals has rested with the European Medicines Evaluation Agency (EMEA).

Surveys have assessed public acceptance of various applications of biotechnology. There is a positive attitude to medical applications of biotechnology in most of the countries in the study. The Austrian public is the most negative towards biotechnology of all the countries in this study. However, medical applications of biotechnology have greater acceptance.

The activities of a variety of public interest non-government organizations (PINGOs) may affect the activities of the pharmaceuticals industry. The UK and The Netherlands have several organizations that review the implications of new treatments, or play a part in public debates on their wider implications. Germany's largest consumer association is concerned about transparency over genetic engineering. Another type of PINGO are patients' associations, supported by the families of patients suffering from specific conditions. Their influence on the pharmaceuticals industry differs between countries. They have very little influence on the pharmaceuticals industry in Spain and Austria, which bases its R&D decisions on international considerations.

Several medical charities in the UK and France fund significant research into specific disease areas and promote the development of new treatments; the French charities focus specifically on genetic disorders. Animal welfare organizations, concerned with the rights of animals and in reducing the number of animal trials and conditions for the animals involved, are the third type of PINGO. They are very active in the UK, The Netherlands and Germany.

3.3.2 Agro-food biotechnology

The market for seeds, pesticides and other agrochemicals or for agro-food products is not controlled *per se*, nor is there any regulation of prices. However, in each country, and at the European level, various agencies control field trials and are concerned with the impact of products on health, the environment and food security. Pesticide regulation is seen as having the potential to create markets for new pesticides, particularly as replacements for older products that are thought to damage the environment or human health (Tait et al. 2001).

Public attitudes, regulation and the response of highly concentrated European food retailers and manufacturers are now having a major impact on European demand for GMOs in crops and food. Citizens associations (Greenpeace, Green political parties and other environmental groups) have played a crucial role in widening public debate about the implications of GM crops and food, and their ideas have been widely disseminated through the media. As a result negative public attitudes to GM crops and food are now widespread, although the strength of antagonism varies from country to country. Three main groups can be discerned. The public has a very negative attitude to GM crops and ingredients in foods in Austria, Germany, Greece and the UK. All these countries have citizens and consumers associations which have taken an active part in stimulating public debate on the topic. Concentrated food retailers in Austria, Germany and the UK have responded to consumers' negative attitudes by removing GM ingredients from their products. Greek food retailers and manufacturers have also adopted this approach. In Austria added opposition to GM agriculture comes from the small scale family-run farm sector, which wants to protect its image as an organic food pioneer.

Public opposition to GM crops and foods is lower in France and The Netherlands although French media interest in the GM debate has increased since 1997. There has been growing public distrust in genetic modification and a great deal of uncertainty about GMO regulations resulting partly from a 'lack of decision-making' by politicians in this domain. Several consumer associations have organized boycotts of products containing GMOs, and the pressure on the concentrated food retailing sector has forced them to avoid GMO food. Public opposition to GMOs, however, does not yet seem to have affected field trials in France or the commercial activities of agro-food biotech firms. This may be explained by French industrial attitudes in the chemical sector which have been described as firms reacting to government plans to gain the best advantage, rather than as taking their own initiatives (Brickman et al. 1985).

The Dutch have developed a tradition for seeking public opinion and this is taken into account by policy-makers. Dutch attitudes towards biotechnology have not changed markedly over time, but the public is well informed and understands both the positive and negative aspects of the technology. Thus, despite negative publicity about genetic modification, the Dutch acknowledge the possible advantages of biotechnology and they judge the risks involved as being acceptable; they remain somewhat negative about applications to food. Nevertheless, the use of GM soy in food products was recently approved by the Dutch government. There is less public opposition to GM food and crops in Spain than in the other countries. The agricultural tradition, the relevance of this sector and limited public reactions against plant biotechnology in Spain have led to agribusiness and public research centres conducting field trials. The Irish public are relatively optimistic about modern biotechnology in general but GM food seems to attract little public support.

The European regulatory framework adopted in the early 1990 (Directives 90/219 and 90/220 on the contained and deliberate release of GMOs) and subsequent modifications to Directive 90/220 on mandatory labelling for GMOs have played an important, if largely negative role, in shaping developments. Gaskell et al. (2000) suggest that regulation and public opinion coevolved in Europe. In the European Union (EU), increased regulatory oversight coincided with growing negative public opinion about agro-biotechnology and diminishing trust in public authorities and regulatory agencies. In this environment, food retailers and manufacturers introduced voluntary standards and labels relevant for their markets in an attempt to gain market share by meeting perceived public demands. With almost no exceptions, voluntary standards have been set with zero (or almost zero) tolerance for products, generating 'no-GMO' or 'GMO-free' claims. Zero tolerance standards have led to reformulation of processed foods to remove biotech products or their derivatives and to identifiable supply chains which ensure the absence of such products. Thus, voluntary GMO-free standards have quickly become the standard, making other standards and regulatory mechanisms irrelevant.

The demand for notification of deliberate field trials under Part B of Directive 90/220/EEC provides information on the extent of field trials which have taken place in EC countries since 2001, as shown in Table 5.2 (countries not covered by this study are shown in italics). This shows that

Table 5.2   Number of field trials in EC countries 1991–2001

Country	No.	%
Austria	3	0.2
Belgium	*110*	*6.9*
Denmark	*39*	*2.4*
Finland	*17*	*1.1*
France	484	30.2
Germany	109	6.8
Greece	19	1.2
Ireland	4	0.3
Italy	*262*	*16.3*
The Netherlands	113	7.0
Portugal	*12*	*0.7*
Spain	167	10.4
Sweden	*61*	*3.8*
UK	203	12.7
Total for the European Union	1603	100

Note: Countries shown in italics did not form part of the study reported in this chapter.

Source: JRC website of deliberate releases at: http://biotech.jrc.it/gmo.asp, accessed July 2001. Data is now found at http://biotech.jrc.it/ deliberate/gmo.asp.

France has conducted one-third of all EU field trials. The other countries in our study with a significant number of field trials are The Netherlands, Spain and the UK.

3.3.3   Equipment and supplies

There is no publicly available statistical information for the biotech research equipment and supplies sector. The lack of information about market size, growth rates and so on makes it difficult to make general statements about the development of this sector. Our study, however, gives strong indications that the market for biotech equipment and supplies within a country is influenced by the amount of relevant research undertaken in the public and private sectors. During the late 1990s, public sector research funding agencies in Germany, The Netherlands and the UK gave biotechnology a high priority thus creating a favourable environment for biotechnology equipment and supplies companies. The rapid growth of dedicated biotech companies in these countries, and the R&D activities of multinational companies using biotech tools and techniques (in agrochemicals, food and pharmaceuticals) increased the market for biotech equipment and supplies. Conversely, the low number of firms in the equipment and supplies sector in Austria, Greece, Ireland and Spain may be explained by the small size of the market, both public and private. Funding for public sector research is limited, and few companies are involved in biotech R&D. Most equipment may be imported and domestic firms may be primarily trading organizations which are not involved in manufacturing.

The low number of equipment and supplies firms in France is rather surprising, given the potential demand provided by high investment in public sector biotech research. There are several possible explanations: the lack of a French tradition in instrumentation development and limited demand from the private sector. First, dedicated biotech firms were slow to develop in France and second, much R&D activity is in subsidiaries of Swiss multinational companies in France. Their equipment and supplies may be sourced from Switzerland.

Compared to the other two fields (biopharmaceuticals and agro-food) the equipment and supplies industry appears to be only slightly affected by numerous European and national laws and regulations which could influence the development of the industry. However, certain regulations may provide the foundation for the business of some companies. One example would be firms that sell equipment to detect GMOs in food or the environment.

Standardization and certification is a major issue for equipment and supplies products. The complexity and systems character of modern biotechnology has led to an increasing number of standards in the industry. There are around 500 standards in Germany alone with a potential impact on product and technology development for companies in this sector. The main benefit of introducing standards into the biotech equipment and supplies sector would be to ensure complementarity between elements within a system. The optimal interface allows for multiple component designs to coexist but, according to industry experts, this is quite difficult in biotechnology. In consequence, it is likely that users will favour systems provided by one supplier rather than a conglomerate of components optimized in isolation.

4. NATIONAL STRENGTHS AND WEAKNESSES BY SECTOR

4.1 Biopharmaceuticals

The knowledge/skills network in France, Germany and the UK far outstrip the other countries in terms of funds allocated to public sector research, the number of university departments and research institutes carrying out research, and in the numbers of postgraduate students being trained. The Netherlands and Spain have broadly comparable, but more modest, achievements in these areas, but the difference in population size between these two countries suggest that per capita expenditure on biotech research is probably higher in The Netherlands than in Spain. The other countries allocate significantly less funds to relevant research and research training.

The industry/supply network appears to be strongest in France, Germany and the UK. These countries are home to multinational pharmaceutical companies and the vast majority of small to medium-sized enterprises (SMEs). These three countries and The Netherlands have attracted foreign pharmaceuticals multinationals to establish research-active subsidiaries and also have strong business interest non-government organizations (BINGOs) to represent the interests of the business community. Subsidiaries of multi-national pharmaceuticals companies involved in R&D can also boost national activity. For instance, they account for most industrial biotech research in Austria and Spain.

The importance of national PSR partners in R&D collaborations with SMEs suggests that countries with a strong knowledge base (the UK, France and Germany) help their SMEs to thrive. However, the importance of the USA for SMEs' private collaborations emphasize the global nature of this sector.

The size of the market and the regimes followed for procuring pharmaceuticals by national health care systems clearly influence the activities of companies in the larger countries. Countries with strong control of pharmaceuticals prices, particularly The Netherlands, Spain and Greece, seem to inhibit the development of national pharmaceuticals companies. Among the smaller countries Austria appears to have attracted inward investment by its previous policy of providing an agreed margin for products manufactured inside Austria. Austria, like Ireland, has a rather strong pharmaceuticals sector based around the activities of foreign subsidiaries. Both countries serve as good entry points to larger markets: Ireland to the EU and Austria to Central and Eastern Europe.

In terms of the factors affecting demand and social acceptability, there is no difference between countries in terms of regulation. The common approach to regulation mainly follows EC Directives and, for the approval of new biopharmaceuticals, rests with the European Medicines Evaluation Agency. Nor could any differences be discerned between countries in terms of the acceptance of biopharmaceuticals.

There has been a long-standing and strong technology policy to support the development of biotech firms in the UK, Germany and Ireland, including mechanisms to encourage technology transfer. Until recently, policy to promote biotech innovation or technology transfer was weak in France, The Netherlands, Austria, Greece and Spain. France and The Netherlands are now doing more to support the creation of small firms. With the exception of Greece, technology policy has also improved in the other countries. Availability of finance capital can affect the creation of start-up firms. Conditions are favourable in the UK, Germany, France and The Netherlands but poor in Austria, Ireland, Greece and Spain.

To sum up: of the leading countries, the UK and Germany have an environment where most factors are supportive or strongly supportive of innovation in biopharmaceuticals and none impede the development of the sector. Most factors in France provide some support to the sector, but there are also several areas which have so far acted as a brake on innovation, although recent policy is now attempting to address these handicaps. Among the other countries, The Netherlands has a mixture of both positive and negative influences on innovation. The negative effect of the small market size is compounded by low expenditure on pharmaceuticals products and this seems to outweigh many positive factors which could encourage pharmaceutical innovation. Austria and Ireland, by contrast, have only a few positive factors. Demand-side factors, particularly the opportunity to provide a launch-pad for access to adjacent markets, however, seems to allow these two countries to play a bigger role in biopharmaceuticals innovation than might otherwise be expected. Spain has certain strengths in knowledge and skills and engages in public–private R&D cooperation. However, the potential for biopharmaceuticals innovation is limited by widespread weaknesses, especially strong control of pharmaceuticals prices. Greece invests in scientific education, but most of the other factors affecting innovation suffer from weakness and impede national innovation. This analysis appears to suggest that though all the factors may have a role to play in contributing to innovation, their significance in the process varies. More importantly, it indicates that demand is one of the strongest promoters of innovation.

4.2 Agro-food biotech

The leading countries in public agro-food research are Germany, France and The Netherlands. Their investments in basic science in the area mainly focus on plant biotechnology. The UK and Ireland also invest in plant biotechnology and plant science but these investments do not generate private investment in agro-food biotechnology. Austria, Spain and Greece are building a scientific base, which may or may not be dedicated to agro-biotechnology. Spain has an improving position in food biotechnology.

The industry/supply network is strongest in Germany, France and The Netherlands, partly due to the activities of large, domestic multinational companies and their subsidiaries. Technology transfer and small business creation was neglected until recently and it now receives increased public policy emphasis. In terms of SMEs, the leading countries are France, The Netherlands and Spain, with Germany in fourth place. Agro-food SMEs dominate Greece's very small number of biotech firms, but there are only a tiny number in the UK. Technology policy is little different to that for bio-pharmaceuticals but there is some indication that venture capital firms are loath to invest in firms in this sector, because of the strength of public opposition.

There is a low demand for, and public acceptance of, GM crops and food, although the strength of public opposition varies from country to country. Despite the EC providing a common regulatory framework, national agencies have sometimes adopted a more stringent approach at the detailed level. Thus there is fragmented regulation and competition between European and national agencies to promote the precautionary principle or other ethical standards. In addition, food retailers and manufacturing have played a major role in eliminating these products from the products they stock and introduced *de facto* regulation by introducing 'zero tolerance' to GM ingredients.

It is very difficult to interpret data on field trials for GM crops, especially since they date back to 1991, when public opposition to GM crops scarcely existed. The concentration of these trials in France, however, confirms that public opposition is lower there than in the other countries in our study. There may also be less public opposition to GMOs in Spain and The Netherlands since a significant percentage of trials also take place in these two countries. The strong agricultural traditions of France, The Netherlands and Spain, together with public attitudes, may explain why these countries also have the most SMEs in the sector.

The results for the UK are difficult to understand. It has a strong science base in the area and there is a national emphasis on commercializing that science base. Although there have been a significant number of field trials, there are very few SMEs. The campaigns of PINGOs, reinforced by media coverage and the response of concentrated food retailers appear to have created an environment where venture capital is loath to invest in these firms. Alternatively those companies which are involved are not prepared to admit that they are active in the area. Another possible explanation is related to the importance of subsidiaries in this sector. Multinationals may choose to locate subsidiaries in countries where public acceptance is higher than in the UK. This hypothesis is partly borne out by data about the countries which appear to have the highest number of field trials for GMO crops.

To sum up, the development of the agro-food sector faces considerable barriers, notably a lack of demand connected to public resistance to GM crops or foods. The countries best placed to develop their competences in the area are France and The Netherlands, based on their science base, the presence of influential multinational companies and less vociferous public opposition to GMOs than in other countries. Spain's fast-growing science base and relative lack of public opposition to GMOs gives it the potential to develop national strength. The main brake on the development of agro-food biotechnology, however, is the weakness of private investment in R&D, together with non-availability of venture capital to support the formation of small firms. Germany has a large number of domestic agro-chemical multinationals; they may choose to use the knowledge developed in the public sector for applications and field trials in other parts of the world where there is less public hostility.

4.3   Equipment and Supplies

The market for the equipment and supplies sector is stable and robust because its products are used in a number of industries and by a wide range of PS organizations and institutions. Unlike other sectors, companies do not appear to have to cope with negative public attitudes to their work. However, the proliferation of standards throughout Europe may hinder the long-term development of the sector.

Although there are some differences, Germany, The Netherlands and the UK form a cluster of countries with a well-developed research equipment and supplies sector. These countries maintain numerous institutions devoted to scientific research and education, have a high number of scientists per capita and a large pool of academics engaged in advanced scientific research. Kenney (1986) considers that a key success factor for a national biotech industry is a highly qualified pool of researchers and this will be more readily available in countries with a strong tradition of scientific research and education. The scope of basic research funding may also serve as an important demand factor for the equipment and supplies industry. There is also considerable investment in research by various multinational chemicals and pharmaceuticals companies in these countries, as well as a growing population of new biotech firms. All these national research actors provide both a market for equipment and supplies and may also stimulate the development of new generations of products (as suggested by Irvine 1991). The domestic market in Germany and the UK is large enough to induce further growth of the sector. The businesses in The Netherlands have strong links to firms in other countries which may compensate for the relatively small domestic market. The strong venture capital markets in these countries nurture the foundation and growth of small start-up companies, mostly established independently. The three countries deviate most in their approach towards public–private and inter-firm collaborations. Whereas firms in The Netherlands strongly emphasize these partnerships, they are less popular in Germany and the UK.

The other cluster comprises Austria, France, Greece, Ireland and Spain. Except for France, these countries allocate low funds to public sector research funding and hamper the development of a large pool of creative scientists. In addition, Austria, Spain, Greece and Ireland have few biotech firms and though Austria, France and Spain host R&D-performing subsidiaries of multinational pharmaceuticals and/or chemicals companies, these companies may source their equipment and supplies from their home countries. In France, academic and industrial research communities may be healthy but they are poorly linked, and institutional mechanisms have failed to exploit these strengths. The weak cluster of countries does not have much of a tradition in engineering or in the development of instrumentation; its academic researchers have a low commercial orientation, and availability of venture capital is poor. As with the other two sectors, demand is an important factor affecting innovation, but as demand in this sector is linked to the strength of the science base, the knowledge/skills network also plays a part. Table 5.3 summarizes the main strengths and weaknesses affecting innovation in the eight countries, by sector.

5. CONCLUSIONS

This chapter has helped in 'setting the stage', by providing an empirical insight into how and why national characteristics and sectoral patterns affect the development of modern biotechnology in Europe. The results confirm the usefulness of using an integrated framework for the study that included contributions from several bodies of theory. Elements of NSI theory are helpful in clarifying why the commercial development of biotechnology is more advanced in some countries than in others, but are not able to throw light on the sectoral pattern of development within a country. This may be partly explained by the existing national structure of production (industrial path dependence). For instance, the strength of France, Germany and the UK in biopharmaceuticals is partly related to existing national strength in the pharmaceuticals sector.

The results of the study show that companies' propensity to innovate is affected not only by supply conditions and the national structure of production. Potential risks, assessed in terms of demand and public acceptance for various applications have been shown to have a significant impact on the national pattern of innovation. This is shown in Table 5.4, summarizing the main differences in the framework conditions for innovation in the

Table 5.3 National innovation strengths and weaknesses by sector

	Biopharmaceuticals	Agro-food	Equipment and supplies
Austria			
Tech policy, etc.	*Policy for biotech innovation and technology transfer recent. No policy for small firm creation. Poor availability of finance capital*		
Knowledge/skills	*Limited*	*Negligible*	*Negligible*
Industry/supply	Modest: Foreign research-active MNCs, few SMEs	*Negligible*	*Negligible*
Demand	**Aided by access to central and E. European markets. Pricing policy favoured local production**	*Strong public opposition incl. important organic farm sector opposition*	*Negligible*
France			
Tech policy, etc.	*Policy for biotech innovation, technology transfer and small firm creation recent. Good availability of finance capital*		
Knowledge/skills	**Strong**	**Good**	**Strong**
Industry/supply	**Strong: domestic/foreign MNCs, many SMEs**	**Strong related MNCs, many SMEs**	Some firms
Demand	**High per capita pharms expenditure** + *control of pharms prices*	Some public opposition to GMOs in food/crops	**Significant pub. and priv. biotech research**

Typographic key: **Strength**   *Weakness*   Fair

125

Table 5.3   (continued)

	Biopharmaceuticals	Agro-food	Equipment and supplies
Germany			
Tech policy, etc.	**Long-standing policy to promote biotech innovation, technology transfer and small firm creation**		
Knowledge/skills	**Strong**	**Moderate**	**Strong**
Industry/supply	**Strong: domestic/foreign MNCs, many SMEs**	**Strong related MNCs, some SMEs**	**Strong**
Demand	**Pharms pricing policy promotes innovation**	*Strong public opposition to GMOs in food/crops*	**Significant pub. and priv. biotech research**
Greece			
Tech policy, etc.	*Limited policy for biotech innovation and technology transfer. No policy for small firm creation. Poor availability of finance capital*		
Knowledge/skills	*Limited*	*Negligible*	*Negligible*
Industry/supply	*Non-existent*	*A few firms*	*None*
Demand	*Strong control of pharms. prices*	*Recent public opposition to GM foods*	*Negligible*
Ireland			
Tech policy, etc.	**Long-standing policy to promote biotech innovation, technology transfer and small firm creation.** *Poor availability of finance capital*		
Knowledge/skills	*Limited*	*Negligible*	*Limited*
Industry/supply	*1 domestic MNC, US mfg subsidiaries, some SMEs*	*A few SMEs*	*A few SMEs*
Demand	**Access to EU Market**	*Little public support*	*Limited*

The Netherlands

Tech policy, etc	*Policy for biotech innovation, technology transfer and small firm creation recent*		**Good availability of finance capital**
Knowledge/skills	Moderate		**Good**
Industry/supply	Foreign research-active MNCs, some SMEs	**Strong related MNCs, many SMEs**	**Large firms and SMEs**
Demand	*Strong control of pharms. prices*	Some public opposition to GMOs in food/crops	**Significant pub. and priv. biotech research + established export links**

Spain

Tech policy, etc	*Policy for biotech innovation and technology transfer recent.*	*No policy for small firm creation. Poor availability of finance capital*	
Knowledge/skills	Moderate	Growing activity	Moderate
Industry/supply	Foreign research-active MNCs, some SMEs	Moderate number of firms	*A few firms*
Demand	*Strong control of pharms. prices*	Less public opposition than the other countries	*Limited*

UK

Tech policy, etc	**Long-standing policy to promote biotech innovation, technology transfer and small firm creation. Good availability of finance capital**		
Knowledge/skills	**Strong**	Good	**Significant**
Industry/supply	**Strong: domestic/foreign MNCs, many SMEs**	*Very few firms*	**Many firms of all sizes**
Demand	**Pharms. pricing policy promotes innovation**	*Strong public opposition to GMOs in food/crops*	**Significant pub. and priv. biotech research**

Typographic key: **Strength** *Weakness* Fair

Table 5.4 Characteristics of framework conditions for innovation by sector

	Biopharmaceuticals	Agro-food	Equipment and supplies
Knowledge/Skills	• Expertise in every country • Major focus of public research funding	• Higher priority for public sector research in Spain and Ireland • Draws on wide science base	• Neglected by public research funding • No specific science base
Industry/Supply	• Commercial activity in all countries • High share of new start-ups • Medium risks and high opportunities for new business creation	• Poor (or hidden) commercial activity in many countries • Diversification rather than new start-ups • Very high risks and limited opportunities for new business creation	• Commercial activity concentrated in countries with large science base and strong pharmaceuticals or chemical MNCs • Diversification and new start-ups • Low risks and high opportunities for new business creation
Demand/Social Acceptance	• High potential demand • High social acceptance	• Unknown customers • Weak demand and exploitation • Strong social opposition	• High actual demand • High potential demand • Demand related to science policy • Not an issue in public debate

three sectors (excluding factors connected to finance and industrial development, as they are largely the same for all three sectors within a given country).

There is a high social acceptance for biopharmaceuticals and strong public opposition to agro-food biotechnology. The equipment and supplies sector does not appear to be affected by public opinion in any way. The nature of demand and the market also varies very strongly by sector, and may be influenced by public attitudes. The market for agro-food biotechnology is characterized by close links between producers, distributors and final consumers (the food chain). Demand has been affected by a combination of public opposition to GMOs, media coverage and the response of concentrated food manufacturers and retailers and led to relatively low [overt] innovation activity by firms. Some entrepreneurs have turned public opposition to their advantage by developing diagnostic kits able to identify GMOs. The market for GM seeds is more global than the food sector, and European MNCs in the agrochemical/seeds sector are pursuing their activities overseas. Demand and innovation activity in the equipment and supplies sector concentrates in those countries where a healthy market is guaranteed by high investment in public and private sector research. National demand for biopharmaceuticals is affected by the size of the country, per capita expenditure on pharmaceutical products and the public procurement regime. Innovation clusters in countries with large markets and procurement policies which guarantee certain profits or in smaller countries which give access to adjacent markets. However, this sector operates at the global level and though innovation may be supported by national demand characteristics, it is also driven by the potential of the global market.

Table 5.4 also emphasizes the importance to innovation of knowledge and skills in terms of the R&D system, the role of the public sector including public policy, and the national education and training system. Some of the difference in national innovation performance in biotechnology is related to public policy for developing the science base (including the date when policy was introduced), but policy also had to be associated with mechanisms to link the science base with industry and to overcome cultural traditions in universities which acted as a barrier to small firm creation.

The results of the study also indicate that while all the framework conditions in the integrated framework are necessary to encourage biotechnology innovation, they do not have an equal impact. Demand and social acceptance together with a well-funded science base appear to be the principal factors which prevail over all the others. National studies in other technologies will reveal whether these conditions have similar impacts on sectoral patterns of innovation in other technologies.

ACKNOWLEDGEMENTS

The author acknowledges support from the EC's Targeted Socio-Economic Research Programme, which funded the project on which this chapter is based through contract SOE1-CT98-1117. The chapter also draws extensively on the work of her partners: Yannis Caloghirou, Stella Zambarloukos and Frangisko Kolisis, Laboratory of Industrial and Energy Economics and Laboratory of Biotechnology, NTUA, Athens, Greece; Christien Enzing and Sander Kern, TNO-STB, Delft, The Netherlands; Vincent Mangematin, INRA, University Pierre Mendès France, Grenoble, France; Renate Martinsen, Institute for Advanced Studies, Vienna, Austria; Emilio Muñoz, Victor Diaz and Juan Espinosa de los Monteros, IESA/CSIC, Madrid, Spain; Seamus O'Hara and Kevin Burke, BioResearch Ireland, Dublin, Ireland; and Thomas Reiss and Stefan Wörner, Fraunhofer Institut für Systems und Innovation Research, Karlsruhe, Germany and many others who contributed to the work.

NOTES

1. European Commission TSER project, 'European biotechnology innovation system', final and other reports available at website www.sussex.ac.uk/spru/biotechnology/ebis/.
2. This includes both biotechnology applied to crop improvement (agro-biotechnology) and to process and product safety in food (food biotech).
3. For a complete review of relevant literature see Senker et al. (1999).
4. It is difficult to identify relevant financial indicators to assess market impact in the biotech sector, because of the length of time needed to take products to market. Most biotech firms do not have products on the market. The only possible indicators to use are number of employees, growth, R&D expenditure and change over time, but these indicators are not relevant to new start-up firms.
5. Data on some aspects is poor. For instance some firms did not provide information on turnover. We do not know whether this reflects lack of turnover or commercial confidentiality. Some gave the amount of turnover, but not the proportion contributed by biotechnology. Others told us the proportion of turnover contributed by biotechnology, but not the amount. Primary information on sector, date of establishment, origin of firm is more comprehensive than other details. The results of the analysis of the database should therefore be regarded as indicative only, and treated with caution.
6. The results are based on 70 per cent of agro-food biotech firms, 60 per cent of equipment and supplies firms and 58 per cent of biopharmaceuticals firms.
7. It appears that approximately 26 per cent of the remaining firms are not yet selling products and we have no information about the remainder.
8. Note for instance that France collects information for 'life sciences' only and does not separately gather statistics on biotechnology. Germany made one attempt only, in 1992, to gather official biotech statistics.
9. In terms of population.
10. For instance, expenditure on public sector research in The Netherlands is not classified in a manner that allows the identification of the amount allocated to biopharmaceuticals-related research.
11. Venture capital companies supply risk capital to new companies set up to exploit scientific research.

12. Based on sales of pharmaceuticals to 'end-consumers' (i.e. patients through pharmacies or other distribution channels) in each country. The data is drawn from a variety of sources, which are not strictly comparable and should be regarded as indicative only.

REFERENCES

Bartholomew, S. (1997), 'National systems of biotechnology innovation: complex interdependence in the global system', *Journal of International Business Studies*, **28**(2), 241–66.

Brickman, R., S. Jasanoff and T. Ilgen (1985), *Controlling Chemicals: The Politics of Regulation in Europe and the United States*, Ithaca, NY: Cornell University Press.

Callon, M., J. Law and A. Rip (1986), *Mapping the Dynamics of Science and Technology*, London: Macmillan.

Carlsson B. (ed.) (1997), *Technological Systems and Industrial Dynamics*, Boston: Kluwer Academic Publishers.

Crowther, S., M. Hopkins, P. Martin, E. Millstone, M. Sharp and P. van Zwanenberg (1999), 'Benchmarking innovation in modern biotechnology', report for the European Commission, DG III – Industry, Brighton: SPRU, University of Sussex.

Danzon, P.M. and L. Chao (2000), 'Cross-national price differences for pharmaceuticals: how large, and why?', *Journal of Health Economics*, **19**(2), 159–95.

Edquist, C. (ed.) (1997), *Systems of Innovation: Technology, Institutions and Organisation*, London: Pinter Publishers.

Edquist C. (ed.) (1995), *Technological Systems and Economic Performance: The Case of Factory Automation*, Boston: Kluwer Academic Publishers.

Enzing, C., J.N. Benedictus, E. Engelen-Smeets et al. (1999), *Inventory of Public Biotechnology R&D Programmes in Europe,* 3 Vols, Luxembourg: Office for Official Publications of the European Communities.

Gaskell, G., N. Allum, M. Bauer et al. (2000), 'Biotechnology and the European public', *Nature Biotechnology*, **18**(9), 935–8.

Gaskell, G., M.W. Bauer and J. Durant (1998), 'Public perceptions of biotechnology in 1996, Eurobarometer 46.1', in J. Durant, M.W. Bauer and G. Gaskell (eds), *Biotechnology in the Public Sphere: A European Sourcebook*, London: Science Museum.

Giesecke, S. (2000), 'The contrasting role of government in the development of biotechnology industry in the US and Germany', *Research Policy*, **29**(2), 205–23.

Hakansson, H. (1987), *Industrial Technological Development: a Network Approach*, London: Croom Helm.

Henderson, R., L. Orsenigo and G. Pisano (1999), 'The pharmaceutical industry and the revolution in molecular biology: interactions among scientific, institutional and organizational change', in D. Mowery and R. Nelson (eds), *Sources of Industrial Leadership: Studies of Seven Industries*, Cambridge: Cambridge University Press, pp. 267–311.

Irvine, J. (1991), 'Government policies for promoting inovation in scientific instruments: innovation in the Australian scientific and medical instruments industry', papers presented at the third meeting of the Prime Minister's Science Council, Canberra: Australian Government Publishing Service.

Kenney, M. (1986), *Biotechnology: The University-industrial Complex*, New Haven, Conn: Yale University Press.

Kivenen, O. and J. Varelius (2003), 'The emerging field of biotechnology: the case of Finland', *Science, Technology & Human Values*, **28**(1), 141–61.

Knorr-Cetina, K. D. (1982), 'Scientific communities or transepistemic arenas of research? A critique of quasi-economic models of science', *Social Studies of Science*, **12**, 101–30,

Latour, B. (1989), *Science in Action*, Milton Keynes: Open University Press.

Lundvall, B.-Å. (ed.) (1992), *National Systems of Innovation: Towards a Theory of Innovation and Interactive Learning*, London: Pinter Publishers.

Lundgren, A. (1995), *Technological Innovation and Network Evolution*, London: Routledge.

Malerba, F. and L. Orsenigo (2002), 'Innovation and market structure in the dynamics of the pharmaceutical industry and biotechnology: towards a history-friendly model', *Industrial and Corporate Change*, **11**(4), 667–703.

Nelson, R. R. (ed.) (1993), *National Innovation Systems: A Comparative Analysis*, New York: Oxford University Press.

Rothwell, R. and W. Zegveld (1982), *Innovation and the Small and Medium-sized Firm*, London: Pinter Publishers.

Saviotti, P. (1998), 'Industrial structure and the dynamics of knowledge generation in biotechnology', in J. Senker (ed.), *Biotechnology and Competitive Advantage*, Cheltenham, UK and Lyme, USA: Edward Elgar, pp. 19–43.

Senker, J., O. Marsili, S. Wörner, T. Reiss, V. Mangematin, C. Enzing and S. Kern (1999), 'Literature review for European biotechnology innovation systems', Brighton: SPRU, University of Sussex.

Tait, J., J. Chataway and D.Wield (2001), 'PITA project: policy influences on technology for agriculture: chemicals, biotechnology and seeds, summary report, SUPRA paper no. 22, TSER project PL 97/1280, Edinburgh: SUPRA.

Walsh, V. (2002), 'Biotechnology and the UK 2000–05: globalization and innovation', *New Genetics and Society*, **21**(2), 149–76.

PART III

Challenging the existing

6. Risk management and the commercialization of human genetic testing in the UK

Michael M. Hopkins and Paul Nightingale

1. INTRODUCTION

While biotechnology offers opportunities for economic development and improved health, it also raises public concerns and fears about its potential risks (see Stirling 2000; Berkhout 2002; Martin et al. 2000). The cases of Monsanto with genetically modified (GM) foods and the UK government with BSE and the MMR vaccine show that organizations that fail to manage these risks and address public fears can suffer serious commercial and political consequences (for example, see Martinson 1999).

Although there is huge diversity within Europe, this concern tends to be more prevalent among Europeans than among their counterparts in the USA (Priest 2000). This is due, in part, to different cultural and political traditions. Industrial lobbies do not have the same power in the European political system as in Washington, and there is a strong environmentally focused 'green' tradition that stresses 'precautionary approaches' to the introduction of new technologies. Consequently, the USA has adopted a largely liberal, *caveat emptores* approach to the introduction of genetics technologies, although this may change as Americans have become increasingly concerned about environmental and social consequences (Priest 2000). European states, on the other hand, particularly after the failure of traditional risk management approaches to address public concern over BSE, have moved towards more participative precautionary approaches. This change in public policy during the 1990s raises questions about how firms can manage the introduction of contentious technologies in environments where the public and policy-makers are concerned about risk.

This chapter attempts to answer some of those questions. It argues that successful exploitation requires different types of risk to be managed by a range of diverse organizations with diverse risk management capabilities.

The explanation is illustrated by exploring how four firms, providing genetic testing services, strategically used charities or the state as complementary assets (see Teece 1986) to disappropriate some of the risks they encountered operating in a market characterized by public and government concern about the social implications of rapid, unpredictable technical change (see Kitcher 1996).

Section 2 briefly outlines the methodological approach for this study while Section 3 explores the literature on risk management and shows how different types of risk can be managed by different kinds of organization. It is followed by a description of human genetic testing in Section 4 and the ethical and social issues raised by such services in Section 5. The context of genetic testing services in the UK is set out in Section 6 and case studies are presented in Section 7. The discussion and conclusion follow in Section 8.

2.   METHODOLOGY AND WEAKNESSES OF DATA

This study analysed the British sector because it has one of the most developed public sector testing services (Harris and Reid 1997), significant private sector activity and a highly contested regulatory environment.[1] The sector is both well established, with public services for rare genetic diseases being delivered since the late 1940s, and innovative with public and private organizations recently offering new and potentially controversial services in novel ways. This provides an opportunity to explore the dynamics of an emerging 'private' sector before problems are resolved and 'black boxed' (Latour 1987, p. 258).

The data was collected as part of a larger project, primarily through 45 semi-structured interviews with companies, NHS clinical scientists and physicians. Interview data was triangulated with information from scientific journals, trade press, newspapers and Security and Exchange Commission (SEC) filings. Statistical analysis was not undertaken due to the small number of firms in the British market, but this small size allowed a reasonably representative analysis of the sector through cases.

Case studies, like all research methodologies, have their weaknesses and the empirical findings are clearly specific to the particular technologies, markets, organizational environments (i.e. the British healthcare system), legislative environment, cultural 'milieu' and contingent strategies that the firms take. The analysis excludes paternity and pharmacogenetic testing and is therefore not comprehensive. The methodology does not, however, rely on generalization of data. Instead, the case studies are used to illustrate and provide context to an 'explanation'. The explanation is based on Teece's (1986) well-established theoretical framework, which is extended to include the work of Moss (2002) and Stirling (2000) and consequently may be generalized to similar contexts.

3. THE STRATEGIC MANAGEMENT OF RISK: MATCHING RISKS TO ORGANIZATIONS

This section explores different types of risk, matches them to appropriate risk management techniques, and examines the organizational capabilities markets, firms and the state require to perform them.

Risk is traditionally understood in terms of its role in 'taming chance' by quantifying and controlling uncertainty (Hacking 1975; Bernstein 1996). It is typically exemplified by a game of dice, where both the likelihoods and magnitudes of losses are well known and defined independently of any subjective framing assumptions. This model and its associated risk management tools (such as cost–benefit analysis) were increasingly criticized in the 1990s following a number of high profile failures, such as underestimating the costs of nuclear reprocessing, and because the public, at least in Europe, no longer trusts the results.

Recent social science, particularly within the fields of environmental policy (Wynne 1992; Stirling 1997) and social choice theory (Arrow 1963, 1974), suggests that the European public is probably right to be dubious about the ability of such a simple model to explain the risks associated with complex technology (Stirling 1998; The Royal Society 1992; see Hayek 1945).

The academic criticisms have focused, first, on the problems of comparing 'apples and oranges' along a single scalar index when dealing with diverse consequences of actions. There is no pre-social, objective way of ranking the relative values of economic growth, human lives or environmental damage when comparing them, particularly when the positive and negative results fall on different sections of the community (Stirling 1998). Second, the process of risk assessment involves selecting a particular outcome as the likely one, but predicting the future is notoriously subjective. The future is 'complex, continuous and often contested' and selecting which one of many predictions about the future will be used for risk assessment involves implicit social choices (Stirling 2000).

Perhaps, most importantly, the implicit framing assumptions (Tversky and Kahneman 1983) underlying the choices about what to include in risk assessments have a major impact on their outcome. Stirling (1993), in a now classic study, explored the risk assessments for a range of energy technologies and found that while the outcomes were presented with great precision, there were order of magnitude differences between the results obtained using different approaches. Moreover, using the assumptions within the published literature, it was possible to rank the different technologies in any order, clearly showing that the framing assumptions rather than the risk assessment process determine the outcomes (Stirling 1993, 2000).

Social scientists now distinguish between different types of risk depending on the soundness of their epistemological foundations (see Table 6.1).

Table 6.1   The different types of risk

		Knowledge about outcomes	
		Well defined	Poorly defined
Knowledge about likelihoods	Firm basis for probabilities	Probabilistic risk	Ambiguity
	Weak basis for probabilities	Uncertainty	Ignorance

Source: Stirling (2002).

The idea that there are different types of risk is not new. Both Knight (1921) and Keynes (1937) made distinctions between risk – where probabilities are known – and uncertainty – where they are unknown; and recent research has distinguished between situations where the potential effects are either well understood or impossible to ascertain (Loasby 1976; Dosi and Egidi 1987; Wynne 1992).

Based on these ideas Stirling (2000) produced a taxonomy of four ideal types of risk characterized by the extent of knowledge about both 'likelihoods' and the magnitude of 'outcomes', (set out in Table 6.1). These correspond to 'probabilistic risk' – where both likelihoods and outcomes are reasonably well understood; 'ambiguity' – where likelihoods are well understood but outcomes are poorly defined; 'uncertainty' – where likelihoods are unknown, but outcomes are understood; and 'ignorance' – where there is little understanding of either the likelihood or impact of a policy choice (Adam and VanLoon 2000; Stirling 2000).[2] Consequently in this chapter, the word 'risk' is used to apply collectively to all the above categories, while the term 'probabilistic risk' is reserved for assessment of outcomes relating to phenomena that are frequently observed and well characterized.

3.1   Diversity of Organizational Capabilities: Markets, Firms and the State

Different types of risk require different risk management procedures: various statistical methods work for probabilistic risks, sensitivity analysis is useful for managing ambiguity, scenario analysis for uncertainty and precaution for ignorance (Stirling 2000). The precautionary approach is not neo-Luddite or anti-technology. It simply involves recognizing that many complex risks cannot be meaningfully reduced to a single percentage likelihood, and therefore will not be well managed by traditional (mathematical) outcome-based methods. Instead they require process-based approaches that address flexible responses to changing circumstances, organizational decision-making structures, and if they are to be trusted by the public, openness and public participation (Stirling 2000; Rose 2000).

Each of the different risk management practices will depend on a range of organizational and technical capabilities which will emerge within organizations in a path-dependent way, structured by their inherent learning capabilities and positions within embedded legal and organizational frameworks (Nelson and Winter 1982; Moss 2002). Consequently as Weber (1934) pointed out, some organizational arrangements will manage the risks and uncertainty of a complex world better than others, and selection pressures may act on organizations to allocate risks and opportunities in different ways (see Nelson and Winter 1982).

3.2 Markets

Insurance and financial markets are extremely effective at reallocating well-defined probabilistic risks, typically through diversification strategies that substantially reduce the volatility and standard deviation of losses by spreading risk among a large number of customers within a market (Nightingale and Poll 2000, Bernstein 1996).[3] Typically this requires markets to be complemented by the technical, organizational and person-embodied capabilities of financial organizations to turn uncertainties into probabilistic risks. However insurance markets may fail due to adverse selection[4] or moral hazard[5] (Moss 2002; Arrow 1963; Rothschild and Stiglitz 1976; Akerlof 1970). Markets also suffer from perception problems, whereby people behave irrationally by not recognizing the level of risk that they face (Weinstein 1989),[6] and commitment problems, whereby insurance companies, regardless of size, cannot convince consumers that they will cover the risk adequately (Moss 2002).[7] Insurance markets also suffer from feedback problems, for example, when concern about the viability of banks causes customers to remove their deposits, making problems worse, and externalization problems whereby the positive and negative effects of an action diverge so that prices do not reflect costs (Moss 2002).[8] In these circumstances, other organizations, particularly the state, typically play a role in risk management (Moss 2002).

3.3 Firms

Because firms use managers to coordinate their actions, they are better able to learn than markets, and are therefore often superior at reducing uncertainties, ambiguities and ignorance. Chandler (1977) for example has argued that the emergence of large firms in nineteenth-century America reflected their ability to organizationally exploit the control techniques developed by the railroads in the 1860s to reduce the inherent risks of operating capital-intensive machinery and systems. For Chandler firms are not 'market failures' but 'organizational successes' that can develop capabilities to dynamically manage risk better than markets (Lazonick 1991).

One area where firms are particularly effective at risk management is research and development (R&D), where they have developed a range of sophisticated methods to turn the uncertainties of investments in research projects into well-defined probabilistic risks for shareholders. This may involve new management practices, such as learning from customers, sourcing external knowledge and integrating specialized knowledge internally (Rothwell et al. 1974). Sometimes internal organizational changes are necessary, such as setting up internal R&D departments (Mowery 1983). In other cases firms need technologies that allow them to better predict and understand the future, such as when banks use information technology (IT) systems to improve their understanding of the probabilistic risks of new financial products, or when pharmaceuticals firms invest in genetics technologies to improve their understanding of the uncertainties of drug development (Nightingale 2000; Nightingale and Poll 2000).

3.4 The State

Similarly governments have sets of capabilities that are unavailable to firms or markets to reduce risks through regulation and enforcement (e.g. moderating environmental pollution) or relocate risks (e.g. from the consumer to the producer in the case of product liability laws, or from the entrepreneur to the creditor in the case of bankruptcy laws) (Moss 2002). These are possible because governments have the power to legislate, investigate and call to account those not in compliance with legislation. This allows them to spread economic risks across the entire taxpaying population by creating 'social contracts' that bind future generations – something that firms cannot do (Moss 2002). The state can also overcome commitment problems because of its ability to, first, print money and deploy resources unavailable to firms, and, second, to act as a trusted third party, reassuring the public about regulatory regimes and risk levels (Moss 2002).[9]

3.5   Complementary Assets for Risk Management

Given the diversity of types of risk, ways of managing them, and differing organizational capabilities, certain organizational structures, or rather combinations and configurations of firms, markets and government bodies, will be better able to deal with specific types of risk than others.[10] We suggest that firms that that are able to use other organizations as complementary assets to cost-effectively disaggregate and disapproapriate some of the uncertain or undesirable consequences of innovation onto third parties can potentially be at a competitive advantage (see Teece 1986). For example, if a complex technology raises public concerns, an endorsement by a well-informed trusted third party would be a complementary asset that could mitigate the commercial risk of public rejection or distrust.

This brief overview of the different capabilities of firms, markets and the state suggests that well-defined, independent probabilistic risks can be managed by diversification strategies that reduce the standard deviation of losses. Firms can either diversify in-house, through joint ventures, or through insurance and financial markets (if substantially more risk needs to be pooled).

Similarly, firms are generally superior to markets and governments at managing the risks of R&D projects because their learning capabilities allow them to reduce technical and market ambiguities and uncertainties and their flexibility allows them to respond to those changes.[11] Ambiguities and uncertainties can be further reduced by diversification strategies that often allow firms to transform them into probabilistic risks through the actions of the capital markets.

Non-independent, systemic probabilistic risks (that cannot be managed by diversification) can either be managed internally or in collaboration with other firms or the state, depending on relative technical capabilities. As Section 4 shows, those risks covered by substantial commitment, feedback, externalization or perception problems are either best left to the state, or will require some form of state intervention.[12] The superior risk management capabilities of the state, particularly the power of legislation and enforcement, allow it to disaggregate risks and disseminate them to different parties. As the state is the 'lender of last resort' when it comes to risk, those most complex and serious risks, particularly those associated with ignorance, are typically covered by the state (Moss 2002).

4. GENETIC TESTING: DEFINITIONS AND DEVELOPMENT

This section defines human genetic testing and describes the clinical emergence of the technology within the National Health Service (NHS). The NHS has been providing clinical services to families with rare genetic disorders since the mid-1940s (Kevles 1985). The early genetic tests were for conditions related to large chromosomal alterations such as aneuploidies (e.g. Down's syndrome). These were detected using cytogenetic techniques that entered widespread clinical use in the early 1960s.[13] Because these disorders tend to be very rare, clinical genetics remained largely distinct from 'mainstream' medicine.[14]

Cytogenetic techniques were followed by biochemical techniques in the late 1960s, and by the mid-1980s so-called 'single gene disorders' with clear patterns of inheritance, such as Duchenne muscular dystrophy (DMD) and haemophilia, could be tested for using molecular genetic techniques.[15] These earliest molecular genetic tests were for conditions where a positive test result indicates a high degree of absolute susceptibility (technically known as high penetrance).[16]

The diffusion of molecular genetic techniques allowed the clinical application of the first gene tests, mainly for prenatal testing, testing for parental carrier status,[17] or to confirm clinicians' observations. As the techniques have developed, more complex, late-onset[18] and lower penetrance genetic associations were analysed, making genetic testing more relevant to mainstream medicine. In the mid-1990s, commercial tests for conditions such as familial breast cancer became available. However, the increasing complexity of the genetic interactions revealed, and the role of environmental factors in these more common disorders, makes genetic test results more difficult to interpret as they only indicate a higher relative susceptibility that often cannot be expressed in a meaningful way as a percentage or quantified likelihood.

These advances in diagnostic technology, and improved scientific understanding of the complexity of genetic influences on disease, have complicated the definitions of 'genetic testing' and 'genetic disease' (Kegley 2000). Furthermore, the technologies used for testing material linked to heritable diseases have wide application, and are used to analyse material from patients, their tumours, or even infectious pathogens (Harper and Clarke 1997).[19] Similarly, while genetic information can be generated by tests of genes and chromosomes, it can also be deduced from studying proteins, metabolic products or from other physiological observation (Zimmern 1999).

These practical complexities mean that the definition of what genetic testing services are is shaped by the activities of firms and NHS laboratories. In this chapter genetic testing relates to tests that produce information about an individual's genetic material (including genes, chromosomes, proteins and other biochemical markers). This would include disease diagnosis, detection of carrier status, disclosure of risk factors for 'complex diseases' (such as heart disease or colon cancer), determination of reaction to medicines, (pharmaco-genetics) and establishing family relationships (i.e. paternity testing) (HGC 2002).

5. THE SOCIAL AND ETHICAL CONCERNS SURROUNDING COMMERCIAL GENETIC TESTING

The scientific complexities and uncertainties surrounding genetic testing raise a range of ethical problems and concerns for policy-makers and the public (Kitcher 1996; BMA 1998; OECD 2000).[20] Within the UK, there is a concern that firms offering commercial testing may offer testing to a larger section of the population than is appropriate, or commence testing before the validity of a procedure is established (Holtzman 1999). Needless or unvalidated testing could generate undue anxiety or give false reassurance (Editorial 1996). In some cases the loss of foetuses to unnecessary termination has been blamed on overzealous use of testing (Concar 2003). Commercial testing also creates worries about patient confidentiality (Lenaghan 1998, p. 81), and the ability of physicians to deal with consumers confronted with disturbing test results.[21] Broader concerns exist about the way in which society will use genetics as a means of avoiding having to address the social problems that are a far more important cause of disease (GeneWatch 2002).

Within the NHS, some clinical scientists are concerned that patented methodologies could restrict public sector testing to older non-proprietary methodologies (CMGS Executive 1998). If private firms 'cream off' simple or profitable services and then drop unprofitable tests in the future, there may be a reduction in the range of tests available (Lenaghan, 1998, p. 81). Were this to occur, public sector capabilities would be difficult to rebuild and the underlying principle in the UK, of access to health care based on need rather than wealth, could be put at risk (CMGS Executive 1998). The quality of private services has also been questioned by clinical scientists (Lenaghan 1998, p. 82).[22]

The market uncertainties of commercial testing are further complicated by the variability in both the predictive power of tests and the treatment options available. Because UK clinicians generally only test when the predictive power is high and treatment is available (see Table 6.2),[23] the commercial

*Table 6.2	When do clinicians favour testing for genetic diseases?*

Effectiveness of treatment	Predictive power of test	
	High	Low
High	(A) Test and treat where economically feasible (e.g. phenylketonuria – PKU)	(C) Test only where preventative measures are economically feasible (e.g. haemochromatosis)
Low	(B) Test only with caution (e.g. Huntington's disease)	(D) Testing not advised (e.g. Alzheimer's disease)

Note:	Table 2 adapted from Burke et al. (2001) with examples added.

viability of a test can radically change if novel treatment possibilities open up.[24] In the UK, public sector clinicians are mainly concerned with the types of test that fall into the categories (A) and (B) shown in Table 6.2. However firms are increasingly interested in providing tests for disorders that fall in categories (C) and (D) where more complex and common diseases such as coronary heart disease, or Alzheimer's disease might be categorized.

The scientific uncertainty that surrounds the aetiology of many conditions, even after genetic associations are found, and the social and ethical problems associated with test results, has led consumer groups to seek changes in market regulation (Copeland 2002a, 2002b). At the same time industry lobbyists are calling for genetic testing to become more consumer-focused and easier to access, with some calling for a 'consumer charter' to give greater opportunities for people to access information about tests (Ledley 2002). However at present there are no statutory regulations governing genetic testing in the UK[25], and the globalization of the market makes legislative control difficult as it can be circumvented by sending samples abroad.[26]

The ambiguity and uncertainty surrounding the technology, ethics, risks and future legislative framework of genetic testing are, for some firms, offset by its lucrative commercial potential. Section 6 explores public genetic testing in the UK, to illustrate the context in which firms operate.

# 6.	PUBLIC GENETIC TESTING SERVICES IN THE UK

The UK has a mixture of public (NHS) and, much smaller, private health services. A public network of 22 regional genetics centres has evolved across the UK to detect and analyse rare genetic diseases (Donnai and Elles 2001).[27] These regional centres can be thought of as three co-located networks of molecular laboratories, cytogenetic laboratories and clinical geneticists/counsellors that are closely linked through their day-to-day work. The centres offer a broad range of genetic tests, diagnosis, risk estimation, surveillance, counselling and support, and are available free through the NHS.[28] The molecular and cytogenetic laboratories situated within the regional centres receive the majority of their samples from specialists such as neurologists, oncologists, paediatricians, and even a number of General Practitioners.

Each molecular genetics laboratory provides tests for 'core' disorders, and most also have specialist knowledge of several additional conditions, often built up through local research interests. For example, Manchester Children's Hospital has expertise in cystic fibrosis (CF), while Guy's Hospital (London) has a well-developed DMD service. This allows the network as a whole to provide tests for many hundreds of rare conditions, some of which might only affect a handful of families in the UK. The range of tests provided, methods used, quality assurance and training of personnel are organised through professional bodies such as the Clinical Molecular Genetics Society and the Association of Clinical Cytogeneticists, which have played a key role in shaping the infrastructure of NHS services. This informal governance system involves frequent assessment and diffusion of 'best practice' technical methods and reporting guidelines.

Despite the stewardship of these professional bodies, the regional genetic testing laboratories, like many other parts of the NHS, are under-resourced and can suffer from bureaucratic inefficiencies. This can produce delays in the processing of samples, particularly for non-urgent referrals and non-routine procedures. This provided the first niche market for a commercial genetic testing service in the UK, which was to supply faster testing services for the fee-paying patients of private doctors. As Section 7 shows, as firms have entered the market, their ability to strategically use NHS services to disappropriate risk has been a major determinant of success. Table 6.3 compares some characteristics of these firms.

7. COMMERCIAL GENETIC TESTING IN THE UK

This section reviews the development of genetic testing services and related activities of the four firms introduced in Table 6.3. The case studies reveal how each firm has approached the problems of market appropriation while at the same time achieving disappropriation of risk associated with the technology.

Table 6.3 Firms with a commercial interest in UK genetic testing services (2002 estimates)

Firm	Year founded	No.of staff	No. of people tested annually (all countries)	Current market focus	Model of service delivery
Cytogenetic DNA Services	1984	25	16000+	International	Through NHS/ private specialists and GPs
Cellmark Diagnostics	1985	100 (in UK)	10–20K	International	Through community-based charity schemes
Myriad Genetics	1991	550 (in USA)	10–100K	International	Mainly through oncologists
Sciona	2000	15	500+ (part year)	UK	Initially direct to public, now via clinicians

CASE STUDY 1: CYTOGENETIC DNA SERVICES LTD – ESTABLISHED TO PROVIDE FAST AND EFFICIENT TESTING OF 'SIMPLE' GENETIC DISORDERS FOR PRIVATE DOCTORS

Cytogenetic Services Ltd. was established in 1984 by Rodney Meredith, a former NHS cytogeneticist with over 15 years experience. He ran the company until 2002 when it was taken over by a private pathology laboratory. Meredith left the NHS after his laboratory at the Middlesex Hospital merged with University College Hospital's laboratory. He had seen a niche for providing cytogenetic testing for private (i.e. non-NHS) doctors in the London area, whose patients would pay for a faster, more efficient service. By focusing on efficiency, the firm's client base grew steadily and the company provided testing for the Middle East and Hong Kong where services could not be found locally. The firm developed capabilities in molecular genetics in the mid-1990s by hiring molecular geneticists and changed its name to Cytogenetic DNA Services Ltd. The company rapidly began to offer services for single gene disorders, partly facilitated by the availability of kits manufactured by Cellmark Diagnostics (see below) and Applied Biosystems, the market leader in instrumentation for DNA analysis.

The company typically invested around 10 per cent of turnover in R&D to improve the processing of urgent samples. For example Cytogenetic DNA Services Ltd. claim to be the first firm to routinely use a technique called QFPCR.[29] This technique allowed them to automate testing of prenatal samples for a range of the most common chromosomal disorders. QFPCR was initially used to complement the slower cytogenetic techniques, which relied on the analysis of whole cells, and required a lengthy preparatory cell culturing stage (Levett, Liddle et al. 2001). QFPCR significantly reduces the sample turn-around time – deemed by client doctors to be a crucial consideration when pregnant mothers are awaiting results. By 1999 they were able to put the technique into routine use, prompting NHS genetic centres to explore the technique further.

The relationship between Cytogenetic DNA Services Ltd. and the NHS genetic services is complex. The clinical scientists employed by both organizations are members of the same

professional bodies, such as the Association of Clinical Cytogeneticists, and they subscribe to the same external quality
assessment and laboratory accreditation schemes. However, they
behave very much as competitors, with client hospitals (i.e. those
that send samples) moving from one to the other depending on
the constraints of the NHS commissioning process. The services
provided by Cytogenetic DNA Services Ltd. rely predominantly on
the same techniques and equipment found in NHS laboratories,
and the company provides many of the same 'core' services.
However, in terms of scope, the molecular genetic tests the
company offers are limited to the more common of the rare single
gene disorders diseases, with other conditions that involve higher
technical, patient and commercial risk left to the NHS laboratories
with specialized services. As Meredith stated:

> [Testing for the rarest disorders] is far better done for the sake of the
> patient by a large institution that sees great variation in the genes,
> [they] can give a very good answer, but if you take something like cystic
> fibrosis or haemochromatosis, very very common ailments [by con-
> trast to other genetic tests], it is perfectly okay for us to do these . . .
> (Interview March 2002).

This reflects the practical problems of molecular genetic testing.
Not all genes are equal and not all mutations are of equally severity or occur with the same frequency. In disorders like cystic fibrosis, novel mutations are rare, and a relatively small number of
mutations account for the majority of affected individuals within a
population. Disorders associated with a small number of common
mutations lend themselves well to identification with diagnostic
kits that are based on simple assay techniques and can be undertaken with a low risk of false findings and consequently a lower
legal and commercial risk to the firm.

By contrast disorders associated with larger genes (such as for
Duchenne muscular dystrophy) can be more complex to analyse.
Mutations in the DMD gene often occur sporadically in previously
unaffected families, rather than being inherited from previous generations. The resulting mutations are more diverse. When the locations of possible mutation sites are distributed throughout the
gene, it requires more laborious testing procedures, such as full
sequencing of the gene. This is slow and relatively expensive compared to tests for specific mutations. Moreover, substantial tacit
knowledge is required to interpret the results in the context of the

specific clinical symptoms of the patient. Because of these clinical interpretation difficulties, some clinicians fear that there is a risk to patients of misleading diagnostic information being given, and consequently, favour confining testing to specialist clinics. Cytogenetic DNA Services Ltd. has therefore made an explicit strategic choice to leave the analysis of complex (and therefore commercially risky) genes to the NHS, and instead concentrate on the efficient testing of more straightforward conditions. They achieve this through a relative lack of bureaucracy and a more extensive internal division of labour between technical staff.[30]

The division of the market between Cytogenetic DNA Services Ltd. and the NHS reflects a clear strategic concern about commercial risk to the company, resulting from the complex nature of some services and social and ethical risks they pose to consumers. The services the company offers are similar to the regional genetic centres, and can therefore avoid unique regulatory concerns. However, Cytogenetic DNA Services Ltd. has avoided undertaking more controversial services, such as breast cancer testing, even though the NHS already provides them. This is to avoid bad publicity and uncertainty as the underlying science is poorly understood. The legal environment in which Cytogenetic DNA Services Ltd. operates means that there is a clear link between the potential risk of harm to patients and commercial risk to the firm should the validity of their test protocol be questioned. For more clearly understood conditions where psychological risk to the patient remains, the firm effectively passes responsibility for these onto the referring physician.

CASE STUDY 2: CELLMARK DIAGNOSTICS – SELLING DNA TESTING KITS, AND EXPLORING MARKETS FOR TESTING SERVICES

Cellmark Diagnostics was founded in 1985 as a subsidiary of ICI to exploit the research of Sir Alec Jeffries (Leicester University, UK) who developed DNA fingerprinting. Cellmark's initial market was in forensic DNA analysis, which they pioneered in the UK and elsewhere. When Zeneca demerged from ICI, Cellmark formed part of Zeneca, but was sold off with Zeneca's other non-core businesses after the merger with Astra. Orchid Biosciences, a US firm interested in diversifying into pharmacogenetics, bought Cellmark in February 2001 for an undisclosed sum.

Throughout its history, Cellmark has attempted to apply its competence in molecular biology to DNA testing. The firm has been awarded various government contracts, first for the Child Support Agency – a government agency that chases absent fathers – (paternity testing), then the National Sheep Scrapie Plan (genotyping sheep for prion disease susceptibility) and most recently the immigration contract for the Foreign Office. Cellmark recently re-entered the forensic business (its first market), and is accredited to supply data to the National DNA Database for Criminal Justice Act samples (generating and cataloguing DNA fingerprints of those suspected of committing criminal offences). However to date Cellmark has not directly been involved in testing British citizens for clinical purposes. Instead Cellmark's commercial interest in health-related genetic testing is mainly through the manufacture of kits for CF (available from 1994) and other disorders which are similarly amenable to the development of a simple assay-based test. The kit provided all the reagents needed to test for the presence of four of the most common CF mutations in northern Europe and North America. Kits have been on sale across Europe for several years, and reagents[31] were launched in the USA in 2001, (see website www.orchid.com/, viewed 12 November 2002).

The kit was based on a proprietary technology called the Amplification Refractory Mutation System (ARMS)[32], and licensed intellectual property rights (IPR) on mutation sites in the CF gene from various patent holders. The advantage of the ARMS is that it is technically simple, requiring only low-cost apparatus. As more CF mutations were characterized and their prevalence in the population assessed by the research community, Cellmark expanded their kits to detect with greater confidence, releasing kits to detect 12, 20 and most recently the 29 mutations. To ensure their acceptability with the user community, Cellmark's kits were assessed independently by clinical scientists at NHS laboratories.

At present Cellmark is the market leader in the supply of kits for CF testing in the UK, and they also supply reagents for diseases affecting the international Jewish community (as discussed further shortly). The provision of kits for DNA testing services is a small market at present but Cellmark's parent company, Orchid, estimates it will grow by 50–100 per cent per year over the next decade (see Website www.orchid.com, viewed 12 November

2002). However opportunity for the development and use of kits is limited by the number of patents held on genes and even mutation sites (Merz, Kriss et al. 2002). Patent holders can make it difficult and often uneconomical for others to produce commercial kits by insisting on impractical licensing agreements (Schissel, Merz et al. 1999).

Apart from problems posed by the current market size and IPR restrictions, kits do offer some commercial opportunities. First, they are relatively lightly regulated: for example, although the European Parliament Directive 98/79/EC on *in vitro* diagnostic medical devices that came into force at the end of 2003[33] requires kits bearing the CF mark to meet standards of manufacturing quality, there is no control over the validity of diagnostic information produced.[34]

Second, the underlying technology generates a low appropriability regime (see Teece 1986) as kits can often be copied and rapidly reach the market. Because much of the information used to develop the Cellmark kits is available in the public domain, the company has not required a strict IPR enforcement strategy to recover large investments in R&D.

Having produced kits and reagents to simplify the process of diagnostic testing, it is unsurprising that Cellmark chose to move on to downstream activities. In 1999 the company began screening the Ashkenazi Jewish population of New York State to identify carriers under a scheme organized and run by Dor Yeshorum, a Jewish charity involved since the mid-1980s in community screening programmes. The charity sources tests from a number of laboratories in the USA and elsewhere, and also arranges genetic counselling. As a result of the scheme, Cellmark currently receives samples for screening against a panel of 10 disease tests, including Tay-Sacks, Gaucher's disease and CF. The samples come from Eastern Europe, the USA and Israel and are predominantly sent to determine the carrier status of individuals. The scheme allows couples to find out before conception whether or not they might have affected children. Cellmark does not offer any similar services in the UK at present.

The Dor Yeshorum scheme is neither wholly public nor private in character, and seems to avoid many of the potential pitfalls associated with purely commercial operations. Cellmark has no access to personal information and does not select its own consumers, reducing the risk of confidential genetic information leaking out. As a result firms participating in such schemes reduce the risk of adverse public perceptions of their services. Interestingly the regulations that New York State enforce on such practices made it necessary for Cellmark's UK site to be inspected by officials from the USA before they could join the Dor Yeshorum scheme. In the UK at present no such regulations apply although a firm offering such a service might be encouraged to join the appropriate British quality assurance scheme. In the case of Cellmark, it already sits on the relevant scheme's advisory panel (Ramsden 2002). The corporate objective of market appropriation is well served by the Dor Yeshorum charity, which plays the role of a 'trusted third party', in Moss' (2002) terms, and uses this position to reduce customer risk through reliable confidentiality, community-based counselling and a clear alignment of interests with patients than might be obvious with a wholly commercial scheme. The charity contract virtually guarantees access to large markets with little need for the firm to search for consumers or gain the attention of communities of physicians. In the UK the firm has to date focused on selling testing kits to public services rather than the downstream activity of testing, where another firm, UDL, proved unsuccessful in recent years.[35]

CASE STUDY 3: MYRIAD GENETICS INC. – TESTING COMPLEX GENES USING PROPRIETARY TECHNOLOGY

By the early 1990s research had begun into the genetics of common diseases such as breast cancer, heart disease and Alzheimer's disease. This research, coupled with higher throughput instrumentation, drawing on increasing automation and more sensitive techniques for DNA analysis, contributed to increasing commercial interest in genetic testing.

Myriad was established in 1991 by Mark H. Skolnick, one of the pioneers of early molecular genetic analytical techniques in the late 1970s and 1980s. Skolnick and former venture capitalist Peter D. Meldrum built up a team with strong credentials including Walter Gilbert, a Nobel laureate and co-developer of DNA sequencing technology. Myriad's initial public offering on the NASDAQ in October 1995 raised close to $50 million through promises to develop innovative pharmaceuticals products and 'predictive medicines' (i.e. diagnostic tests). The key to this dual strategy relied on finding genes that can provide both targets for drugs and predictive information on disease susceptibility. From inception up to March 2002, the company had accumulated a deficit of $67 million, mainly through R&D spending. It is difficult to identify Myriad's spending on the development of genetic testing products because R&D provides leads for both the pharmaceuticals and the predictive businesses, although an internal estimate suggests around 20–40 per cent of R&D is not pharmaceuticals-related. An advantage of this dual strategy is that, while Myriad is yet to launch a pharmaceuticals product, its predictive medicines are providing a rapidly growing revenue stream ($17.1 million in the year to June 2001). Table 6.4 contains details of the predictive medicines Myriad has released to date.

Table 6.4   Myriad's predictive medicines

Product	Disease	Release date
COLARIS AP	Familial adenomatous polyposis and attenuated FAP	May 2002
MELARIS™	Melanoma and pancreatic cancer	September 2001
COLARIS™	Hereditary non-polyposis colorectal cancer	September 2000
CardiaRisk®	Risk factor for cardiovascular disease	January 1998
BRACAnalysis®	Hereditary breast and ovarian cancer	October 1996

Each test is based on genes for which Myriad holds patents or licensed proprietary knowledge. Almost all the products rely on DNA sequence analysis rather than screening for specific mutations. Reliable full sequence analysis is complex and expensive for competitors to produce in high volume, and would remain so even assuming Myriad granted licences for such an activity. By using customized sequencing technology and IPR, Myriad considers itself well protected from major commercial competition. Indeed, even leading US providers of genetic tests such as Genzyme Inc. (who carry out 350,000 genetic tests a year) only compete with Myriad in testing for one condition, colon cancer, and here the testing strategies are different. Myriad provides a full screen, while Genzyme only looks for specific mutations.[36]

Myriad has been widely criticized for its DNA patenting activities, for example see Westphal (2002). Nonetheless at present Myriad provides a two week service, compared to the public service that can take many months to process samples for BRCA 1&2 testing. Furthermore, Myriad provides comprehensive analysis, routinely sequencing 17 000 to 34 000 base pairs of DNA per patient, whereas NHS laboratories sequence only portions of the gene most frequently associated with mutations.

Despite the technological and proprietary advantages Myriad have secured over other potential providers in its chosen disease areas, Myriad have had very limited success in exploiting these in the UK. In March 2000 they attempted to link with a British company, Rosgen Ltd. However Rosgen went into liquidation shortly afterwards and Myriad has been without a British operation ever since. As a result Myriad receives only a trickle of customers from the UK (Cancer Clinical Geneticist 1 – personal communication 2 September 2002). However they continue to pursue other international markets such as Germany and Japan with more success. The majority of its customers continue to be from North America. In the meanwhile the NHS is offering its own unlicensed version of the BRCA1&2 tests.

Negotiations between the Department of Health (UK) and Myriad regarding a licence for NHS BRCA 1&2 testing are ongoing, but in the meantime there has been no pressure from Myriad for the NHS to cease such testing (Nuffield Council on Bioethics 2002). Myriad's present route of service provision is through medical professionals, generally oncologists at cancer centres. Myriad has reduced legal and consumer risk by not supplying tests directly to consumers. This is particularly important because some of Myriad's tests, such as the BRCA 1 & 2 tests, are for lower penetrance conditions than the early tests for rarer genetic disorders, and may prove to have complex environmental components. Consequently, the test cannot give patients clear answers about susceptibility, that is that they definitely are or are not going to get a particular type of cancer. None the less, the NHS is confident enough in the utility of BRCA1&2 tests to offer their own test protocol in the NHS laboratory network.

In Europe the most publicized aspect of Myriad's strategy is probably its patenting of high profile genes (so-called 'DNA patents'), rather than the utility of their tests. DNA patents are often regarded as insufficiently inventive to deserve legal protection.[37] Whether or not these patents stand, Myriad claims the patents are justified to ensure protection of their significant research investment that has allowed the optimization of analytical protocols and technologies necessary to ensure high standards in the testing process. It is worth considering that if services for genes requiring complex analysis are to be offered to the public, whether through clinicians or otherwise, the protocols used will require comprehensive and therefore expensive validation. Myriad is prepared to spend millions of dollars on such work. The question remains as to whether or not NHS genetic testing services could afford similar validation processes if the government were to establish a common regulatory framework.

Myriad is another example of a firm that avoids selling tests direct to the public so as to reduce its commercial risks. Instead it relies on physicians to act as 'gatekeepers' of the service. Myriad also only pursues investments where high levels of market appropriation can be achieved through patenting. At present, the DNA patents that are central to this strategy are being largely ignored by the UK's clinical scientists, and at least one of Myriad's patents is actively being challenged in Europe (Thomas, Hopkins et al. 2002). Should these patents fall, Myriad may have to rely on its technical capabilities as a means of market appropriation, focusing on attracting consumers through the speed and quality of their analysis. While these technical capabilities are currently 'state of the art', the degree to which they can be imitated is yet to be tested. Myriad does, however, have a large sales force and an international presence, ensuring market appropriability strategies in this firm are multifaceted.

CASE STUDY 4: SCIONA – PROVIDING CONSUMERS WITH DIETARY ADVICE BASED ON PERSONAL GENETIC PROFILES

Sciona was the first British company to offer genetic testing services for complex diseases directly to the public. It was established by Chris Martin in September 2000 and has so far raised £3 million in venture capital funding. Sciona's vision is to make genomic information available to both clinicians and consumers, especially where such information could allow individuals to take action to

improve their long-term health. Before Sciona, the mass consumer market was untapped as the majority of firms exploiting
genomics had targeted their services towards drug development
rather than clinicians or the public.

Sciona has been able to rapidly develop a commercial service
through assembling and combining externally produced datasets,
often already in the public domain, together with proprietary technology gained through the acquisition of a small private firm,
Genostic Pharma Ltd., and proprietary data analysis processes
developed through Sciona's in-house R&D programme. Sciona's
first service focuses on 'nutritional genomics',[38] although future
products may focus on other areas such as drug metabolism and
the avoidance of adverse drug reactions. The service is based on
linking variation in the general population in genotypes associated
with the poor metabolism of alcohol and certain toxins, vitamin utilization, and oxidative stress, with data on disease susceptibility.
The consumer's DNA is examined to determine their genotypes at
around 20 loci. The test results are not advertised as predictive of
particular diseases and Sciona makes no medical claims. Rather,
they give advice tailored to the consumer's genotype in the form
of a report that focuses on maintaining health and well-being.

Sciona's initial marketing strategy was aimed at the general
public and invited clients to apply for the service online or to purchase a kit from a high street retailer such as The Body Shop (kits
contain a swab for collecting cheek cells which are sent to Sciona
for analysis). This approach attracted considerable media interest
and resulted in the pressure group GeneWatch strongly condemn-
ing Sciona for supplying 'meaningless tests' and 'misleading the
public' (GeneWatch 2002; Hawkes 2003). As a result of growing
adverse publicity, The Body Shop withdrew the kits (Cookson
2002). Sciona invited the Human Genetics Commission to visit
their facilities in an attempt to ensure the company retained its reputation by using a government body as a trusted third party to
increase public trust and reduce commercial risk.

However the HGC first raised certain concerns about the
accreditation of Sciona's laboratories and, second, having no statutory powers, could only conclude that British guidelines were not
broad enough to cover Sciona's services.[39] As a consequence,
Sciona no longer has any retail distributors, but continues to offer
its services to consumers through dieticians and physicians.[40]

Sciona claim to be in favour of some form of regulation that would establish the legal position of genetic information, encourage consumer confidence, protect against disreputable suppliers damaging the market, and allow an educated public to make informed choices. Sciona's strategy relied on the regulative and investigative power of the state to reallocate the firm's commercial risk through changes in the regulatory environment to place onus on the consumer (*caveat emptor*). Following pressure group resistance, and without the state acting as risk reallocator, Sciona was unable to manage the risks inherent in the field of human genetic testing as effectively as the three firms discussed in the previous case studies.

It is too early to fully assess the effectiveness of their market appropriation/risk disappropriation strategy. As with Monsanto, an initial problematic strategy may damage consumer trust and make subsequent strategies more difficult to successfully implement.

DISCUSSION AND CONCLUSION

These case studies of human genetic testing services in the UK, while limited to a specific geographic and technological market, have shown that firms strategically use third parties as complementary assets to disappropriate risks. The relocation of risks onto non-commercial organizations appears to be an important step in stabilizing highly uncertain markets such as those for genetic testing services. Table 6.5 provides a summary of the market appropriation and risk disappropriation strategies undertaken by the case study firms. The NHS regional genetic testing centres have been included for comparison.

The participants' main technical competences are outlined to indicate the sophistication of their services. The IPR held by each participant has been separately described to the other strategies for market appropriation, to highlight the finding that although DNA patents have been a highly controversial and much-discussed part of the commercialization debate, it appears firms in this market rarely rely primarily on IPR. Only the firm that invested significantly in R&D (i.e. Myriad) has claimed patenting to be a central part of its market appropriation strategy, and they claim this was necessary because of the complexity of the genes involved. Indeed, the firm has partly based its market appropriation strategy on developing mastery of this complexity that would be hard for other firms to replicate. Although the other firms hold patents, their patents on methods are likely to be less enforceable (Teece 1986, p. 187), and in at least one case (Cytogenetic DNA Services Ltd.) it is claimed they serve a primarily defensive purpose.

Table 6.5 Strategies for selected UK market participants in genetic testing

Testing organization	Technical competences	Intellectual property market appropriation	Non-IPR strategies for appropriation of risk	Strategy for Disappropriation
Regional genetic testing laboratories (public)	Cytogenetics, simple and complex genes	Negligible	First and major incumbent. Control referrals for majority of population	Professional bodies establish/run training and QA schemes
Cytogenetic DNA Services Ltd.	Cytogenetics, simple genes	Negligible	Efficiency of routines through division of labour/automation	Uncontroversial tests, uses independent gatekeeper
Cellmark	Simple genes	On methods	Makes upstream products which could reduce costs of services	Uses independent gatekeeper
Myriad	Complex genes	On genes and methods	IPR, automation, large sales force	Uses independent gatekeeper
Sciona	Complex genes (multi-factorial conditions)	On methods	First mover advantage in an untapped market	Change in strategy and reliance on future regulation

The evidence shows that the four case study firms have followed diverse strategies to meet their objectives, particularly with respect to market appropriation. Yet all have found that the main commercial risk of genetic testing relates to public access to novel services. Each of these firms, whether through deliberate choice or external pressure, have found it necessary to relocate a portion of their commercial risk to a third party, whether a charity or a public body. This ultimately places control of the technology, and potentially some of the financial benefits, in the hands of these organizations.

In terms of the risk taxonomy presented in Section 3, the relationship between the presence of particular genes and the penetrance and severity of the disorders to which they are associated is highly complex. Diagnostic test results can only be ascribed a meaningful probabilistic risk for a very few conditions, which tend to be particularly severe and require patient counselling. While there is clearly a market for genetic tests for more complex disorders the uncertainties involved in diagnosis and the ignorance we have over their implications mean that they can generate personal risk for consumers. This risk generates public concern and attempts to regulate a very fast-moving technology, which, in turn, generates considerable commercial risk to firms supplying the services. However, firms can dissapropriate such risks and appropriate the markets for these tests if they are able to enlist support from trusted third parties. In particular, it appears firms are best able to exploit genetic testing technologies through a combination of organizations wherein firms offer a technical procedure to which they can apply economies of scale, and a trusted third party offers counselling services.

This may not be suitable for all or even a majority of genetic tests, but none the less this approach has offered a toehold for firms in what was previously a public monopoly. Such an arrangement may not please those who see the state as ultimately meeting the cost for the possible increase in demand for counselling services. Nor will it please those concerned that firms may cherry-pick the best markets and expand services into areas where benefits to the consumer are more questionable; (see Table 6.2 C & D rather than A & B).

The analysis is based on only initial moves of firms in this new market, and in the cases of Myriad and Cellmark, their main human disease testing activities remain abroad, and so beyond the scope of detailed examination here. Furthermore, these two firms have substantial programmes to develop products or services outside the area of human disease testing, and so cannot be regarded solely as human genetic testing firms, unlike Cytogentic DNA Services Ltd. or Sciona. Diversifying into these alternative services perhaps represent the greatest risk-reducing strategy, but these activities have not been explored substantially here.

From a policy perspective, the firms' technological capabilities (particularly in R&D and high throughput services), and the state, or charities' capabilities for patient management suggests a likely division of labour with the private sector concentrating on high throughput services, while non-profit organizations focus on rare disease screening and counselling. Such arrangements would limit unequal access to testing (see Holtzman 1998). Interview data suggests that the prospect of private sector testing is acceptable to many clinicians, provided it is to a high standard and is cost-effective.

Whether private testing expands or not, the existing regional genetics services are in need of reform to address the increasing demand for genetic counselling that will result from growing public interest in genetic testing services.

The division of labour between private and public genetic testing services and their implications are yet to be fully worked out in the UK. However, it seems clear that organizational risk reallocation will play a key role in the economic development of these markets in the UK, and potentially within Europe as a whole.

ACKNOWLEDGEMENTS

The authors are grateful for the helpful comments received in early drafts of this paper from Ben Martin, Andrew Stirling, Paul Martin, Jackie Senker and Annemiek Nelis. We thank our interviewees for their time and cooperation, and the UK's Economic and Social Research Council for supporting this work through a PhD studentship and the CoPS Innovation Centre.

NOTES

1. For example see websites www.genewatch.org or www.hgalert.org.
2. With these undefined and uncalculable risks, simply producing a number from a risk assessment procedure does not make it 'objective' in the sense that the process still involves subjective framing assumptions, subjective assessments of the likelihoods of particular uncertain future outcomes, and normative assumptions about the relative ranking of various good and bad consequences, that fall on diverse subsets of the population.
3. However, markets often have difficulty dealing with catastrophic or systemic risk that cannot be managed by diversification because they share a common cause. This is why insurance houses do not sell policies against falling house prices (Moss 2002).
4. As when someone has a medical examination, finds out they are susceptible to an illness and acquire health insurance without informing the insurance company of the known risk.

5. As when the act of insuring against a risk increases its likelihood by making people less concerned about its consequences, for example, where having fire insurance stops house-holders being concerned about putting out house fires. See Baker (1996) for a review of moral hazards.

6. For example, certain workers at high risk tend to assume that they will lead a 'charmed life' and legalisation is often needed to cover their safety (Moss 2002). This was a partic-ular problem of worker health and safety insurance in nineteenth-century America (Moss 2002).

7. Similarly people will not buy flood insurance if they are convinced the government will bail them out in the event of a flood (Moss 2002).

8. For example, before the US government acted to ensure that companies were responsible for their own clean up costs, toxic waste was a major environmental problem.

9. At least in theory. In practice, it easy to over-indulge and regulatory capture may be a problem.

10. This differs from the traditional evolutionary economics approach of Nelson and Winter (1982) in that firms' fitness is not derived from selecting routines within a fixed techno-logical trajectory, but rather by firms reconfiguring their environment and their relation-ship to it by strategically managing their interactions with the state and with markets.

11. Where individual firms lack these capabilities, they may consider outsourcing those risks, but should recognize the difficulties involved (Mowery 1983).

12. Clearly, there will be situations where the shifting of risk is socially undesirable, as when firms bribe government officials to reduce the risk of losing procurement contracts, or contribute to political lobbying in an attempt to bypass the commercial risks of due dem-ocratic process or antitrust legislation. There are however major social and economic benefits from shifting risks between institutions as those claiming against house, car or life insurance can demonstrate.

13. Cytogenetics is the study of whole cells using microscope-based techniques to ascertain the condition of the chromosomes.

14. However, testing for the hereditary metabolic disease PKU, became a routine part of neo-natal care in the UK in the early 1960s; see Pollitt, Green, et al. (1997).

15. Unlike cytogenetics, molecular genetic and biochemical genetic techniques use a variety of platforms to conduct indirect analysis from which genetic change may be inferred. These require more complex methodologies but can reveal results beyond the resolution of cytogenetic tests.

16. The notion that the presence or absence, and severity of a disorder can be attributed to a single gene is increasingly understood to be an oversimplification. For example see Badano and Katsanis (2002), Moss (2003) and Morange (2001). For a fuller description of genetic disorders, their modes of inheritance and clinical features see Rose and Lucassen (1999).

17. For many single gene disorders to be inherited, a child must receive a copy of a faulty gene from both parents. Carriers of faulty genes may not be affected as they have a working copy of the gene; however knowing their carrier status may affect a couple's reproductive choices.

18. The term 'late onset disorder' refers to genetic diseases that are not apparent at birth or in childhood, but manifest themselves in later life, often many years into adult-hood.

19. For example, cytogenetic techniques used for identifying foetal abnormalities can diag-nose forms of non-inherited cancer, and molecular genetic techniques for single gene dis-orders can also identify pathogens.

20. One only has to consider the implications of a false positive result for a gene associated with cancer, or the impact on a family of a member finding out the person they thought was their father is not, to recognize the ethical issues that are involved. There are also problems associated with 'the right not to know'. If a person whose father died of a highly penetrant dominant late onset genetic disorder (e.g. Huntington's disease) decided that they did not want a test, but their son decided that they did, then a positive result for the son would reveal the father's probable fate.

21. HGC meetings minutes 5 February 2003, website, www.hgc.gov.uk/business_meetings.htm. viewed 6 October 2003.

22. These sorts of policy debates clearly indicate the potential conflicts of interest at work. The concerns of NHS clinical geneticists, as a professional group, could be as easily seen as attempting to protect a monopoly as to protect the public, while similarly, commercial enterprises have an obvious incentive to push for increased consumer choice. We take no position on these debates, but use them to highlight the problematic nature of the assessments of both risks and opportunities – not only are they highly technical and complex, they are also framed by a range of social assumptions.

23. In the USA there is less concern over genetic testing and private firms, such as Athena Diagnostics, offer tests from each of the quadrants in Table 6.2 provided a physician requests them – highlighting both the national specificity of our study and the dubious nature of 'Anglo-Saxons' as an analytical category. See website www.athenadiagnostics.com/site/content/testing/list_profiles.asp, viewed 18 October 2002. Athena's website only recommends the Alzheimer test for those showing symptoms of the disease rather than predictive testing.

24. Equally, improved scientific understanding of the role a gene plays in other diseases might make testing less desirable. For example a test associated with the metabolism of folic acid, potentially helpful in predicting spina bifida, may also provide a weak correlation with colon cancer in the mother, because genes often play complex roles in multiple or interrelated systems (see website www.hgc.gov.uk/business_meeting.htm, viewed 6 October 2003.)

25. In 1997 the UK Government's Advisory Committee on Genetic Testing (ACGT) issued *guidelines* for monitoring testing for single gene disorders (ACGT 1997). But they were acknowledged to be inadequate for 'over-the-counter' or 'retail' genetic tests by the ACGT's successor, the Human Genetics Commission (HGC). The HGC has been asked by a Department of Health official to consider all options (ranging from a ban on retail tests to regulation-free tests). The government was under considerable pressure and a number of parliamentary questions had been raised on the subject (see HGC meeting minutes 20 November 2002, website www.hgc.gov.uk/business_meeting.htm).

26. HGC meetings minutes 5 February 2003, website www.hgc.gov.uk/business_meetings.htm, viewed 6 October 2003.

27. The UK government has recently started to initiate more coordinated approaches; see Milburn (2001).

28. The genetic counselling services are a tertiary service accessed by referral from medical specialists, although similar services have been established by haematologists, offering testing and counselling for common blood disorders in a distinct network of clinics and laboratories.

29. The technique was originally developed at Guy's Hospital in London (see Adinolfi, Pertl et al. 1997) and involves the quantitative analysis of fluorescence produced by dye-labelled DNA to screen for multiple chromosomal abnormalities simultaneously.

30. 'There have been [several] attempts by people in the UK [to form] this type of laboratory and it has failed simply because it has not been viable to them. We have been able to undercut them in price and in efficiency' Interview – Rodney Meredith 18 March 2002.

31. In some cases separate reagents rather than the full kits are sold for licensing reasons.

32. See Newton et al. (1989).

33. See website www.mdevices.net/directv.htm, viewed 14 January 2003.

34. HGC meetings minutes 5 February 2003, see at website www.hgc.gov.uk/business_meetings.htm, viewed 6 October 2003.

35. The very first company in the UK to offer genetic tests direct to the public was University Diagnostics Ltd. (UDL). They offered carrier testing for CF in the late 1990s; however UDL withdrew its tests after a change of policy by the Cystic Fibrosis Trust (a trusted third party who had provided endorsement). UDL withdrew their CF to focus on paternity testing (personal communication with former UDL Director 13 November 2002).

36. One industry interviewee suggested that these services were differentiated as one offered 'quality', and the other affordability.

37. See Wadman (2001) and Nuffield Council on Bioethics (2002) for a fuller analysis.
38. Nutritional genomics is based on the application of genomic information to discover how nutrition may affect gene expression in the individual. For a review see Elliot and Ong (2002).
39. In their August 2002 newsletter, the HGC stated 'During the process it became clear that the [1997] Code, which had been written with single gene disorders in mind, did not cover all necessary aspects for other types of tests'. Whether or not regulation on these tests can be effective is questionable. There are many unregulated products available to consumers that purport to predict risk factors to future health, such as bone density scans, blood cholesterol tests and blood pressure measurement.
40. Website www.sciona.com/, viewed 3 September 2002.

REFERENCES

ACGT (1997), 'Code of practice and guidance on human genetic testing services supplied direct to the public', London: Department of Health, Advisory Commitee in Genetic Testing.

Adam, B. and J. VanLoon (2000), 'Introduction: repositioning risk; the challenge for social theory', in U.B.B Adam and J. VanLoon (eds), *The Risk Society and Beyond*, London: SAGE Publications, pp. 1–31.

Adinolfi, M., B. Pertl et al. (1997), 'Rapid detection of aneuploides by microsatellite and quantitative florescent polymerase chain reaction', *Prenatal Diagnosis*, **17**, 1299–311.

Akerlof, G.A. (1970), 'The market for lemons: qualitative uncertainty and the market mechanism', *Quarterly Journal of Economics*, **84**(3).

Arrow K.J. (1963), 'Uncertainty and the welfare economics of medical care', *American Economic Review*, **53**(5), 961.

Arrow K.J. (1964), 'The role of securities in the optimal allocation of risk-bearing', *Review of Economic Studies*, **31**(2), 91–6.

Arrow, K.J. (1974), The Limits of Organisation, New York: W.W. Norton & Company.

Baker, T. (1996), 'On the geneology of moral hazard', *Texas Law Review*, **75**, 237.

Badano, J.L. and Katsanis, N. (2002). 'Beyond Mendel: an evolving view of human genetic disease transmission', *Nature Reviews Genetics*, **3**, 779–89.

Berkhout, F. (2002), 'Novel foods, traditional debates: framing biotechnology sustainability', *New Genetics and Society*, **21**(2) 131–48

Bernstein, P.L. (1996), *Against the Gods: The Remarkable Story of Risk*, New York: Wiley and Sons.

Bijker, W.E. (1995), *Of Bicycles, Bulbs, and Bakerlite: Towards a Theory of Social Change'*. Cambridge, MA: The MIT Press.

BMA (1998), *Human Genetics: Choice and Responsibility*, Oxford: Oxford University Press.

Burke, W., L. Pinsky et al. (2001), 'Categorizing genetic tests to identify their ethical, legal and social implications', *American Journal of Medical Genetics*, **106**(3), 233–40.

Chandler, A.D. jr. (1977), *The Visible Hand*, Harvard, MA: Belknap Press.

CMGS Executive (1998), 'Gene patents and clinical molecular genetic testing in the UK: threats, weaknesses, opportunities and strengths', Clinical Molecular Genetics Society, 2002.

Cookson, C. (2002), 'Company to stop direct selling of genetic tests', *Financial Times*.

Copeland, E. (2002a), 'All in the genes', *Health Which?* (June), 10–13.

Copeland, E. (2002b), 'Healthy genes.' *Health Which?* (December), 22–4.

Concar, D. (2003), 'Test blunders risk needless abortions', *New Scientist*, 4 May, 4–7.

Donnai, D. and R. Elles (2001), 'Integrated regional genetics centres: current and future provision', *British Medical Journal*, **322**(28 April), 1048–52.

Dosi, G. and M. Egidi (1987), 'Substantive and procedural uncertainty: an exploration of economic behaviours in complex and changing environments', SPRU DRC discussion paper no. 46, Brighton: SPRU, University of Sussex.

Editorial (1996), 'Keep an eye on genetic screening', *Nature*, **384**, 93.

Elliot, R. and T. Ong (2002), 'Nutritional genomics', *British Medical Journal*, **324**, 1438–42.

GeneWatch (2002), 'Genetics and predictive medicine: selling pills, ignoring causes', GeneWatch, UK.

Hacking, I. (1975), *The Emergence of Probability: A Philosophical Study of Early Ideas about Probability, Induction, and Statistical Inference*, New York: Cambridge University Press.

Harris, R. and M. Reid (1997), 'Medical genetic services in 31 countries: an overview', *European Journal of Human Genetics*, **5** (supplement 2), 3–21.

Harper, P. and A. Clarke (1997), *Genetics, Society and Clinical Practice*, Abingdon, UK: Bios Scientific.

Hawkes, N. (2003), 'Genetic tests must not go on sale, government told', *The Times*, London, 10 April.

Hayek, F. von (1945), 'The use of knowledge in society', *American Economic Review*, **35**, 519–30.

Hayek, F. von (1978), *New Studies in Philosophy, Politics, Economics and the History of Ideas*, Chicago: Chicago University Press.

HGC (2002), 'The supply of genetic tests direct to the public: a consultation document', London: Department of Health, Human Genetics Commission.

Holtzman, N. (1998), 'The UK's policy on genetic testing services supplied direct to the public – two spheres and two tiers', *Community Genetics*, **1**(1), 49–52.

Holtzman, N. (1999), 'Are genetic tests adequately regulated?', *Science*, **286**, 409.

Kegley, J. (2000), 'Confused legal and medical policy: the misconceptions of genetic screening', *Medicine and Law*, **19**, 197–207.

Kevles, D. (1985), *In the name of eugenics: Genetics and the Uses of Human Heredity*, New York: Knopf.

Keynes, J.M. (1937), 'The general theory of employment', *Quarterly Journal of Economics*, **51**(2), 109–23.

Kitcher, P. (1996), *Lives to Come: The Genetic Revolution and Human Possibilities*, New York: Simon and Schuster.

Knight, F. (1921), *Risk Uncertainty and Profit*, Boston, MA: Houghton Mifflin.

Latour, B. (1987), *Science in Action*, Cambridge, MA: Harvard University Press.

Lazonick, W. (1991), *The Myth of the Market Economy*, Cambridge: Cambridge University Press.

Ledley, F. (2002), 'A consumer charter for genomic services', *Nature Biotechnology*, **20** (August), 767.

Lenaghan, J. (1998), *Brave New NHS?*, London: IPPR.

Levett, L., S. Liddle et al. (2001). 'A large-scale evaluation of amino-PCR for rapid prenatal diagnosis of fetal trisomy', *Ultrasound Obstetrics and Gynaecology*, **17**, 115–18.

Loasby B. (1976), *Choice, Complexity and Ignorance: An Inquiry into Economic Theory and the Practice of Decision Making*, Cambridge: Cambridge University Press.

Martin P., S. Crowther et. al. (2000), *Mammalian cloning in Europe: prospects and public policy*, Brighton: SPRU, University of Sussex.

Martinson, J. (1999), 'Monsanto pays GM price', *The Guardian*, 21 December.

Merz, J., A. Kriss et al. (2002), 'Diagnostic testing fails the test', *Nature* **415** (February), 577–9.

Milburn, A. (2001), 'Speech by the Secretary of State for Health', International Centre for Life, Newcastle, 19 April.

Morange, M. (2001), *The Misunderstood Gene*, London: Harvard University Press.

Moss, D. (2002), *When All Else Fails*, Cambridge, MA: Harvard University Press.

Moss, L. (2003), *What Genes Can't Do*, Cambridge, MA: The MIT Press.

Mowery, D.C. (1983), 'The relationship between intra-firm and contractual forms of industrial research in American manufacturing 1900–1940', *Explorations in Economic History*, **20**, 351–74.

Newton, C.R., A. Graham, L.E. Hepinstall, S.J. Powell, C. Summers, N. Kalsherer, J.C. Smith and A.F. Markham (1989), 'Analysis of any point mutation in DNA. The amplification refractory mutation system (ARMS), *Nucleic Acids Research*, **17**, 2503–16.

Nightingale, P. (2000), 'Economies of scale in experimentation: knowledge and technology in pharmaceutical R&D', *Industrial and Corporate Change*, **9**(2), 315–59.

Nightingale, P. and R. Poll (2000), 'Innovation in investment banking: the dynamics of control systems within the Chandlerian firm', *Industrial and Corporate Change*, **9**(1) 113–41.

Nelson, R.R. and S. Winter (1982), *An Evolutionary Theory of Economic Change*, Harvard, MA: Belknap Press.

Nuffield Council on Bioethics (2002), 'The ethics of patenting DNA: a Discussion Paper', London: Nuffield Council on Bioethics.

Organisation for Economic Co-operation and Development (2000), *Genetic Testing: Policy Issues For The New Millennium*, Paris: OECD.

Pollitt, R.J., A. Green et al. (1997), 'Neonatal screening for inborn errors of metabolism: cost, yield and outcome', London: Health Technology Assessment.

Priest S. (2000), 'US public opinion divided over biotechnology?', *Nature Biotechnology*, **18**, September, 939–42.

Ramsden, S. (2002), 'Participants Manual', National External Quality Assessment Service for Molecular Biologists, Manchester, UK.

Rose, P. and A. Lucassen (1999), *Practical Genetics for Primary Care*, Oxford: Oxford University Press.

Rose, H. (2000), 'Risk, trust and scepticism in the age of the new genetics', in U.B.B. Adam and J. VanLoon (eds), *The Risk Society and Beyond*, London: Sage Publications, pp. 63–77.

Rothschild, M. and J. Stiglitz (1976), 'Equilibrium in competitive insurance markets', *Quarterly Journal of Economics*, **90**, 629–49.

Rothwell, R., C. Freeman, A. Horsley, V.T.P. Jervis, A.B. Robertson and J. Townsend (1974), 'SAPPHO updated', *Research Policy*, **3**(3), 372–87.

Schissel, A., J. Merz et al. (1999), 'Survey confirms fears about licensing of genetic tests' *Nature*, **402**, 118.

Stirling, A. (1993), 'Environmental valuations: how much is the emperor wearing?', *Ecologist*, **23**, 15–32.

Stirling, A. (1997), 'Multicriteria mapping: mitigating the problems of environmental valuation?', in J. Foster (ed.), *Valuing Nature? Economics, Ethics and Environments*, London: Routledge, 186–210.

Stirling, A. (1998), 'Risk at a turning point', *Journal of Risk Research*, **1**(2) 97–110.

Stirling, A. (2000), 'On "precautionary" and "science based" approaches to risk assessment and environmental appraisal', in A. Klinke, O. Renn, A. Rip, A. Salo and A. Stirling (eds), *'On Science and Precaution in the Management of Technological Risk'*, vol. II, case studies, European Science and Technology Observatory, Sevilla, report EUR 19056/EN/2, European Commission Joint Research Centre.

Teece, D. (1986), 'Profiting from technological innovation: implications for integration, collaboration, licensing and public policy', *Research Policy*, **15**, 285–305.

The Royal Society (1992), 'Risk: analysis, perception and management', London: Royal Society.

Thomas, S.M., M.M. Hopkins et al. (2002), 'Shares in the human genome – the future of DNA patenting', *Nature Biotechnology*, **20** (December), 1185–8.

Tversky A. and D. Kahneman (1974), 'The framing of decisions under uncertainty', *Science*, 185, 1124–31.

Tversky, A. and D. Kahnemen (1983), 'Extensional versus intuitive reasoning: The conjunction fallacy in probability judgments', *Psychological Review*, **90**, 293–315.

Weinstein, N.D. (1989), 'Optimistic biases about personal risks', *Science*, **246**, 4935.

Wynne, B. (1992), 'Uncertainty and environmental learning', *Global Environmental Change*, June, 111–27.

Wadman, M. (2001), 'Testing time for gene patent as Europe rebels', *Nature*, **413**, 443.

Weber, M. (1934), *Essays in Economic Sociology*, New York: Bernsmann.

Westphal, S. (2002), 'Your money or your life', *New Scientist* (13 July), 29–33.

Zimmern, R. (1999), 'Genetic testing: a conceptual exploration', *Journal of Medical Ethics*, **25**(2), 151–6.

7. Networks and technology systems in science-driven fields: the case of European food biotechnology

Finn Valentin and Rasmus Lund Jensen

1. INTRODUCTION

This chapter has a theoretical as well as a practical aspect. From the latter perspective we study a representative field of biotechnology – microorganisms applied in food processing – where Europe innovates at orders of magnitude higher than the USA, much in contrast to most other fields of biotechnology (Henderson et al 1999). The enormous economic potential of biotechnology makes it an issue of considerable practical relevance to understand why Europe performs particularly well in this field. At the same time this area of food science, like all research and development (R&D) in biotechnology, relies heavily on network formation between organizations (Saviotti 1998). Leads are derived primarily from relationships between organizations, less from their advantages as separate units. The better we theoretically understand these inter-organizational arrangements, the more precisely may we extract the lessons from this particular niche of European success.

Inter-organizational advantages have been theorized in more versions than can be considered in a single chapter. Theories on systems of innovation, however, offer a conceptual strategy particularly pertinent for understanding the European case of food biotechnology, and two versions of this theory are principally relatable: (1) national particularities are emphasized in the version known as national systems of innovation (NSI); (2) technological interdependencies are considered essential for system formation in technology systems (TS) theory. Their different answers to the question of how innovations may benefit from relationships between organizations also make them useful ways of considering inter-organizational advantages in this study of food biotechnology.

The research issue of this chapter, consequently, concerns key differences between NSI and TS approaches. Empirically we examine to what extent patterns of network formation in our representative field of food biotech R&D are shaped by NSI effects, respectively by TS effects. This issue gives rise to an additional, methodological objective of the chapter. We cannot fully appreciate the role of research networks without first understanding what their research is about. A classic dilemma in innovation studies has been restrictions in the possibilities for characterizing large quantities of R&D activities in terms of their thematic content. R&D content is accessible primarily in the form of qualitative data, producing a trade-off in research designs between quantity versus depth and richness. Innovation studies, for this reason, have a considerable appetite for methodologies and tools alleviating precisely this trade-off. In this spirit, this chapter tries out novel text-mining tools to bring out dimensions in the text sections of patents in quantitative forms. In this way we build a clearer structure of the R&D agenda for which inter-organizational networking is the requisite organizational form.

We take as our example the specific field of food science and technology that relates to lactic acid bacteria (LAB). This family of microorganisms is used widely in existing food product and process technologies, and also has implications for the emerging partial fusion of food and pharmaceuticals (Valentin and Jensen 2002). LAB has been quite intensively targeted with the tools of biotechnology as they have migrated into food science from their origin in the pharmaceuticals-related discovery chain (Salminin and Wright 1998). Consequently LAB-related research and inventions offer an attractive and well-delimited window on the exploitation of the new biotech science regime in food R&D. The 180 biotech-related LAB patents claimed until year 2000 provide rich information on that exploitation, and they are the key source of the data analysed below (patent search and processing procedures are presented in Valentin and Jensen 2003).

The chapter is structured with a presentation of theory on inter-organizational R&D in Section 2, which also discusses theoretical dissimilarities and complementarities between NSI and TS theories. Sections 3–5 each present an empirical analysis of a separate dimension of the technology system, working their way through the cognitive dimension (3), the actor network dimension (4) and the dimension of economic competence (5). Sections 6 and 7 summarize results and discuss their implications.

2.   THEORY

The literature analyses inter-organizational R&D at three different levels: projects, networks and systems. These separate levels are visualized in Figure 7.1 in a form that invites us also to consider inter-level relationships.

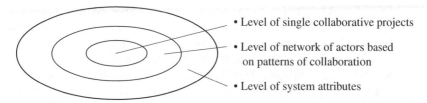

• Level of single collaborative projects

• Level of network of actors based
  on patterns of collaboration

• Level of system attributes

Figure 7.1   Levels of analysis in theories on systems of technology creation

At the project level the literature is focused on collaborative R&D, examining primarily how organizations benefit from participation. Collaborative R&D allows companies to pool their skills in specific processes of problem-solving, hence also to go through learning processes more effectively (Dodgson 1993). As a hybrid governance mode it offers advantages for temporary and partial integration of assets from different companies (Llerena and Matt 1999; Leveque et al. 1996). It also has been argued to be particularly well suited for the type of 'integrative R&D' typifying the science technology interface (Iansiti and West 1999; Valentin 1999).

The network level shifts the centre of attention to the patterns of actors contributing to multiple overlapping collaborative arrangements. Directly and indirectly these contributions make actors part of wider structures, the properties of which impact back on single participants, not least regarding their exposure to relevant information (Powell 1990). For R&D collaborations networks become a repository of distributed knowledge and experience (Küppers 2003). In the fast-learning regime of biotechnology, requiring the integration of diverse skills and experience, the access of single actors to this distributed knowledge significantly affects their subsequent performance (Powell et al. 2001; Liebeskind et al. 1996). Benefits of 'advantageous search positions' become self-perpetuating (Valentin and Jensen 2002), until new architectures in problem-solving downgrade the relevance of previous relational advantages (Orsenigo et al. 2001).

The system level conceptualizes inter-organizational R&D as part of the broader issue of how technological and economic change is affected by its system context. System effects have been theorized in several versions (referred to as e.g. national system, regional systems, technology systems or sectoral systems) (Edquist 1997). They share the non-atomistic point of departure that economic behaviour to some extent is shaped by higher-order effects, which are not adequately conceptualized merely as aspects of single transactions, collaborations or networks (Nelson 1993). Common ground among system theories also includes the views that:

1. Systems are reinforced by the behaviour they shape. Actors build routines and develop micro-level socio-economic patterns of interactions that reinforce system characteristics. Path-dependency and cumulative effects therefore are also essential aspects of their development.
2. Systems are multidimensional, comprising incentives, regulation, norms, cognitive characteristics and so on. Inconsistency between dimensions translates into tensions, and their resolution plays an important part in driving system change.
3. So while systems define contingencies for economic behaviour at the levels of actors and networks, this behaviour, in turn, may accumulate into patterns that modify or resolve system level contingencies, forming inter-level interactions as it were.
4. The explanandum refers to competitiveness of techno-economic systems. Fields in which systems are economically specialized have a central, mediating role in explanatory structures because they typify consistency and congruence between various system dimensions. These are the fields where economic behaviour in particular benefits from and at the same time reinforces system characteristics.

Beyond these and a few additional common assumptions important differences distinguish specific versions of system theories. Among these versions we consider only TS and NSI theories, of which NSI is the more inclusive, while TS has the more rigorous conceptual structure.

Technology system theory defines system in terms of three dimensions and their mutual congruence and interactions, as visualized in Figure 7.1. What distinguishes TS theory from other system theories is primarily its point of departure in the cognitive dimension of technologies, which has particularly rich implications for other parts of the system during the emerging phase of new technologies (Carlsson 1997). Technological knowledge is modified and expanded through (a) addition of new capabilities, (b) integration and restructuring of knowledge components and (c) accumulation of application specific know-how (Carlsson et al. 2002; Stankiewicz 2000; Stankiewicz 2002). These three processes translate into continuous shifts and modifications of the R&D agenda, calling for ongoing modifications of the dimension referring to actor networks. R&D problem-solving combines internal and external skills and information flows, whereby each unit becomes part of a wider network. Shifts and expansions of R&D issues call for requisite reconfigurations and extensions of the network, giving rise to new patterns of collaborations engaging firms and universities, government laboratories and so on. Cognitive shifts that redefine the boundaries of the R&D agenda are particularly detectable at the level of actor organizations where networks expand and form new areas of density and actor centrality.

The economic dimension affects the system by providing incentives and disincentives for formation of actor relationships, by allocating the R&D funds that directs innovation efforts, and by identifying and organizing the commercial exploitation of resultant new technologies. This conversion of R&D opportunities into business activities is based on economic competence (Eliasson 2000) that resides in firms, entrepreneurs, venture capitalists and so on. Economic competence therefore is specific to markets and application domains, and several blocks of economic competence may interact with problem-solving actor networks and its underlying cognitive dimension (compare current overlapping relationships into biotech research networks from both pharmaceuticals and food companies).

TS is first and foremost a theory on the translation of a cognitively defined opportunity set into economic activity. This translation takes place through formation of various types of inter-level correspondence and congruence (in Figure 7.2 symbolized by vertical lines connecting the triangular configura-

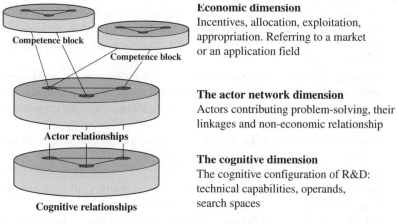

Economic dimension
Incentives, allocation, exploitation, appropriation. Referring to a market or an application field

The actor network dimension
Actors contributing problem-solving, their linkages and non-economic relationship

The cognitive dimension
The cognitive configuration of R&D: technical capabilities, operands, search spaces

Figure 7.2   The three dimensions in the theory of technological systems

tion recurring in all three dimensions). The notion of congruence between dimensions remains to be developed into more rigorous theoretical form. But TS theory already has spurred empirical studies clearly demonstrating effects of differential congruence on performance of systems operating within the same cognitive opportunity set (Carlsson 2002).

Causes of differential congruence, however, are exogenous to TS theory, rendering shortcomings in adjustments to shifting opportunities on part of specific actors a matter of empirical observation. To structure such empirical inquiries, the NSI approach offers a useful complementary perspective.

NSI theory was the first version of system theories on innovation system that in the 1980s emerged as an integration of several strains of previous literature (Freeman and Soete 1997), and it also embraced arguments and approaches of some epistemological heterogeneity (Nelson 1993a). NSI theory on the one hand includes stylized attributes of single nations (e.g. their styles in science or labour market policies, or their specific industrial profiles) and offers on the other hand several novel theoretical arguments on much more general characteristics of innovation dynamics.

The theoretical substance of user–producer relationships, to take a well-known component of NSI thinking (Lundvall 1992), is concerned with interactive learning, arguably not a national but a universal aspect of technological innovation. Of course actors embodying the roles of users and producers in exact interactions bring their national specificities to the setting, but these specificities do not provide the core of the theoretical argument. Real world interactive innovation processes in this way become the combined effect of general attributes of interactive learning and the impact of national specificities, but the NSI framework lacks conceptual distinctions between the two. Along the same lines key arguments of TS theory was introduced as components of the NSI framework, with unclear distinctions between national particularities and more general theorizing of technology dynamics.

NSI thinking became an immense inspiration for subsequent work, some of which necessarily addressed the disentanglement of general innovation and technology dynamics from the sphere of national specificities, as exemplified in, for example, TS theory.

What is left of NSI thinking after these general theories have spun out? While a thorough deliberation of that question is still missing it seems obvious that the issues of national specificities will remain the core of a NSI framework subjected to further improved specification. NSI as a separate body of thinking is essentially concerned with effects – direct as well as cumulative – of specific national institutions and regulations as they intertwine with the specific history of national techno-economic contexts. As such NSI brings out aspects of technology dynamics that are needed for any exhaustive understanding of specific cases.

In the empirical analysis below we take our point of departure in a complementary view on TS and NSI theories. In line with the TS argument we expect actor networks to iterate towards alignment with the opportunities and requirement for problem-solving coming out of the cognitive dimension of their technology system. However, iterations towards alignment of specific actors and their networks are affected by their previous history as part of national particularities, exactly as argued in NSI thinking. Adjustment and response of actors grows out of path-dependency caused by their previous coevolution with a specific national institutional and regulatory setting, that is the core factors in NSI analysis. NSI effects, in other words, may delay, accelerate or modify the way specific networks in the end become aligned with the dynamics of a specific technology system. In this way TS and NSI theories may be seen as defining two complimentary sets of influences, pretty much like wave patterns originating from two different sources of energy. Any specific system evolves in the interference formed by 'waves' originating from both sources. Combinations may be mutually reinforcing, or NSI effects may noticeably impair or delay interdimensional congruence required for full exploitation of TS dynamics. This view essentially sees TS as the system driver and NSI as the response modifier.

If TS and NSI are useful theories we should expect their effects to be detectable as attributes of empirical actor networks. Looking for these attributes is the main empirical task of this chapter, which brings us back to Figure 7.1 above. Relationships between the two inner circles have been fairly well researched (e.g. Pyka and Küppers 2003). Relationships inwards from the outer system circle are essential in systems theories, but have attracted much less rigorous research.

This chapter examines how the intermediate level of networks relates to the outer circle of system attributes. This focus is selected not only because empirical studies of system networks relationships are scarce, but also because precisely this relationship, we submit, is a useful way to uncover core differences between the two versions of system theory we are considering in this chapter. We structure empirical analysis along the lines of the three TS dimensions. The remaining parts of this chapter works itself upwards through the three dimensions of TS analysis, as visualized in Figure 7.2, beginning with the cognitive dimension.

3.   THE COGNITIVE DIMENSION

3.1   Food R&D and the Evolution of Biotechnology

Turning first to the cognitive dimension of food biotechnology this section identifies its origin in the broader revolution of molecular biology and its basic pattern of evolution over the past 25 years. This section focuses on the issues and the cognitive attributes of this R&D agenda. Building on the science of molecular biology and genetics accumulated over the 1950s–60s several interrelated discoveries and inventions provided the breakthrough in biotechnology in the 1970s. They included discovery of reverse transcriptase (1970), the first recombinant plasmids (1973–74). The second half of the decade saw developments of cloning, genetic libraries and DNA sequencing. In 1982 the fundamental techniques of genetic engineering were collected and presented in *Molecular Cloning: A Laboratory Manual*, marking the end of the breakthrough stage of the new science and technology (S&T) field (Morange 1998; Judson 1979). Commercial activities remained sporadic, but included the invention of genetically engineered insulin (1978) and human growth hormone (1979).

Through the 1990s a number of complementary technologies began to align into a coherent, effective set of technologies. High throughput screening techniques allowed the first sequencing of entire genomes (including that of lactobacillus). Bioinformatic tools in the form of DNA chips, data translation tools, protein structure prediction and modelling all combined to make biotech R&D far more cost-effective (Daniell 1999).

This gradual accumulation of biotech insights, instrumentation and tools is largely driven by the search for new opportunities related to pharmaceuticals, but their implications are straightforward for food science. However, the two fields form separate research environments not only in terms of specialized corporate research laboratories, but also in terms of infrastructure provided by university departments, academic degrees and government research institutes. Lags will occur before food science absorbs and utilizes advances coming out of pharmaceuticals'-related research.

Our focus on biotech R&D in the food industry leaves out the vast research effort directed at genetic modification of crops and animals. These agro-products supply the basic raw materials that provide the proteins, fats, carbohydrates and so on, which feed into the value chain of the food industry. Biotech R&D in food processing is concerned with (1) controlling and modifying the basic ingredients to improve their performance (2) production of novel ingredients and (3) processing systems for the incorporation of ingredients into finished products (Jeffcoat 1999; Cheetham 1999).

We use LAB as a representative case of the impact of biotechnology because it plays a crucial role in many areas of food technology. It is a crucial operand in modern dairy technologies, including their increasing attention to functional (probiotic) foods. As such LAB has been a natural focus for application and further development of new biotech tools as they gradually became available over the last 20 years (Margolis and Duyk 1998).

The actual time patterns in food biotech patenting appear in the curves in Figure 7.3 where a dotted line plots the180 LAB patents by year of application. The full line, plotted against the right vertical axis, shows all 3425 food-related biotech patents.

During the 1980 only few food biotech patents appear (less than 100 per year), and LAB patenting is sporadic. A move to a moderately elevated

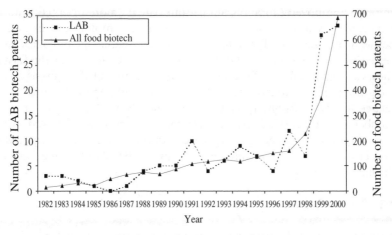

Figure 7.3   Patents in food biotechnology and in the sub-group specifically referring to lactic acid bacteria

plateau begins around 1990. The early 1990s appear to have brought notable changes to LAB R&D. Towards the end of the 1990s a steep increase occurs[1].

The two indicators presented in Table 7.1 bring out a shift in R&D in LAB biotechnology before and after 1992. The first indicator uses the main International Patent Classification (IPC) to distinguish patents

Table 7.1   Indicators of shifts in LAB biotech R&D before and after 1992

Indicators		Periods		All patents
Analytical dimensions	Definitions	1980–91	1992–2000	
Orientation of patents towards tools of genetic engineering versus application	Ratio of method orientation over application orientation	0.7	4.3	2.8
Complexity of research skills	Average number of inventors per patent**	2.3	4.1	3.8

Notes:
* 'Orientation towards tools of genetic engineering' combines patents that are oriented towards enabling technologies with those introducing novelties or modifications at more generic levels. The classification of patents, based on their main IPCs, into these groups is presented in Valentin and Jensen 2002.
** ANOVA (analysis of variance) test: $p < 5\%$.

emphasizing novelty in application from patents more oriented towards biotechnology novelties in tools and enablers, resembling the distinction between 'co-specialised' versus 'transversal research technologies' suggested by Orsenigo et al. (2001). The ratio of the latter over the former shows a steep rise from 1992 onwards, reflecting an added emphasis on the integration of an expanding set of techniques and approaches into LAB-related R&D. This interpretation corresponds well with the growing number of inventors per patent over the three periods, observed in the second indicator, reflecting an increase in the specialized skills required to master the diversity of techniques and approaches that become available through the 1990s.

3.2 Themes in R&D

Turning now to a closer look at the cognitive dimensions of LAB biotech R&D, we draw on data extracted from text sections – titles and abstracts – on each patent front page, identifying the novelty claimed and the key principles of how it is brought about. The standard conversion into quantitative representation of this information starts out by selecting keywords in the text sections of patents. Frequent co-occurrences within patents of keywords signify some common dimension of meaning, and patents may be characterized quantitatively on the basis of their affiliation with various dimensions. Data mining tools are now becoming available that allow large numbers of co-words to be considered, enhancing effectiveness in testing outcomes of different ways of bundling multiple co-words into higher-order dimensions. This chapter uses BibTechMon software in a procedure accounted for in the Appendix, which also explains the principles behind the keyword map presented in Figures 7.4a, 7.4b and 7.4c.

To characterize recurrent themes in the 180 LAB patents we use the 973 keywords that are identified by single circles in the map. Proximity between keywords reflects intensity of co-occurrences. Lines represent the strongest 5 per cent of all 46410 co-occurrences that were used in the iterations to identify 23 themes (statistics on which also are presented in the Appendix).

Not all of the 23 themes lend themselves to meaningful interpretation, and a few of the meaningful themes appear quite randomly across the 180 patents, as indeed they should on the basis of their content. The 12 themes reported on are those that offer both meaningful interpretation and some level of discriminatory effect between interesting categories of patents.

The 12 themes are summarized in Table 7.2 and they fall into three main groups. Two groups refer each to broad areas of application, respectively: (1) food process and quality; (2) pharmaceuticals and probiotic functions; and (3) the third group – for lack of a better term – is referred to as 'enablers'. Without having the status of generic tools, enablers are techniques that allow effects to be achieved and controlled across multiple applications of LAB-related genetic engineering.

Figure 7.4(a) positions these themes in a visualization of the keyword map. Each theme is seen to appear as a 'slice' of the map where co-occurrences of its constitutive keywords are particularly dense, indicated by the lines delimiting each theme. Themes in food process and quality (starter cultures and fermented foods) are positioned on one side of the map, opposite the group of pharmaceuticals and probiotic functions (intestinal infections, probiotics). The four enablers are located apart from each other, in proximity to the areas of applications for which they are particularly pertinent.

Lines delimiting each theme are merely indicative. Even though themes are concentrated in one particular area of the map, their exact scope may vary. For obvious reasons we would expect areas defined by a field of application to materialize with a more narrow focus than will the enablers. Exemplifying this point, Figures 7.4(b–c) give a more detailed presentation of one application theme and one enabler theme, with the keywords absorbed into each theme highlighted in the map. Theme 6 – covering pharmaceuticals applications referring to intestinal infections – has a focused appearance in the upper 'north-eastern' corner of the map. Theme 18 – enablers relating to nucleotide cloning and transfer – has a more distributed pattern, indicating particular affiliation between these enablers and applications in fermented foods and in starter cultures.

On the basis of these findings we tested for differences in the configuration of skills mobilized within each of the above 12 R&D themes. We identified in each patent the scientists coming from public research organizations (PROs) and calculated their share of the entire inventor team into an index referred to as 'PRO intensity'. Correlations were examined between the PRO intensity index and the intensity with which each of the 12 research themes were present in patents (based on occurrence of their lead keywords). Clearly significant correlations were found, positive with some themes, negative with others. There are indications, in other words, of variability in the configuration of skills mobilized in different patents, some themes requiring stronger involvement of public research, other themes requiring less.

To sum up what we learn from the above findings about the cognitive dimension of LAB-related biotech R&D:

- Biotechnology has been applied to LAB in a number of R&D themes, ranging from processing issues (e.g. starter cultures) to food functionalities (e.g. preservation) and further on to pharmaceuticals and probiotic effects. Innovations also include a set of enabling technologies that augment analysis and problem-solving across multiple

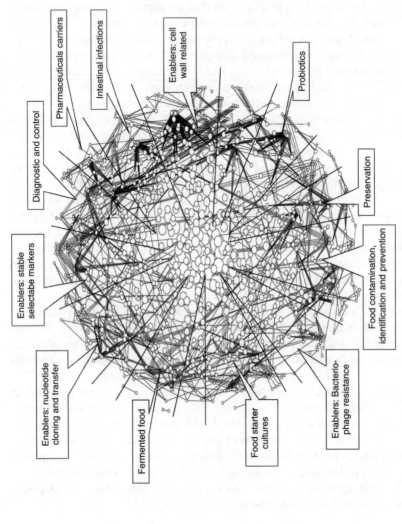

Figure 7.4(a)   R&D themes in 180 patents of LAB-related biotechnology

Probiotics

Enablers: cell
wall related

Intestinal infections

Pharmaceuticals carriers

Diagnostic and control

Enablers: stable
selectabe markers

Enablers: nucleotide
cloning and transfer

Fermented food

Food starter
cultures

Enablers: Bacterio-
phage resistance

Food contamination,
identification and prevention

Preservation

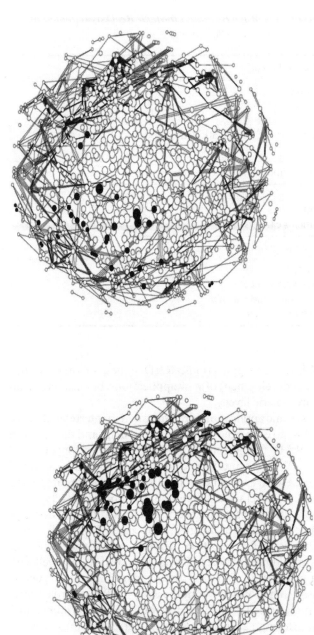

Figure 7.4(b) Keywords referring to R&D theme 6:
intestinal infections

Figure 7.4(c) Keywords referring to R&D theme 18:
enablers; nucleotide cloning and transfer

Notes:
Circles: 973 separate keywords, size indicating number of occurrences for each keyword.
Lines: co-occurrence of keywords within patents.
Number of lines: only 2320 out of a total of 46410 co-occurrences included in this visualization, representing the 5% strongest Jaccard intensities of co-occurrence.

Table 7.2   Selected themes in LAB-related biotech R&D as reflected in titles and abstracts of 180 patents

Theme	Food process and quality
14	Starter cultures
15	Fermented foods
17	Diagnostics and control
19	Food contamination (identification and prevention)
2	Preservation

Theme	Pharmaceutical and probiotic functions
1	Probiotics
6	Intestinal infections
20	Pharmaceuticals carriers

Theme	Enabling genetic technologies
4	Cell wall related
16	Bacteriophage resistance
18	Nucleotide cloning and transfer
21	Stable, selectable markers

areas of LAB-related applications. R&D issues, in other words, involve not only development of new applications, but also problem-solving at more generic levels.

- Enablers and specific applications are innovated in interrelated forms in positions close to each other in the keyword map, and the keyword mix in each R&D theme reflects an integration of process and product issues. Whereas products and process applications would draw on in-house local experience in food and ingredients companies, research skills to develop enabling biotech approaches typically would draw on outside research expertise. Application and development of science takes place as integrated processes. This is a key indication of the necessity for firms to collaborate with outside expertise, as it typically would be available in PROs. This corresponds well with observations reported in Table 7.1 that enablers and more generic issues increased in importance from the early 1990s onwards, bringing with it a pronounced growth in 'distributed forms of innovation' (Coombs and Metcalfe 2000) with a steep increase of inventors per patent.

- Although separate R&D themes are detectable (co-occurences of multiple keywords positioned close to each other in the map) they also overlap (single keywords having co-occurences into several themes). These thematic overlaps signify common elements in the R&D behind the patent and hence also paths for various types of cognitive connections between patents, be they spillovers, cumulative learning, repeated use of same skills across different applications and so on.

- R&D themes differ in the configuration of skills required to carry out innovations, specifically in their involvement of researchers from public science, that is as firms shift their R&D agenda from one theme to another they also will have to reconfigure their networks.

These findings offer a richer and more detailed appreciation of a research agenda compared to what we normally obtain from co-word analysis. It tells us why single R&D projects would often require a multiplicity of skills, ranging from process experience, insights into the molecular biology of raw materials, abilities to develop new research tools and concepts and so on. And it informs us on the need for flexible reconfiguration of these skills and experiences from project to project.

4. THE ACTOR NETWORK DIMENSION

4.1 Industry Structure and R&D

To what extent do networks formed by the actors involved in the R&D problem-solving that was profiled in the previous section develop congruent organizational forms? This section examines attributes of the global organization of this R&D actor network, starting out with a brief summary of the way R&D is structured in food processing.

The vertical structure of the food industry gives rise to a particular distribution of R&D across its sub-sectors and across different types of firms. Each of the sub-sectors in food processing (dairy products, meat products etc.), source not only raw materials that are specific for its final products. They also source a complex mix of ingredients that are essential in process regulation and in modifying tastes, structures and other product functions. Producers of ingredients deliver these inputs based on quite intensive R&D into process and product technology issues across a broad scope of downstream food products. Firms in food processing, on the other hand, traditionally have undertaken R&D on a limited scale, often narrowly focused on particular parameters of quality and hygiene of raw materials and final products. On this basis, the ingredients sector has come to play a growing role in advancing the knowledge frontier in food technologies

Very large multinational corporations (MNCs) like Nestlé and Unilever form a third type of company. Within their global brand name products they are active throughout their value chains, including highly advanced R&D to support each of their links (Boutellier et al. 1999). As a result their R&D agendas stretch from plant biotechnology, to process and ingredients technologies, and further on to food consumption (e.g. taste experience, nutrition etc.) (Pridmore et al. 2000).

Reflecting their different roles, these three types of firms (i.e. food processing, food ingredients and global brand firms based on vertical integration) are highly heterogeneous not only in the intensity but also in the scope of the R&D they undertake. This heterogeneity fundamentally shapes the R&D response of the food industry to the opportunities offered by biotechnology.

4.2   The Global Distribution of R&D

Tables 7.3(a–b) present the geographic distribution of patents in food biotechnology and in LAB specifically. The lead role of the USA in pharmaceuticals-related biotechnology is less pronounced in food, where almost 60 per cent of patents are assigned to non-US organizations. And in the specific field of LAB, roles are completely reversed. The USA has

Tables 7.3(a) and (b)   Geographic distribution of patents in food biotechnology, and the segment thereof relating to LAB

(a) Global distribution			(b) European countries			
Global regions	All food biotech %	All LAB %	European countries	All food biotech %	All LAB %	N
Europe	27	71	NL	13	24	31
Japan	21	14	UK	23	13	17
USA	41	9	FR	16	20	25
Others	11	6	CH	7	22	28
Total	100	100	DK	11	13	16
N	3425	180	Others	30	9	11
			Total	100	100	
			N	925		128

Notes: Nationality of patents refers to the organization(s) to which the patent is assigned. Twenty-eight patents have multiple assignee organizations coming from the same country. The three patents with assignees from different countries were categorized on the basis of predominant nationality of the inventor team.

merely 9 per cent of patents, Japan has 14 per cent, but Europe clearly has a dominant position with 71 per cent of all patents.

The dominant position of Europe in LAB biotechnology is seen in Table 7.3(b) to be concentrated in a few countries. The five most active European countries have a global share of LAB patents of 65 per cent, equivalent to 91 per cent of all LAB patents coming out of Europe.

This geographical distribution of LAB patenting not only deviates markedly from the overall pattern of US dominance in biotechnology; it also brings out the important role played in this field by small European countries. Some specific factors in the five countries help explain their performance. They are countries with a strong performance in biology and food science (Salter et al. 2000), and quite active S&T policies have supported this position. For decades France and Holland have contributed significantly to advances in molecular biology, including specific programmes in sequencing the lactobacillus genome (Morange 1998; van der Meulen and Rip 1998; Roseboom and Rutten 1998). Denmark has operated a series of S&T programmes in biotechnology, with a specific sub-programme in LAB research (Valentin 2000).

More importantly, however, all five countries are also home to one or several MNCs with significant positions in foods and ingredients: Danone and Rodia in France, the Chr. Hansen Group and Carlsberg in Denmark. Nestlé gives strong priority to this field of S&T, including notable activity in fundamental research, concentrated particularly in its headquarter laboratories in Switzerland (Boutellier et al. 1999). Unilever has a dual base in Great Britain and in The Netherlands, where we also find major companies like Quest (a subsidiary of Unilever until it was acquired by ICI in 1996) and DSM. We demonstrate below that the influence of Unilever on the organization of collaborative R&D has been pervasive enough to combine research organizations in The Netherlands and in Great Britain into one coherent network, which we refer to as the 'NUG-system' (Netherlands–Unilever–Great Britain).

4.3   Patterns in the Global Organization of R&D

To examine more closely the organizational attributes of LAB biotech R&D we now turn to the composition of the networks formed by the organizations collaborating in various configurations in the 180 LAB biotech patents.

The inventors behind each patent are listed in its front page. In all 180 LAB patents to a total of 437 inventors are listed. Using a variety of bibliometric sources and Internet searches we identified the host organizations of all but 11 per cent of all inventors and labelled them Inventor

Organisations (IOs). One hundred and eighteen different IOs were identified, to which may be added 32 organizations that have patents assigned to them without having own scientist contributing to the R&D behind the patent ('inventor passive organizations'). A total of 597 inventor participations and 320 organizational participations[2] (OPs) were recorded. Each patent, in other words on the average mobilizes 3.3 inventor participations, distributed on 1.8 different organizations.

Collaboration between scientists coming from different host organizations becomes apparent as a key characteristic of the R&D behind the 180 patents. Intra-patent collaboration, in turn, may be seen as building blocs of a much wider network, each scientist and each organization having overlapping participation in multiple collaborative projects.

In this way the R&D underlying the 180 patents form a network, the global structure and evolution of which may be examined in a graph analysis. To examine how the global network evolves over time, we split the total 1978–2000 period into three six-year intervals and one initial five-year interval, based on year of patent application. These patterns of collaboration are presented for each interval in Figures 7.5(a–d) in the form of network graphs (UCINET software) (Borgatti et al. 1999), while their statistical profiles are presented in Tables 7.4(a–b).

Each node in Figure 7.5(a–d) represents a single IO, graphs signifying collaborations between them. The network graphs for the four periods convey immediately that nationalities of IOs influence how networks evolve and get structured, and that the lead LAB-patenting countries identified in Table 7.3B play a key role in this nation-based order. Yet the exact impact of nationality varies across time and across nationalities in ways that make its role less transparent. We bring out the role of nationality more clearly using the graph analytical concept of 'cut points' (Wasserman and Faust 1994) in the following procedure:

1. When a fully connected graph in a specific period includes several distinct national concentrations, very few OPs may connect these concentrations. One or more OPs form cut points when their removal produces a complete separation into two unconnected sub-graphs. We activate such separations when a single cut point produces sub-graphs characterized by a distinct 'predominant nationality', meaning that at least 50 per cent of its OPs are located in the same country. In this way the share of OPs required as cut points to bring out disconnected national graphs indicates the tendency in networks to form separate systems[3].

2. It follows that a graph with a predominant nationality may include IOs from other countries. Such IOs are referred to as 'non-nationals'. In other words, OPs from other countries may be affiliated with a national

Figure 7.5(a–d)   Networks of R&D collaboration of 180 LAB biotech patents 1978–2000, divided into four periods

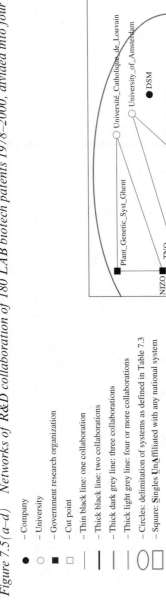

● – Company
○ – University
■ – Government research organization
□ – Cut point
| – Thin black line: one collaboration
| – Thick black line: two collaborations
| – Thick dark grey line: three collaborations
| – Thick light grey line: four or more collaborations
○ – Circles: delimitation of systems as defined in Table 7.3
□ – Square: <u>S</u>ingles <u>UnA</u>ffiliated with any national system

185

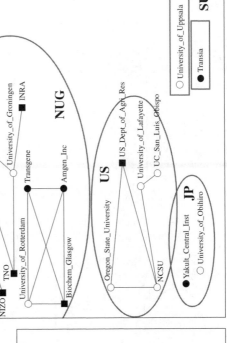

Figure 7.5(b)   Period II: 1983–1988

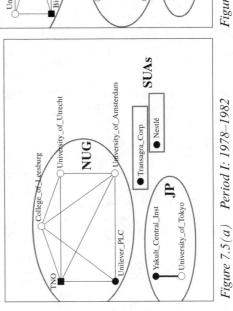

Figure 7.5(a)   Period I: 1978–1982

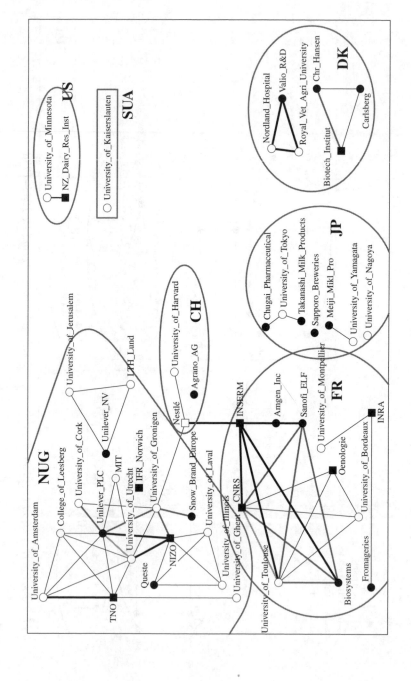

Figure 7.5(c)   Period III: 1989–1994

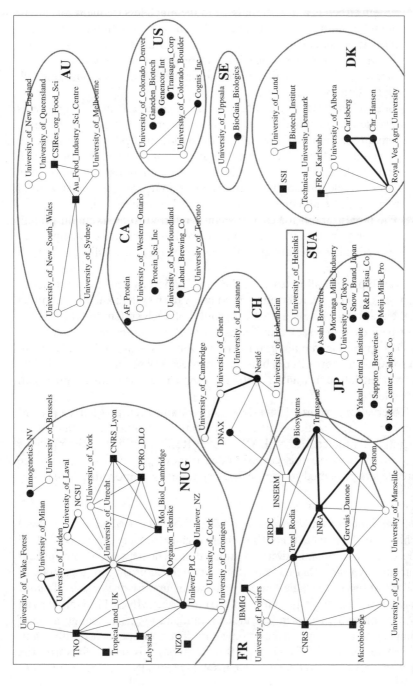

Figure 7.5(d)   Period IV: 1995–2000

187

system either as non-nationals, or in the much less frequent role as cut points to another system. The inclusion of non-nationals emphasize that the present procedure is designed to identify the quality of systems not in terms of geographical borders, but rather as a gravitational force capable of attracting the collaboration also of non-national research organizations.

3. 'Singles' are IOs that are not connected to any graph, indicating patents invented by scientists coming from only one IO, which furthermore remains unconnected throughout the period in question. When located in a country that also classifies the predominant nationality of a graph they are defined as 'singles' affiliated with that national system. In a few cases a single is the only recorded activity in its country, and is referred to a 'single unaffiliated' with any national system (SUA).

4) The largest graph for each system is referred to as its 'main graph'. Smaller graphs may connect additional IOs within each system. They are defined and delimited by the same criteria as are main graphs[4].

All participations affiliated with the same nation as defined in the above steps and criteria are referred to as an innovation system and are delimited in the graph by a circle. Statistical profiles of each system for each of the four periods are presented in Tables 7.4(a–b), giving the percentage share of OPs appearing in main graphs, as cut points, as single OPs and as non-nationals. Numbers of collaborations between organizations, indicated by the nomenclature in Figures 7.5(a–d) complement the statistics presented in Table 7.4.

Observations recorded in Tables 7.4(a–b) and Figures 7.5(a–d) present number of organizational attributes of actor networks, that relate to the cognitive characteristics of LAB biotech R&D identified in the previous section. We analyse this correspondence moving outwards through the three layers presented in Figure 7.1, beginning with individual collaborations.

The predominant architecture of R&D collaboration Most of the R&D behind the 180 patents is collaboratively organized. Only 9.7 per cent of all OPs come from organizations inventing alone. The network graphs make it immediately visible that the most frequent architecture in this large body of collaborations has a single company partnering up with one or several public research organizations, primarily universities. Large MNCs, like Unilever and Nestlé, appear as centres in several such configurations. The configuration of company-to-company collaboration is rare. The correspondence with cognitive characteristics of R&D in LAB biotechnology is straightforward. The previous section showed how problem-solving typically involves a confluence

Table 7.4(a)   Patterns of organizational participations in innovation system during four periods 1978–2000

Org'al participation (OPs) and Inv. Org.[1]		1978–82	1983–88	1989–94	1995–2000	SUM
NUG	% in main graph[2]	100.0	37.5	78.7	93.9	
	% cut points	0.0	0.0	1.1	0.0	
	% single OPs[3]	0.0	6.3	6.4	2.0	
	% non-national [4]	20.0	31.3	12.8	14.3	
	All OPs (N)	5	16	47	49	117
	All organizations (N)	5	12	19	21	
Japan[5]	% in main graph	100.0	0.0	45.5	45.5	
	% cut points	0.0	0.0	0.0	0.0	
	% single OPs	0.0	100.0	18.2	54.5	
	% non-national	0.0	0.0	0.0	0.0	
	All OPs (N)	2	4	11	11	28
	All organizations (N)	2	2	8	9	
USA	% in main graph		100.0	100.0	50.0	
	% cut points		0.0	0.0	0.0	
	% single OPs		0.0	0.0	50.0	
	% non-national		0.0	50.0	0.0	
	All OPs (N)		8	2	6	16
	All organizations (N)		5	2	6	
France	% in main graph			65.0	92.3	
	% cut points			5.0	3.8	
	% single OPs			5.0	7.7	
	% Non-national			5.0	0.0	
	All OPs (N)			20	39	59
	All organizations (N)			11	14	
Switzerland	% in main graph			73.3	100.0	
	% cut points			6.7	5.6	
	% single OPs			13.3	0.0	
	% non-national			26.7	37.0	
	All OPs (N)			15	27	42
	All organizations (N)			3	6	
Denmark	% in main graph			53.8	57.1	
	% cut points			0.0	0.0	
	% single OPs			0.0	9.5	
	% non-national			30.8	14.3	
	All OPs (N)			13	21	34
	All organizations (N)			6	9	

Table 7.4(a) (continued)

Org'al participation (OPs) and Inv. Org.		1978–82	1983–88	1989–94	1995–2000	SUM
Australia,	% in main graph				81.3	
Canada,	% cut points				0.0	
Sweden	% single OPs				18.8	
consolidated	% non-national				0.0	
	All OPs (N)				16	16
	All organizations (N)				15	
Inventor org. without system affiliation		2	2	1	1	6
All OPs without system affiliation		3	2	3	1	9

Table 7.4(b)   *Average shares for all patents and supplementary descriptive statistics*

	1978–82	1983–88	1989–94	1995–2000	%	N
Total % in main graph	77.8	36.7	67.6	84.0	73.4	235
% cut points [5]	0.0	0.0	5.4	3.6	3.8	12
% single OPs	0.0	16.7	7.2	9.2	9.7	31
% non-national	10.0	16.7	14.4	11.8	13.1	42
All OPs (N)	10	30	111	169		320
All organizations (N)	10	21	50	80		118[6]
All inventors	16	46	134	241		437
All inventor participations	15	45	206	331		597
All inventor-passive organizations[7]	1	5	9	17		30
Inventors with unidentified host organization[8]	6	9	12	22		49
Total number of patents	7	17	60	96		180

Notes:
[1] The national affiliation of a graph is defined by the predominant location of its constituent organizations. To qualify as a national graph, at least 50 per cent of its OPs must have identical national location.
[2] The main graph of a system is its largest connected graph in terms of number of OPs.
[3] Not part of a connected graph in the system, but having the same national affiliation.
[4] Part of the connected graph, but located outside the country defining the graph.
[5] Each cut point will appear in two or more of the above system, but counts only once in this summation.
[6] Sum excludes multiple appearances of same inventor organization in different periods.
[7] Organization appearing as assignee only, without having own inventors in any of the 180 patents.
[8] Japan is the most incompletely identified system. Host organizations have been identified for only 28 Japanese inventors, 50 per cent of the total.

of application-specific product and process insights from firms being combined with biotech tools and concepts that would be provided in the most updated version from public science. This is the cognitive confluence that we see reflected in company–university partnerships as the most frequent architecture of collaboration.

Ubiquity of networks Furthermore, organizations have overlapping participations in multiple projects, so that not only collaborations but also wider networks growing out of their overlaps become ubiquitous. In a minority of cases this takes the form of smaller 'satellite' networks, but most of the activity (73.4 per cent of all OPs) appears as part of 'main graphs' that tie together sizeable amounts of organizations and numbers of participations. This ubiquity of networks reflects two of the cognitive characteristics that were brought out in the previous section: (1) in the keyword 'map skills' was observed to reoccur and to be accumulated and exploited across several applications (same theme being expressed – in various versions – in multiple patents). At the actor level that is translated into single organizations contributing to multiple collaborations, thus bringing about an extension of the network; and (2) that also makes the network an effective medium for spreading awareness of specific skills and experience that easily would remain invisible in systems based on arm's length transactions. This argument reverses the direction of causality, networks now perceived not as growing out of innovations, but rather innovations growing out of the particular ability of a network to carry information of such richness that combinatorial novelties more likely come to life. Networks, in other words, are at one and the same time the effect of and the cause of innovations, and that is what endows them with a certain self-perpetuating quality. The present findings confirm that networks once established have continued existence in subsequent periods where they grow (in terms of both size and partially also in density), albeit at different rates.

Networks combine into a few, mutually unconnected innovation systems While networks build on lower-order project collaborations, they become themselves part of higher-order architectures that we refer to as 'innovation systems'. They are systems in the sense that their constituent organizations have frequent internal collaborations, but have very few connections across boundaries. Boundary spanners in the form of cut points are rare, in fact constituting only 3.8 per cent of all participations. The global collaborative R&D in LAB biotechnology, in other words, is segmented into innovation systems that are mutually unconnected. Somewhat surprisingly, there are not numerous innovations systems, each comprising small networks. On the contrary, all activity is concentrated into a small number of systems. Furthermore this concentration is almost all-inclusive. Only six inventor organizations operate as singles, unaffiliated with any of the systems identified in Table 7.4A. The six largest systems account for 289 (90 per cent) of all 320 OPs. As the largest single system the combined NUG system absorbs more than one-third of all OPs. These few, mutually disconnected networks exhibit high levels of internal connectivity. Main graphs tend to connect the vast majority of OPs in each system, accounting for 72.3 per cent of all 320 OPs. Inventor collaborations tend to form a connected chain within each system.

Cognitive aspects of R&D provide a rationality why systems are few and unconnected. Small networks bring only few benefits to collaborative R&D, because their advantages in transmitting rich information is not exploited if actors are few and cognitive variety is limited. Too large networks, on the other hand, make distances between actors exceed the transmission capacity of network linkages. A network offers little advantage to a company looking for complementary external experience to solve a specific R&D problem if that experience is located in another organization several network degrees away. Judging from the architecture of the systems presented in Figures 7.5(a–d) advantages based on skill combinations, spillovers and so on tend to disappear for connections exceeding three degrees. Segmentation into unconnected systems, in other words, shape networks that are neither too small nor too large to offer advantages to collaborative R&D. This argument, of course, gives no indication of where networks would be formed or how they would differ in shape and size within the range where they offer advantages. Looking at their national dimension takes us further in that direction.

Nationality   Innovation systems appear to be segmented along several dimensions. Nationality is one such dimension, but its significance varies across systems. National delimitation is outspoken for the Japanese system in which only domestic organizations participate, confirming the introvert orientation identified also in other studies of Japanese biotechnology (Lynskey 2001). France is similarly self-enclosed, although French organizations on a few occasions contribute to other systems (e.g. INRA in Figure 7.5(b), and DNAX in Figure 7.5(d)). The USA is a small and introvert system in terms of its internal network, but has a number of participations into other systems, particularly in collaborations with Unilever and Nestlé. However, nationality matters less where technology creation has its highest intensity, that is in the two systems with the highest concentration of patent assignments – NUG and Switzerland with their combined share of 44 per cent of all LAB biotech patents (Table 7.3(b)) and 45 per cent of all OPs. The Netherlands and Great Britain are completely intertwined in a single configuration, largely orchestrated though collaborations with Unilever (Figures 7.5(c–d)). This NUG system comprises one-third of all 320 OPs. The Swiss system is in fact a 'Nestlé -system'. Apart from a few appearances of Agrano and the University of Lausanne all other organizations in this system are geographically located in other countries (the UK, Belgium, France, USA) and are referred to as part of the Swiss system only because all their participation occurs in collaboration with Nestlé. The Swiss system, in other words, predominantly consists of Nestlé's activity as a single innovator plus its shifting collaborations with a handful of non-domestic universities.

Clearly nationality partly shapes the way networks coalesce into innovation systems. But other factors are equally important. To better understand this mix of impacts we turn to the third level of TS theory, the dimension of economic competence.

5.   THE DIMENSION OF ECONOMIC COMPETENCE

Economic competence combines market insights with the ability to shape and align actor networks so as to render its R&D problem-solving effective for commercial opportunities and results. Status in the patent as assignee and/or inventor host organization indicates where among the R&D collaborators we find a concentration of economic competence.

In the majority of the 180 LAB patents, assignee organizations have their own scientists participating in a collaborative R&D project involving also other organizations. Assignment presumably has gone to the party who recognized the commercial potential, who conceived and defined the R&D issue, and who orchestrated and financed the project, based on ability to assess potential returns. For these reasons assignment only to a selection of R&D contributors signify their economic competence. The concentration of economic competence in specific organizations and across different types of actors is presented in Table 7.5 in the form of the rate of assignments per organization (AR).

The three most prevalent assignees – Unilever, Nestlé and Chr. Hansen – have an average AR of 25, reflecting their combined share of 37.5 per cent of all 200 assignments. At the next level we find a group of four firms with medium level of AR (6), while all other 31 firms average 1.1 assignments per organization. By far the most frequent type of organization is in public research with an average AR of 0.5. In other words, the overall AR average of 1.4 covers a pronounced polarization. In the majority of organizations economic competence by this indicator is either absent or very thin, and they contribute primarily to problem-solving in collaborative projects

Table 7.5   Organizations and their share of patent assignments

Organizations by type and by assignment rate (AR)	No. of organizations*		No. of patent assignments		No. of patent assignments per organization
	%	N	%	N	
3 firms with highest AR { Unilever**	0.8	1	19.0	38	38.0
Nestlé	0.8	1	13.0	26	26.0
Chr. Hansen	0.8	1	5.5	11	11.0
SUM	*2.4*	*3*	*37.5*	*75*	*25.0*
Firms with medium AR	2.7	4	12.0	24	6
Other firms	35.1	52	27.5	55	1.1
Public research organizations	60.1	89	23.0	46	0.5
All organizations	100.0	148	100.0	200	1.4

Notes:
 * Including all inventor-active and inventor-passive organizations.
** Including subsidiaries, one of which is Quest Int., which in 1996 was sold to ICI. All
    patents assigned to Quest have application dates prior to 1996.

orchestrated by only a handful of firms. Economic competence, as indicated by the AR, is heavily concentrated in very few firms, the three top assignees in particular.

If these three MNCs stand out as particularly competent in spotting market opportunities, do they also represent economic competence by actively shaping and aligning the actor network in R&D problem-solving? That role may be carried out in multiple ways, many of which would not be observable through the present data. But these data do offer an indication of their *direct role* in organising collaborative R&D within their respective systems. For the period 1992–2000 Table 7.6 examines the share of organizations involved in direct collaboration with the three lead MNCs of their respective systems, that is being in 1-degree distance in the terminology of network analysis. This distance is calculated separately for domestic and foreign organizations that are part of the system (as defined by the above criteria).

The direct collaborations of Unilever and Nestlé extend to at least half of all organizations in their respective systems, in the case of Nestlé even including a complete reach in a clearly centric formation (compare the Swiss graph in Figure 7.5(d)). The scope of Unilever's collaboration in actual numbers is by far the largest, and is particularly effective towards foreign organizations in the NUG system. The scope of Chr. Hansen's collaborations includes a comparatively lower share of 33 per cent of the

Table 7.6   Share of all inventor organizations with direct collaboration with lead MNC in each of three systems, 1992–2000

| MNC | System | % of organizations with direct collaboration | | | MNC's % share of all patents assigned to the system | Total no. of organizations in the system |
		% of domestic org.	% of foreign org.	% of all org.		
Unilever	NUG	54.9	100.0	67.7	63.3	31
Nestlé	CH	100.0	100.0	100.0	100.0	7
Chr. Hansen	DK	33.3	20.0	25.0	58.8	8

Danish system. (Chr. Hansen, however, has an indirect (2-degrees) reach of most of the Danish system through its key PRO partner, The Biotechnology Institute.)

To summarize, economic competence is particularly concentrated in three MNCs as indicated by AR. The same companies reveal strong economic competence by their significant role in shaping and aligning the actor networks involved in their respective innovation systems. Ninety-two LAB biotech patents, 52 per cent of the total, come out of these three systems (the combined share for NL, GB, CH and DK in Tables 7.3(a–b). The transparency of these systems provided by the network graphs and the statistics in the previous Section 4 reveal that systems do not coincide with national borders. At the same time nationality unquestionably matters.

6.   RESULTS

The innovation systems observed in this study conform to a technology system model in several fundamental ways.

At the cognitive level the emergence and growth of R&D in LAB biotechnology clearly reflects its origin in the wider biotech revolution, that is it materializes and grows over two decades in rates that are driven from the cognitive level. The global occurrence of patents (Table 7.1) mirrors the general adoption of biotechnology in food R&D, reflecting how biotech tools and approaches gradually become increasingly accessible, steadily expanding the opportunities for profitable R&D. These cognitive dynamics set the scene for organization and networking, as conceptualized in TS theory. Single national systems cannot substantially alter this pattern, but they may adapt to it at different rates and in different ways.

The keyword map brought out key cognitive attributes of LAB-related biotechnology. It showed multiple fields of applications, some closely aligned, others (like food starter cultures and pharmaceuticals) mutually unrelated. Main areas of application each include development of enablers allowing biotechnology tools and solutions to be applied to a product or process domain. LAB-related innovations grow out of these ongoing confluences between application knowledge and biotech research. At the same time thematic overlaps signify common elements in the R&D behind patents in diverse application domains, lending themselves as various types of cognitive connections between patents, be they spillovers, cumulative learning, repeated use of same skills across different applications and so on.

The actor network has key characteristics that are clearly congruent with these cognitive attributes of R&D. Not only is the bulk of all R&D carried out in collaborations combining skills from different organizations. The recurrent basic architecture of these collaborations bring together application specific insights of companies with the most recent advances in biotech skills and knowledge as conveyed by scientists from public research organizations.

Furthermore collaborations overlap into networks, and do so in a telling pattern. Instead of having one interconnected global network or, at the other extreme, having a plethora of small disconnected networks, R&D collaborations form networks that are segmented into a few mutually unconnected systems. Each of these systems includes actors of sufficient numbers, variability and heterogeneity to offer combinatorial advantages. At the same time they are not too large to allow the network to channel opportunities towards effective exploitation. Segmentation into these systems, in other words, make networks particularly instrumental for combinatorial exploitation of opportunities, experience pools and spillovers, generated as cognitive outputs of R&D.

Economic competence materializes as a key driver of system formation. It is particularly concentrated in three MNCs, and the networks branching out from their geographical home bases comprise more than half of all R&D activity (52 per cent of all OPs). The systems associated with these lead MNCs have networks comprising also non-domestic partners. Indeed Unilever's dual base in both The Netherlands and Great Britain brings about a complete integration of the two nations into one internally coherent actor network. Nestlé mobilizes inventor partners independently of their national affiliation, in effect developing its own transnational system.

Indirectly the case of the US innovation system in LAB biotechnology offers information of the role of economic competence. The USA has a number of different organizations involved in LAB biotechnology. But they are either absorbed into collaborative configurations mobilised by Unilever or Nestlé, or they invent sporadically as single organizations, or as small sub-systems. These are indications of the potential for a strong US innovation system in this field matching its overall impressive standing in biotechnology. This potential remains unrealized apparently because the USA in this specific field lacks economic competence at the level offered to their respective systems by Unilever and Nestlé.

NSI effects, nevertheless, also are reflected in the data, particularly in the form of revealed preference in collaborative R&D for domestic partners. This preference appears as an underlying propensity in most systems, the US and the Swiss systems being the most notable exceptions, and Japan being the most pronounced example (all 28 OPs observed since 1978 being domestic). The French pattern also is highly introvert, although French inventor organizations sporadically appear as partners in collaborations instigated from other systems (NUG and Nestlé). For these countries the innovation system does indeed follow national borders. And even Unilever with all its capabilities and global reach has most of its collaborations with domestic (particularly Dutch) public research organizations and so does Chr. Hansen in Denmark. To the extent economic competence finds relevant complementary skills among domestic research organizations, they become their preferred partners.

Furthermore public science has an initiating and anticipatory role that is best understood as a NSI effect. Countries enter this field of R&D in a timing that cannot be attributed to the economic competence of large MNCs. Around 1990 there is considerable French activity with public research organizations playing a leading part (compare Figure 7.5(b)). Similarly the Dutch system 1983–88 is not yet integrated with the British system and seems largely to be driven by public research, which also plays a significant role in the Danish system 1995–2000. Public research might have been induced into this anticipatory role for example by national research programmes or by public scientists simply being more attentive to the broader research needs of their national contexts. Both explanations are consistent with NSI theory.

Clearly both NSI and TS effects are at work in the dynamics of innovation systems, but in the specific science technology field studied in this chapter TS appears to be the more fundamental of the two. NSI effects may explain how networks initially emerge, and how, as in the cases of France and The Netherlands, they may anticipate subsequent commercial requests for advanced research skills and insights. However, the same two examples also clearly demonstrate the difference made by the presence of powerful economic competence. It remains below critical levels in France where no company emerges as a strong organizer of network formation. Its presence in the form of Unilever and Nestlé, on the other hand, is what shapes network growth and architectures in NUG and the Swiss system and what propels them into very high levels of innovation in the latter half of the 1990s. TS effects emerge as the driver, NSI filling primarily a modifying role.

7. DISCUSSION

The implication of these results is that TS and NSI are not just two separate causal mechanisms. In the long run they also affect each other. As an example consider the preference for partnering in collaborative R&D with domestic research organizations observed for the most economically competent MNCs. This preference has a number of causes, but its *sine qua non* is that the company in domestic public research finds skills and experience of particular relevance for the R&D problem at hand. Finding this match in domestic public science to a large extent rests on co-specialisation of public and corporate R&D that precedes the emergence of the new biotech-based TS. It is no coincidence in this context that The Netherlands, France and Denmark are countries with strong research traditions in food and agricultural sciences, particularly in research relating to dairy products where LAB has been a key issue over the past 100 years (Salminin and Wright 1998; Leisner 2002). This history of previous public–private R&D relationships now materializes as a set of national specificities, in that sense constituting a NSI rendering domestic partnering a particular advantage for the launching of a new biotech-based TS. Once domestic public science organizations adjust their R&D agendas to their roles as preferred partners in the new TS they in effect rejuvenate the pattern of public–private co-specialization within the R&D agenda of the new biotech regime. This co-specialization will be fuelled as long as the scientific opportunities offered by the biotech revolution replenishes the opportunity pool (Klevorick et al. 1995), and as long as the economic competence of key actors is not eroded. Once this new TS for one reason or another loses its energy and dissolves it will leave behind its mark on national co-specialization between public science and industrial technology. The co-specialization thus left behind may, or may not, fit into the needs of some future new TS, making the specific NSI either an asset or a liability for the formation of that future TS.

Generally nations exhibit considerable co-specialization of their industrial and public research profiles (Pavitt 1998; Laursen and Salter 2002), and it is maintained and formed over long historic developments (Nelson

1993), clearly spanning the lifetimes of several TSs. Several exchanges between NSI and TS mechanisms, in other words, have revitalized this co-specialization over consecutive rounds. That indicates an ability on part of public science to adjust to the quite dramatic shifts associated with the rise of new TSs and their radically new R&D agendas. That makes the responsiveness of public science – its mechanisms and its limits – a key issue for a deeper understanding of the formation of TSs as motors of science-driven growth. An emerging research agenda is beginning to take a closer look at this endogenous role of science in techno-economic development (Rosenberg 2000; Geuna et al. 2003; Larédo and Mustar 2001). This new agenda also has rich implications for further research into the relationships between TS and NSI effects on science-driven competitiveness.

NOTES

1. The time pattern in Figure 7.1 fits well into a more generalized theory of the dynamics of science-driven technologies (Grupp 1998; Valentin and Jensen 2002).
2. Multiple inventors from the same organization appearing in a single patent is counted as one OP.
3. It should be noted that cut points refer to specific participations, not to entire organizations.
4. Table 7.4 returns the share of inventor organizations in secondary graphs only indirectly as the residual not accounted for by the combined share of the main graph plus 'singles'. For example in the case of Period 3 in France, 65 per cent of organizations are in the main graph, 5 per cent are singles, 5 per cent are involved in cut points; this implies that 25 per cent are connected in separate sub-graphs.

REFERENCES

Borgatti, S.P., M.G. Everett and L.C. Freeman (1999), 'Ucinet 5 for Windows: software for social network analysis', Natick, Analytic Technologies.

Boutellier, R., O. Gassmann and M. von Zedtwitz (eds) (1999), *Managing Global Innovation*, Berlin: Springer.

Carlsson, B. (1997), 'Introduction', in B. Carlsson (ed.), *Technological Systems and Industrial Dynamics*, Boston/Dordrecht/London: Kluwer Academic Publishers, pp. 1–22.

Carlsson, B. (ed.) (2002), *Technological Systems in the Bio Industries*, Boston: Kluwer Academic Publishers.

Carlsson, B., M. Holmén, S. Jacobsson, A. Rickne and R. Stankiewicz (2002), 'The analytical approach and methodology', in B. Carlsson (ed.), *Technological Systems in the Bio Industries*, Boston: Kluwer Academic Publishers, pp. 9–33.

Cheetham, Peter S.J. (1999), 'The flavour and fragrance industry', in V. Moses, R. E. Cape and D.G. Springham (eds), *Biotechnology: The Science and the Business*, Amsterdam: Harwood Academic Publishers, pp. 533–562.

Coombs, R. and S. Metcalfe (2000), 'Organizing for innovation: co-ordinating distributed innovation capabilities', in N.J. Foss and V. Mahnke (eds), *Competence, Governance, and Entrepreneurship*, Oxford: Oxford University Press, pp. 209–31.

Dachs, B., T. Roediger-Schluga, C. Widhalm and A. Zartl (2001), 'Mapping evolutionary economics – a bibiometric analysis', paper presented at the EMAEE Conference, Vienna University, 13–15 September.

Daniell, E. (1999), 'Polymerase chain reaction: development of a novel technology in a corporate environment', in V. Moses, R. E. Cape and D.G. Springham (eds), *Biotechnology: The Science and the Business*, Amsterdam: Harwood Academic Publishers, pp. 147–54.

Dodgson, M. (1993), 'Why collaborate?', in M. Dodgson (ed.), *Technological Collaboration in Industry*, London and New York: Routledge, pp. 25–56.

Edquist, C. (1997), 'Systems of innovation approaches – their emergence and characteristics', in C. Edquist (ed.), *Systems of Innovation: Technologies, Institutions and Organization*, London: Pinter Publishers, pp. 1–35.

Eliasson, G. (2000), 'Industrial policy, competence blocs and the role of science in economic development', *Journal of Evolutionary Economics*, **10**, 217–41.

Freeman, C. and L. Soete (1997), *The Economics of Industrial Innovation*, Cambridge, MA: The MIT Press.

Geuna, A., A.J. Salter and W.E. Steinmueller (eds) (2003), *Science and Innovation: Rethinking the Rationales for Funding and Governance*, Cheltenham, UK and Northampton, MA, USA: Edward Elgar.

Grupp, H. (1998), *Foundations of the Economics of Innovaton*, Cheltenham, UK and Lyme, USA: Edward Elgar.

Henderson, R.M., L. Orsenigo and G.P. Pisano (1999), 'The pharmaceutical industry and the revolution in molecular biology: interactions among scientific, institutional and organizational change', in D.C. Mowery and R. Nelson (eds), *Sources of Industrial Leadership: Studies of Seven Industries*, Cambridge: Cambridge University Press, pp. 267–307.

Iansiti, M. and J. West (1999), 'From physics to function: an empirical study of research and development performance in the semiconductor industry', *Journal of Product Innovation Management*, **16**, 385–99.

Jeffcoat, R. (1999), 'The impact of biotechnology on the food industry', in V. Moses, R.E. Cape and D.G. Springham (eds), *Biotechnology: The Science and the Business*, Amsterdam: Harwood Academic Publishers, pp. 515–32.

Judson, H.F. (1979), *The Eigthth Day of Creation: Makers of the Revolution in Biology*, London: Penguin Books.

Klevorick, A.K., R.C. Levin, R.R. Nelson and S.G. Winter (1995), 'On the sources and significance of interindustry differences in technological opportunities', *Research Policy*, 185–205.

Küppers, G. (2003), 'Complexity, self-organisation and innovation networks: a new theoretical approach', in A. Pyka and G. Küppers (eds), *Innovation Networks: Theory and Practice*, Cheltenham, UK and Northampton, MA, USA: Edward Elgar, pp. 22–54.

Larédo, P. and P. Mustar (2001), *Research and Innovation Policies in the New Global Economy: An International Comparative Analysis*, Cheltenham, UK and Northampton, MA, USA: Edward Elgar Publishing.

Laursen, K. and A. Salter (2002), 'The fruits of intellectual production', DRUID working paper.

Leisner, J. (2002), 'Mælk og Bakterier', Erhvervshistorisk Årbog, 51.

Leveque, F., C. Bonazzi and C. Quental (1996), 'Dynamics of cooperation and industrial R&D: first insights into the black box II', in R. Coombs, A. Richards, P.P. Saviotti and V. Walsh (eds), *Technological Collaboration: The Dynamics of Coorperation in Industrial Innovation*, Cheltenham, UK and Lyme, USA: Edward Elgar, pp. 180–200.

Liebeskind, J.P., A.L. Oliver, L.G. Zucker and M.B. Brewer (1996), 'Social networks, learning and flexibility: sourcing scientific knowledge in new biotechnology firms', *Organization Science*, 7(4), 428–43.

Llerena, P. and M. Matt (1999), 'Inter-organizational collaboration: the theories and their policy implications', in A. Gambardella and F. Malerba (eds), *The Organization of Economic Innovation in Europe*, Cambridge: Cambridge University Press, pp. 179–201.

Lundvall, Bengt-Åke (1992), 'User-producer relationships, national systems of innovation and internationalisation', in B.-Å. Lundvall (ed.), *National Systems of Innovation*, London: Pinter Publishers, pp. 45–67.

Lynskey, M.J. (2001), 'Technological distance, spatial distance and sources of knowledge: Japanese "new" entrants' in new biotechnology', *Comparative Studies of Technological Evolution*, Amsterdam: JAI. Elsevier Science Ltd., pp. 127–205.

Margolis, J. and G. Duyk (1998) 'The emerging role of the genomics revolution in agricultural biotechnology', *Nature of Biotechnology*, 16(4), 311.

Morange, M. (1998), *A History of Molecular Biology*, Cambridge, MA: Harvard University Press.

Nelson, R.R. (ed.) (1993a), *National Innovation Systems*, New York: Oxford University Press.

Nelson, R. (1993b), 'A retrospective', in R. Nelson (ed.), *National Innovation Systems*, New York: Oxford University Press, pp. 505–23.

Noll, M., D. Fröhlich, A. Kopcsa and G. Seidler (2001), 'Knowledge in a picture', paper presented at the R&D Managements Conference Leveraging Research and Technology, Wellington, New Zealand, 7–9 February.

Orsenigo, L., F. Pammolli and M. Riccaboni (2001), 'Technological change and network dynamics: Lessons from the pharmaceutical industry', *Research Policy*, 30(3), 485–508.

Pavitt, K. (1998), 'The social shaping of the national science base', *Research Policy*, 27(8), 793–805.

Powell, W.W. (1990), 'Neither market nor hierarchy: network forms of organisation', *Research in Organizational Behavior*, 12, 295–336.

Powell, W.W., K.W. Koput and L. Smith-Doerr (2001), 'Interorganizational collaboration and the locus of innovation: networks of learning in biotechnology', *Administrative Science Quarterly*, 41(1), 116–45.

Pridmore, R.D., D. Crouzzillat, C. Walker, S. Foley, R. Zink, M.-C. Zwahlen, H. Brüssow, V. Pétiard and B. Mollet (2000), 'Genomics, molecular genetics and the food industry', *Journal of Biotechnology*, (78), 251–8.

Pyka, A., G. Küppers (eds) (2003), *Innovation Networks: Theory and Practice*, Cheltenham, UK and Northampton, MA, USA: Edward Elgar.

Roseboom, J. and H. Rutten (1998), 'The transformation of the Dutch agricultural research system: an unfinished agenda', *World Development*, 26(6), 1113–26.

Rosenberg, N. (2000), *Schumpeter and the Endogeneity of Technology: Some American Perspectives*, London: Routledge.

Salminin, S. and A. von Wright (1998), *Lactic Acid Bacteria: Microbiology and Functional Aspects*, New York: Marcel Dekker.

Salter, A., P. D'Este, K. Pavitt, P. Patel, A. Scott, B. Martin, A. Geuna and P. Nightingale (2000), 'Talent: not technology: the impact of publicly funded research on innovation in the UK', Brighton: SPRU, University of Sussex.

Saviotti, P.P. (1998), 'Industrial structure and dynamics of knowledge generation in biotechnology', in J. Senker and R. van Vliet (eds), *Biotechnology and Competitive Advantage*, Cheltenham, UK and Lyme, USA: Edward Elgar, pp. 19–43.

Stankiewicz, R. (2000), 'The concept of "design space",' in J. Ziman (ed.), *Technological Innovation as an Evolutionary Process*, Cambridge: Cambridge University Press, pp. 234–47.

Stankiewicz, R. (2002), 'The cognitive dynamics of biotechnology and the evolution of its technological systems', in B. Carlsson (ed.), *Technological Systems in the Bio Industries*, Boston: Kluwer Academic Publishers, pp. 35–52.

Valentin, Finn (1999), 'Knowledge and uncertainty in R&D: does inter-firm collaboration make a difference?', paper presented at the 8th International Conference on Management of Technology, IAMOT.

Valentin, F. (2000), *Danske virksomheders brug af offentlig forskning: En casebaseret undersøgelse*, Copenhagen: Danmarks Forskningsråd.

Valentin, F. and R.L. Jensen (2002) 'Reaping the fruits of science', *Economic Systems Research*, **14**(4), 363–88.

Valentin, F. and R.L. Jensen (2003), 'Discontinuities and distributed innovation: the case of biotechnology in food processing', *Industry and Innovation*, **10**(3), 275.

van der Meulen, B. and A. Rip (1998), 'Mediation in the Dutch science system', *Research Policy*, **27**(8), 757–69.

Wasserman, S. and K. Faust (1994), *Social Network Analysis: Methods and Applications*, Cambridge: Cambridge University Press.

APPENDIX: APPLYING BIBTECHMON SOFTWARE TO GENERATE RESEARCH AND TECHNOLOGY THEMES ACROSS 180 LAB PATENTS

To identify common themes in the research and technology issues addressed in the 180 patents we examine co-occurrences of key terms, using the BibTechMon data-mining software (for an introduction see e.g. Noll et al. (2001).

Our initial entry of titles and abstracts of 180 patents into the database generates more than 10 000 separate terms, counting one appearance only of each keyword in each patent. All redundant terms (like 'and, or, else, if' etc.) are deleted, as are terms that in our particular sample would be incapable of generating differences in meaning or final interpretation. The latter include terms like 'contain, mol, concentration, solution'. This reduced set of 2095 terms is standardized to handle differences in spelling, abbreviation, synonyms and so on, bringing us to a final set of 1045 terms, hereafter referred to as keywords.

The network is generated by a co-word analysis, calculating the intensities of all relations between keywords. The intensity of a relationship between any two keywords reflects how frequently they appear together in different patents. This generates a co-occurrence matrix that is normalized using the Jaccard Index given by the equation. The Jaccard Index generates a normalized co-occurrences matrix by the equation:

$$J_{ij} = \frac{C_{ij}}{C_{ii} + C_{jj} - C_{ij}}$$

where C_{ij} are co-occurrences of keywords i and j, and C_{ii} is the total number of occurrences of keywords i.

Using the procedure presented in (Dachs et al. 2001) the two-dimensional map presented in Figure 7.4 is generated, with proximity between keywords reflecting the Jaccard intensity of their relationship. In the keyword network we identify 'themes', defined as configurations of keywords connected by their highest Jaccard intensities. The formation of each theme takes its point of departure in a core of highly connected keywords. Keywords are added successively on the basis of their Jaccard intensities with the core configuration (applying the Shell facility offered by BibTechMon) down to an intensity level of 0.20 (thus leaving 72 keywords (out of a total of 973) unaffiliated with themes). Each keyword contributes to one theme only. Through this procedure a total of 973 keywords are categorized into 23 different themes.

Table 7.A.1 shows themes comprising an average of 42 keywords, each of which has an average occurrence of 3.15 (3). The 23 keywords ranking

Table 7.A.1 Statistics on 23 themes

Theme	Keyword characteristics				Patent characteristics		
	1 Sum of kw per theme	2 Sum of kw occurrences in all pts.	3 Average occurrence	4 Occurrence of most frequent kw	5 Max. occurrence of kw in a single pt.	6 Pts. with 0 kw occurrences	7 Pts. in top medium of 2
1	37	149	4.03	16	26	116	20
2	31	68	2.19	9	20	149	10
3	55	115	2.09	9	23	138	15
4	56	255	4.55	18	37	92	26
5	62	123	1.98	7	32	136	13
6	58	166	2.86	16	34	115	15
7	39	162	4.15	19	20	106	20
8	42	130	3.10	10	18	118	21
9	33	94	2.85	8	11	130	11
10	53	124	2.34	8	25	129	14
11	53	127	2.40	8	11	125	13
12	42	168	4.00	12	23	112	33

13	40	123	3.08	18	14	121	21
14	52	136	2.62	15	19	119	20
15	33	129	3.91	10	20	122	20
16	40	122	3.05	8	17	124	20
17	42	127	3.02	16	19	123	24
18	37	111	3.00	11	10	127	18
19	51	147	2.88	12	10	117	16
20	42	169	4.02	21	12	111	25
21	20	79	3.95	14	8	132	11
22	31	107	3.45	13	19	136	13
23	24	72	3.00	8	13	139	10
SUM	973	3003	72.53				
Average	42	131	3.15	12.64	19.17	123.35	17.8

as the most frequently occurring in their respective themes has an average occurrence of 12.6 (4).

Each patent may be characterized by its number of 'hits' among all keywords or among the subset of keywords within single themes. The most 'keyword-intensive' single patent in each theme average 19 keyword hits. On the average themes are totally unaffiliated (keywords scoring = 0) with 123 patents (6), that is 57 patents having some level of positive scoring within single themes. Column 7 gives patents located in the top median of this distribution, referred to as 'Theme carriers'. For all 23 themes, the average size of the Main Carrier group is 18 patents. The average for the 12 themes included in the analysis in this paper is 24 patents.

8. Future imperfect: the response of the insurance industry to the emergence of predictive genetic testing*

Stefano Brusoni, Rachel Cutts and Aldo Geuna

1. INTRODUCTION

This exploratory study of knowledge investment in the life insurance industry aims at examining the impact of increasing medical knowledge on the actuarial practice of life insurance companies. A huge literature exists on the innovative dynamics of a number of service industries. In the service industries, knowledge is often taken to be highly embedded in day-to-day operations or to be received through the purchase of equipment from other sectors. Indeed, most of the available studies of innovation in the service industries focus on the introduction of information and communication technologies (ICTs) in banking and insurance. Most of these studies focus on the impact of ICTs on the way recipient organizations do traditional things, and how this leads them to introduce new products (see Hecht 2001 and the seminal work of Barras 1990).

In contrast to these studies, this chapter does not consider the diffusion of new or improved equipment. Its emphasis is squarely on the diffusion of new information and knowledge and their impact on the accumulation of intangible capital by firms, for example, the process by which a firm acquires specific new knowledge. Specifically, our analysis is directed to firms' responses to the emergence of a particular body of knowledge: genetics and the related development of genetic screening techniques. The innovative character of this study lies in the examination of how externally generated knowledge is acquired by a service industry. Genetics was chosen because of its potential impact on one of the key activities of the insurance industry – risk assessment. ICTs may enable firms to assess the risk attached to an increasingly greater number of policies, but they do not necessarily bring about any major change in firms' risk assessment capabilities.

This study is necessarily exploratory in nature given the lack of indicators for understanding the extent to which (and in what ways) the service industry approaches knowledge generated by other parts of the economy. An industry study approach is adopted in order to analyse how insurance companies are adapting and developing their knowledge base to keep pace with the breathtaking changes that are occurring, particularly in the scientific understanding of the genetic causes of an increasing number of diseases. In other words, we look at what kind of competence-building strategies (e.g. in-house R&D versus joint research activities) insurance companies are deploying to react to the emergence of genetic screening techniques.

This chapter focuses on a sample of UK-based European firms. The UK focus was adopted for a number of reasons. First, the regulatory framework related to access to and use of genetic information varies widely from country to country. It is likely that such differences impact dramatically on firms' decisions. Therefore, given the limited scale of this study, it became necessary to control for this variable. Second, the UK (life) insurance market is the largest in Europe and all major players are active in it. Thus, our sample of UK-based firms includes all the major European Union (EU) and Swiss firms. Third, the UK industry association, the Association of British Insurers (ABI) has developed a specific code of conduct that goes well beyond the traditional moratoria and bans introduced in other countries. Although a moratorium is now in place in the UK, insurance firms appear to be taking steps to develop competences related to the use of genetic tests, as opposed to the 'wait-and-see' strategy being adopted by their EU-based competitors. In short, the UK is a reasonable setting for an exploratory study of a set of activities that have just started attracting the attention and efforts of the industry.

This chapter is organized as follows. Section 2 reports on the rationale for this study, the key research questions and the methodology applied. Section 3 describes the current state of the technology and the regulatory response of the UK government and the ABI. Section 4 summarizes the main issues raised by the use of genetic information for insurance purposes. Section 5 concludes.

2. INSURANCE AND GENETIC TESTING: THE TECHNOLOGICAL AND INSTITUTIONAL FRAMEWORK IN THE UK

2.1 Genetic Testing

The possibility of using genetic information to assess individual risks for insurance purposes has gained enormous attention in the past few years, particularly in the UK and the USA. Insurance is an indispensable part of life for the vast majority of people in the UK. For instance, mortgages are generally only granted if life insurance is in place. Increasingly, people rely on personal insurance schemes to gain access to various health services.

There are fears among the industry that allowing people to hide the results of genetic tests from insurers could result in the breakdown of the private insurance market. This is a phenomenon known as adverse selection. In this scenario, those who know they are at high risk will be more likely to purchase insurance than those who know their risk is low. Insurers, if unaware of this, assume their risk pool is unchanged and thus underwrite premiums which are too low to cover future claims.[1] This would result in insurers being forced to raise premiums, making insurance cover a less attractive option for those with normal life expectancy. Eventually only those who are classed as high risk will want to buy insurance, leading to the collapse of the pooling system and the insurance market itself.

Genetic information in the form of family history has long played a part in insurance underwriting. The difference between family history and genetic testing is the apparent accuracy of a genetic test in predicting the probability of an individual succumbing to a disease. This would allow insurers to match individual insurance policies and premiums with the predicted diseases. Thus, those individuals with a 'clean' genetic record would be able to obtain insurance at lower rates, while those with 'bad' genes might be unable to obtain insurance at all, and would constitute a 'genetic underclass'.

The clash between these two opposing positions (adverse selection versus a genetic underclass) has generated a rather heated debate about the consequences of the commercial exploitation of genetic information. At this stage, this debate clearly does not anticipate any immediate large-scale application of genetic tests for insurance purposes. As an example of the current size of the problem, the Co-operative Insurance Society (CIS) saw 14 test results out of 460 000 applications for insurance between April 1997 and January 2001 of which only one test affected the terms of the policy (Department of Health 2001: 18).

As discussed later in this chapter, there are still limitations on what the technology can actually deliver. Indeed, the predictive capacity of current genetic tests is limited to identifying those who carry a certain gene, which may predispose them to a particular disease. The timing and likelihood of the disease developing and its severity cannot be specifically forecast. It is argued that because the timing of a genetic disease is not predictable from a test, it is possible that an applicant could be denied cover but might not develop the disease until the period of insurance cover has expired, that is, when they would no longer be a risk to the insurer (O'Neill 1997). It may be unreasonable (and commercially unsound) to deny insurance to certain applicants who may pose no greater risk than others during the period of insurance. Besides, the UK industry, also in response to the pressure of consumers' associations, has established a rather strict 'code of conduct' that further limits the use of those few genetic tests that are currently available.

There has also been discussion about the reliability of genetic tests and their results (Human Genetics Commission 2001; Meek 2001). Positive test results raise uncertainties about the course of disease development. In addition, it may be the case that several different genetic mutations are responsible for a single disorder, and a negative test result for one mutation does not guarantee freedom from the disease. While this scenario does not violate the insurance principle of equal knowledge, as neither the insurer nor the applicant is aware of the error, it highlights the uncertainties surrounding genetic testing.

Besides, not all genetic tests are relevant for insurance purposes, and some categories of insurance may be more affected by genetic tests than others. The disorders most pertinent for insurance purposes are the late-onset dominant mono-genic disorders, of which the classic example is Huntington's disease (Macdonald 2001).[2] The late-onset nature of a genetic disorder is particularly important in discussions about insurance, because it is the predictive nature of the genetic test that is of concern in assessing an individual's risk. There are only a small number of genetic diseases that fall within this category, primarily due to the rarity of late-onset diseases. Currently, few of these diseases can be treated. These diseases are the ones that most concern insurers, because a positive test result will greatly increase an individual's risk, and that risk cannot currently be reduced by treatment or other preventative measures (see Table 8.1).

Some of these diseases may already be included in underwriting considerations based on family history. It thus follows that a negative test result will be positively beneficial in terms of obtaining insurance cover to someone with a family history of a disease. As an example, an applicant who has a parent who suffers from, or has died from, Huntington's disease and has not taken a test themselves is likely to be classed as 'uninsurable' (strictly speaking, insurable, but at an unacceptably high premium rate). If the applicant were to take a test and it proved negative, they would then be able to get insurance cover at a reasonable rate. If the test were positive they would remain uninsurable, but, from an insurance point of view, would be no worse off than if they had not taken the test. The implication, at least for those with a history of Huntington's disease, is that 50 per cent of those refused insurance based on family history could be accepted for insurance following a genetic test.

Finally, it is worth noting that genetic testing may impact on different

Table 8.1   Tests that may be used by insurance companies

Condition	Type of insurance
Huntington's disease	Critical illness
	Income protection
	Long-term care
Early onset familial Alzheimer's disease	Life insurance
Amyloid precursor protein gene	Critical illness
Presenilin 1 gene	Income protection
	Long-term care
Hereditary breast and ovarian cancer	Life insurance
BRCA1 gene	Critical illness
BRCA2 gene	Income protection

Source: Adapted from ABI News Release, 3 April 2001.

insurance products in different ways. Life insurance is a very extensive, mature market and therefore may be able to absorb a certain amount of adverse selection. The newer markets, for long-term care, critical illness and income protection, are much smaller than the life insurance market, and thus may be more vulnerable to the impacts of genetic testing.

2.2   Regulation of the Use of Genetic Tests in Insurance

Genetic testing and its potential use in insurance entered the UK political arena fully in 1995 when the House of Commons Select Committee on Science and Technology published their report entitled 'Human genetics: the science and its consequences' (Department of Health 2001). This report recommended that the industry be given one year to develop a suitable solution to the potential problems of adverse selection or creation of a genetic underclass. A second recommendation was for the creation of a statutory body to oversee developments in human genetics. The government felt that the insurance industry should draw up its own code of practice but, in response to a follow-up report from the Committee, the government set up the Human Genetics Advisory Commission (HGAC), a non-statutory body, in 1996.

The HGAC was in operation from December 1996 to December 1999. In March 1997 it published its first report entitled 'The implications of genetic testing for insurance' (HGAC 1997). This report recommended, first, that a two-year moratorium on the use of genetic tests in insurance should be put in place and, second, that a mechanism should be established whereby the actuarial relevance and reliability of genetic tests could be assessed. The notion of a legislated moratorium was rejected by the Government, but the Genetics and Insurance Committee (GAIC) was established in April 1999 to provide the assessment mechanism recommended by the HGAC. The GAIC also has responsibility for reporting on the level of compliance with its decisions within the industry. By June 2000 the GAIC had published the criteria by which genetic tests would be assessed and approved. These covered issues of technical, clinical and actuarial relevance. The first application was submitted to the GAIC by the ABI in October 2000.

In May 1999 the government established a new body to look at the wider social and ethical issues, as well as the scientific issues, relating to genetics. This Human Genetics Commission (HGC) took office in December 1999 taking on the role previously filled by the HGAC. Following a period of consultation in May 2001 the HGC published its interim recommendations on the use of genetic information in insurance (HGC 2001). This report called for a three-year moratorium on the use of genetic tests in insurance, with an exception being made for policies over £500000. Similarly, the report of the House of Commons Select Committee on Science and Technology on 'Genetics and insurance' which was published in March 2001, called for a voluntary moratorium on the use of positive test results for at least two years (Department of Health 2001, p. xxvi). Both the HGC and the House of Commons committee reports recommended that the moratorium be backed by legislation unless the insurers were able to prove they could regulate themselves.

2.3   The ABI and the Genetic Testing Code of Practice

The ABI represents over 400 insurance companies in the UK, covering about 96 per cent of the market (Department of Health 2001, p. 51). By 1995 the ABI was aware of the issue of genetic testing, and set up the ABI Genetics Committee in 1996. That same year the ABI appointed Professor Raeburn of Nottingham University as its Genetics Advisor. In December 1997, following consultation with its members and the public, the first version of the Genetic Testing Code of Practice was published (ABI 1999a). The Code prevents insurance companies from asking applicants to undergo a genetic test in order to obtain insurance, and companies may only consider the results of tests which have been approved or are being considered by the GAIC. The Code also covers provision for an independent appeals system to consider complaints under the Code and requires companies to certify their compliance annually. Companies that do not comply with the Code could face possible withdrawal of membership of the ABI, and its accompanying bad publicity.

Originally, tests did not have to be disclosed in applications for mortgage-related life assurance up to £100000 for a house to be occupied by the applicant. Following the publication of the House of Commons Select Committee on Science and Technology report in March 2001 and the HGC interim report in May 2001 the ABI extended this moratorium to include all classes of insurance up to a total value of £300000 (ABI 2001a). The Code sets out additional responsibilities for the Chief Medical Officer (CMO) and the Nominated Genetics Underwriter (NGU) in an insurance company. The CMO must be consulted over every application containing a genetic test result, and must provide expert medical advice and judgement. The role of the NGU is also of interest, as it requires a senior underwriter to become the central reference point within the company for matters relating to genetic testing in insurance applications. The responsibilities of the NGU include keeping up to date on genetics and assisting with the training of appropriate staff, and the role provides a 'hub' for company activities relating to genetic testing. While members of the ABI are required to comply with the Code, it could be viewed as a 'minimum standard', with some insurance companies ignoring genetic test results for all applications, or considering only negative results from the tests received (Department of Health 2001).

Currently only one test for one type of insurance has been approved by the GAIC and that is the test for Huntington's disease in relation to life insurance applications. However, until GAIC reaches a decision on applications submitted by the ABI in December 2000, companies may use test results relating to the following conditions and types of insurance which were the subject of those applications.

These tests and four others (for myotonic dystrophy, familial adenomatous polyposis, multiple endocrine neoplasia and hereditary motor and sensory neuropathy) were originally included in a list of tests issued by the ABI in 1998, which could be used by insurers. It has been agreed that, for the four tests which were not submitted to the GAIC and for any submitted which are subsequently rejected by the GAIC, insurers who made use of these tests must refer back to November 1998, when the list was first published. Any applications which received less favourable treatment as a result of the tests being used from that point must be re-underwritten as if the tests had not existed (Department of Health 2001, p. 80).

3. RATIONALE AND METHODOLOGY

3.1 Rationale and research questions

Given the potential implications of genetic testing for the industry, it is no surprise that a substantial body of journalistic and academic literature has accumulated in the last few years focusing on the appropriateness of alternative policy responses to the emergence of predictive genetic testing. This chapter looks at yet another aspect: alongside the policy debate, and the efforts that both insurance firms (directly and through their associations) and customer associations are devoting to influencing the design of the policy framework, are insurance companies actually developing the 'technical' capabilities they need to be able to assess the actuarial and economic implications of genetic testing? And, if so, how?

There is a huge literature that focuses on the processes through which firms monitor, absorb, develop and commercially exploit new information and knowledge. Classic studies on the emergence of large, innovating firms stress the pivotal role played by the in-house R&D department (Chandler 1990). Mowery (1983) also pointed out that in-house and contract research actually complement each other, although contract research tends to focus on the easier, and more predictable, steps of the research process. More recently, both practitioners and academics have highlighted the emergence of new, highly specialized bodies of scientific and technological knowledge that are pushing firms to increase their reliance on external sources of knowledge. Building on the seminal work of Cohen and Levinthal (1989), which introduced the notion of absorptive capacity, a large body of empirical literature stresses the importance of networks of innovators as the locus of learning and innovation (Powell 1990; Freeman 1991).

This research aims to look at whether there is evidence of the development of competences related to genetic screening in the insurance industry. If so, we are interested in understanding whether these competencies are developed in-house, and by what kind of firms, or relying on external suppliers of relevant skills. In doing so, several assumptions have been made. First, that firms benefit from competence building because it improves their competitive advantage and firms that exhibit dynamic capabilities are able to improve or change their competences. Second, insurance companies dealing in long-term businesses, such as life, health and critical care insurance, will benefit from improving their competences in dealing with genetic testing in insurance applications. Third, some genetic tests are more relevant for insurance purposes than others. Examples would be predictive tests over diagnostic tests, and dominant over recessive genetic disorders. A rise in the use of relevant genetic tests will increase the pressure on insurance companies to obtain the competences needed to deal with the tests.

The research hypotheses are that companies will be adopting competence-building strategies in response to genetic testing, and that there will be a difference in the efforts being made by different types of firms to improve their competences in relation to genetics. Moreover, we explore the issue of whether and how firms in the insurance industry are building linkages with external sources of specialized knowledge and whether these networking activities involve all firms in the industry or only some of them.

3.2 Method, data and unit of analysis

There are two main categories of insurance, general insurance (including motor, property and liability policies), and long-term insurance, which includes life assurance, critical illness and income protection insurance, and pensions. This chapter is focused on the long-term insurance sector (excluding pensions) because it is the long-term products such as life and critical illness insurance that are most likely to be affected by genetic testing. For the purposes of this study the terms 'critical illness' and 'income protection' will be collectively referred to as 'health insurance'. This should not be confused with private medical insurance, often provided by employers, but which is not included in this research as it is annually renewable and generally insures groups rather than individuals. Genetic tests are not, at the moment, relevant to this type of policy (Interview I). See Appendix, Table 8.A.1, for a full list of interviewees.

In 1999 (ABI 1999b) there were 829 insurance companies authorized to transact business in the UK, of which 233 were authorized to transact life assurance. This number includes both those companies which deal only in life insurance, and those which deal in both general and life insurance, the composites. In practice, the number actually writing insurance is lower than this (Howitt 1999). The ABI estimates that there are about 120 companies writing long-term policies for life insurance, critical illness insurance and income protection insurance (Interview I).

Table 8.2 shows that the volume of insurance business in the UK generally and in the life insurance market in particular has generally increased, while the number of companies active in that market has fallen. The decrease in the number of companies licensed to operate in the UK echoes the decline seen across Europe during the same period (Eurostat 1999). In 1999 the UK life insurance market was worth £91 775 million and made up 72.5 per cent of the insurance market in the UK. Table 8.2 shows that the life insurance market has almost doubled in value since 1993 when measured by premiums written. During the same period, life insurance has

Table 8.2   The size of the UK insurance market

Year	Total premiums (£ million)	Life premiums (£ million)	Life as % of total	Life companies	Composite companies	Total companies writing life
1989				206	64	270
1990				203	64	267
1991				202	64	266
1992				196	62	258
1993	76778	50207	65.4	194	59	253
1994	76405	48867	63.9	191	57	248
1995	81091	46075	56.8	174	58	232
1996	87867	53999	61.5	177	59	236
1997	96408	61929	64.2	177	65	242
1998	109319	74880	68.5	176	62	238
1999	126648	91775	72.5	171	62	233

Source: Data from Sigma World Insurance series 1994–1999 and ABI yearbook
1989–1999.

increased its share in the total insurance market in the UK from 65.4 per
cent to 72.5 per cent. The UK life insurance market is the largest in Europe,
followed by France and Germany, and is the third largest in the world after
North America and Japan, having 10.5 per cent of the world market in 1999
(Sigma 2000).

The largest 20 insurers in the UK account for most of the life insurance
market. This figure has increased in recent years from 67 per cent in 1994
to the 1998 figure of 80 per cent (Sigma 1999), suggesting that the market
in the UK is becoming increasingly concentrated. In 1998, 35 per cent of
market share (by premiums) was held by the top five companies. The top
10 companies accounted for 54 per cent of the market. Compared to certain
European countries, the UK market is not particularly concentrated. For
example, in France the top five companies hold 53 per cent of the market,
while in Germany the top three have a market share of 36 per cent.

Several sources of information were used in the attempt to study the
competence-building strategies of insurance companies, and to look for
differences between companies. The primary information on firm activities
was gained through an in-depth review of the existing technical and policy
literature, from interviews and questionnaires. It was important to interview both large and small insurance companies, and the organizations they
may interact with. Interviews in the pilot stage were conducted with senior
staff from three large insurance companies and one small insurance

company. Due to the limited availability of comparable financial information on insurance companies in terms of premiums written or turnover, large companies were defined as those included in the ABI Top 20 long-term insurers in 1999 (ABI 1999b). Three were publicly quoted companies and one was a mutual. Two of the companies were independent, and two were subsidiaries of larger groups. Following these interviews it became clear that both the ABI and the reinsurance companies had an important role to play in the way insurers learn about genetics. An interview was carried out with the Chief Underwriter of a reinsurance company to learn more about reinsurers'[3] interactions with insurance companies and the activities of the reinsurers themselves. The Assistant Manager of Life and the Head of Health at the ABI were also interviewed. Finally, the Genetics Advisor to the ABI and the Director of the Genetics and Insurance Research Centre at Heriot-Watt University were interviewed to gain a better understanding of the research work that is going on outside the industry, and the possible impacts of genetics on insurance and individuals.

The semi-structured interviews with companies asked about the different ways in which internal competences are changed, and the use of external and internal bodies to provide information. Specific questions dealt with staff training and employment, internal research efforts, external information sources and activities and the level of genetic testing penetration in the company. Interviews with insurers were anonymous both for the company and the person, as the purpose was to identify some general activities and efforts rather than individual corporate strategies. The non-company interviews were less structured, with simply a brief list of points to be covered prepared beforehand. A number of the interviews were carried out face to face, but time constraints meant that several interviews were conducted over the telephone.[4]

Semi-structured interviews provided a flexible way to identify key areas of activity within companies and external organizations which it was important to focus on in the questionnaires. Information gathered in this way has the advantage that it is up to date, which is often not the case with published information, and is important in an area where changes may occur rapidly. However, the relatively small sample of interviews possible in the time available may have resulted in the information being unrepresentative, but the choice of fairly diverse sample of companies it was hoped would minimize this problem. Furthermore, the interviews did not aim to be exhaustive; they were carried out to ensure that no highly important areas of activity were not excluded from the questionnaire. In addition to these methodological reasons, they were used to depict a background against which we could interpret the results of the survey.

The main areas of activity identified by the interviews were addressed in a short structured questionnaire directed to either the NGUs or CMOs in a sample of insurers. The initial choice for contact was the NGU or the deputy NGU. This is often the chief underwriter in a firm, who has the additional responsibility of ensuring that the firm keeps up to date with developments related to genetics. Thus, it was felt that their knowledge of activities in relation to genetic testing was likely to be the most complete of any individual in the company. The names and contact details (either fax or email) were obtained by telephoning the companies on the ABI list of life and composite insurers. Where the NGU (or deputy) could not be contacted the questionnaire was directed to the CMO, who sees all applications containing a genetic test. Prior to wider distribution the questionnaire was piloted on two senior members of staff from the ABI, and the chief underwriter of one insurance company. Following their comments some amendments were made and the final version achieved. The questionnaire contained stratification questions, to allow comparisons to be made between different types of companies. The main part of the questionnaire consisted of six questions covering the companies' internal and external activities to see if firms were placing greater importance on external sources of information, and thus making use of a network outside the company.

The sample of companies to whom the questionnaire was sent was created from the lists of life and composite insurers (selling both long-term and general insurance) held by the ABI. A total of 32 composite and 144 life companies were listed some of which figured on both lists and so were removed from one. Research on the structure of the industry allowed us to identify a large number of entries that were for subsidiary companies, with a single underwriting department covering all the companies in that group. It also became apparent that several of the companies listed were not carrying out their own underwriting, but were outsourcing it to another insurance firm. Firms which sold only pensions were excluded from the survey. Questionnaires were finally sent to a total of 15 composite and 47 life companies. In our estimation, these firms included all the independent company members of ABI, which represent about 96 per cent of the UK industry. The questionnaire was sent out by fax or email in mid-July 2001, and a reminder was sent two weeks later to those who had not responded. Companies still not responding were contacted a third time and offered the option of completing the questionnaire over the telephone. The response rate to the questionnaire was 32 per cent, which included one response that could not be used in the quantitative analysis, but which provided additional background information. The completed questionnaire response rate was 30.6 per cent, and it is these that are included in the following analysis. We estimate that the number of respondents, although small, covers over 50 per cent of the UK industry in terms of premiums, and includes the vast majority of firms with in-house underwriting capabilities.

4. RESPONSE OF FIRMS TO THE EMERGENCE OF GENETIC TESTING

4.1 Empirical analysis

This section analyses and interprets the results of the empirical research which was undertaken. In this exploratory study the aim is to ascertain whether companies are making efforts to build competences and, if so, whether they are using networks in order to do this. The research tries to identify some characteristics of these networks relating to the role of different participants, the flow of knowledge and the efforts of smaller and larger firms. The insurance industry is involved in a variety of activities in relation to genetics and genetic testing. These include the ABI's Genetics and Insurance Forum (GIF), the UK Forum for Genetics and Insurance (a multi-disciplinary discussion group), and the Genetics and Insurance Research Centre (GIRC) at Heriot-Watt University. This last organization is funded by a group of 12 insurance companies through the ABI, though the centre itself is independent and all its work is available in the public domain. The aims of the centre are primarily to develop models of single and multi-gene disorders, and their effects on insurance, and to develop links with other groups such as geneticists, epidemiologists and health economists (GIF 2001, p. 1).

Research into the mortality and morbidity of people with long-term insurance is undertaken on a continuous basis by the Institute and Faculty of Actuaries, whose Continuous Mortality Investigation Bureau (CMIB) uses data from the insurance companies which are no longer commercially sensitive (Interview I). The CMIB does not at the moment include genetic testing in its research, perhaps due to the small number of policies currently affected by genetic testing (Interview K). As a minimum requirement, those firms that are members of the ABI have to keep up to date with developments and disseminate new information to relevant personnel within the firm. It was important, therefore, to interview a number of companies and other actors in the field to find out how this requirement was being manifested in specific firm activities and whether other activities, not identified by the literature review, were also taking place.

The insurance company interviewees agreed that at present, reliable genetic tests are so few that there is no problem with either adverse selection or a genetic underclass likely at this time. Internally, the NGU has a central role to play in the transfer of information within the company, and the collation of information about genetic testing from both internal and external sources. The responsibility of the NGU to scrutinize all applications including genetic information was confirmed by all the companies interviewed. In the companies interviewed, only the reinsurer was involved in a programme of in-house research, where published medical data are collated by in-house doctors and used for the creation of mortality and morbidity statistics by in-house actuaries (Interview G). In order to access this data subscriptions to a variety of medical journals are taken out by the company. The statistics obtained are used to compile manuals, which are used by the client insurance companies to underwrite all but the most unusual or expensive policies. One insurer was carrying out internal literature reviews, though of an unspecified nature (Interview B).

The ABI is an important source of information for insurance companies, providing regular circulars to the industry. In addition, ABI committees and forums bring representatives of the industry together and 'provide an opportunity to meet with and discuss with peers at other companies' (Interview C). Reinsurers also have an important role to play for the insurance companies. As well as the production of underwriting manuals, reinsurers hold club meetings for underwriters on a regular basis for the discussion of a range of topics, including genetics (Interviews A and D). The participation of the Institute of Actuaries in the collection and analysis of data by the CMIB was mentioned (Interviews H, I and J), as was their role in pricing (Interview C), and in holding actuarial conferences which have recently been opened to underwriters (Interview A). The use of medical publications and other scientific literature in keeping up to date with changes involving genetic testing was mentioned specifically by three of the interviewees (Interviews B, C and G).

Joint activities revealed by the interviews appeared to be limited to involvement in the conferences and committees run by the ABI, reinsurers and Institute of Actuaries. The emphasis of the companies on the small number of applications affected by genetic tests and the lack of data available due to the rarity of both the diseases and the tests has not apparently encouraged them to pool their available data. As one interviewee said: 'There is informal interaction between companies to integrate approaches, . . . but there is a lack of co-ordination between companies' (Interview B). The Genetics and Insurance Research Centre (GIRC) is a joint venture funded by a group of insurance and reinsurance companies coordinated by the ABI. Company involvement is restricted to funding the centre and having a place on the steering committee. There is no formal mechanism for data sharing between the companies and the centre, with research being based on published epidemiological data (Interview K).

The information obtained by the literature review and interviews revealed that companies *are* adopting competence-building strategies in relation to genetic testing. These strategies include both internal activities such as staff training, and external activities such as conferences, seminars and committee membership. The differences in these strategies between large and small firms were not discernible from the interviews. To address this issue, and the role of networks in learning about genetics, the responses to the questionnaire were analysed.

As stated, the response rate to the questionnaire was 32 per cent (20 returned out of 62 sent). All of these firms have in-house underwriting capabilities. Their composition is described in Table 8.3.[5] The breakdown

Table 8.3   Composition of respondents

Composition	Large companies	Small companies	Reinsurers
	42%	42%	16%
GIRC Funders	50%	0%	33%

Source: GIRC: Generic and Insurance Research Centre.

by those companies which are funding the GIRC shows that half of the large companies and a third of the reinsurers which responded contribute to funding the centre: no small firms did. This allows us to claim that the sample of respondents is not biased towards those that tend to cooperate (funders of GIRC).

Further confirmation of the reliability of the sample comes from the differing characteristics of respondents. Thirty-seven per cent of respondents were mutual or cooperative firms, with the remaining 63 per cent being either public quoted, or public limited companies. Thirty-seven per cent were independent firms and 26 per cent were subsidiaries of UK companies, while 37 per cent were subsidiaries of international companies and had their head offices outside the UK. An interesting, and thought-provoking, result of the questionnaire is that only 84 per cent of the firms that responded were training their staff in connection with genetic testing, as required by the ABI Code of Practice. That 16 per cent of the firms were not training their staff should perhaps be of concern, as all the firms who received the questionnaire are members of the ABI, and should thus comply with the ABI Code of Practice.

On the basis of these results we proceed to analyse first the importance of sources of information, to ascertain the role of external sources of information in firms' learning, and thus the potential for the use of networks. This is followed by a discussion of the importance of different sources of

internal and external information and methods of communication for all
the firms. A comparison of these data for large firms, small firms and reinsurers is made, to identify whether differences in learning activities exist
across the groups, as suggested by the research hypothesis.

From Table 8.4 it is clear that external sources of information are significantly more important than internal sources and there is a difference

*Table 8.4   The importance of internal and external sources of information
              for all companies*

Sources	% of companies
Internal most important	5
External most important	58
Both the same	37

between the percentage of companies that rated external as most important
and those that rated internal and external as being of equal importance. It
is not possible to state whether this difference is significant.

A majority of both large and small companies stated that external
sources of information were most important. Far fewer felt that internal
and external sources were of equal importance, and very few felt that internal sources were the most important. This suggests that insurers are relying
more on external information than on information generated inside the
firm. While both large and small insurance companies rely predominantly
on external sources of information, reinsurers stated that internal and
external sources were equally important. Thus, it would seem that in-house
sources of information are much more important to reinsurers than to
insurers. Reinsurers differ greatly from insurers in this respect.

The importance of both internal and external sources has increased in the
last three years for more than half the companies and only 5 per cent believe
that there has been any decrease in the importance of sources about genetics. While there has been an increase in the importance of both internal and
external sources of information for nearly two-thirds of the large insurance
companies, all the reinsurers reported that both internal and external
sources had become more important. In marked contrast to this, the majority of small insurance companies stated that there had been no change in the
importance of either source of information (see Tables 8.5–8.7).

As shown in Figure 8.1, the most important source of information within
the insurance firm is the NGU, with 89 per cent of respondents rating this
as 'extremely important'. However, 5 per cent of firms said that the NGU
was not used as a source of information. Staff other than the NGU, and
seminars and courses were ranked equally as the second most important

Table 8.5   Large insurance companies

Sources	Increased	Decreased	Stayed the same
Internal	63%	0%	38%
External	63%	13%	25%

Table 8.6   Small insurance companies

Sources	Increased	Decreased	Stayed the same
Internal	25%	0%	75%
External	37%	0%	63%

Table 8.7   Reinsurance companies

Sources	Increased	Decreased	Stayed the same
Internal	100%	0%	0%
External	100%	0%	0%

sources of information within the firm with 42 per cent of the respondents rating them as 'extremely important'. Some firms may have included the CMO in the 'other staff' category, rather than listing them in the 'other' category. A fairly small percentage (22 per cent) of all firms rated in-house research as a source of extremely (11 per cent) or moderately (11 per cent)

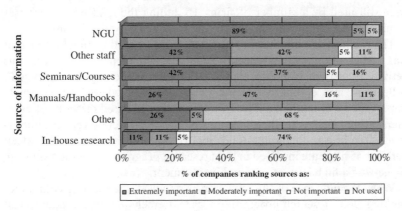

Figure 8.1   How important is each of the internal sources of information to all companies?

important information and this decreased to just 7 per cent when the rankings given by reinsurers were excluded. This suggests that in-house research is of greater importance to reinsurers than insurers. The importance of other internal sources is not greatly affected by removing the influence of the reinsurers.

Table 8.8 shows that, in fact, in-house research is important to all the reinsurers who responded to the questionnaire, in marked contrast to the insurance companies. Only 13 per cent of the large companies rated in-house research important, while none of the small companies made use of in-house research.

Table 8.8   Importance of in-house research by company type

Company Type	Extremely important	Moderately important	Not important	Not used
Reinsurers	33%	67%	0%	0%
Large firms	13%	0%	13%	64%
Small firms	0%	0%	0%	100%

Moving to external sources of information (see Figure 8.2), the ABI is overwhelmingly the most important. Ninety-five per cent of firms ranked the ABI as extremely important, the remaining 5 per cent ranking it as moderately important. All companies claim to make use of the ABI as a source of information. Insurers rely heavily on reinsurers for information with half of the firms stating that they were an extremely important source of information. The other sources, such as other firms, universities, patient groups, consumer groups and so on, were ranked as less important, but were still used by a number of firms. Excluding the reinsurers' responses resulted in no significant changes to the ranking of external sources.

The method of communicating which firms ranked as most important was participation in conferences or seminars, with 68 per cent ranking this activity as extremely important (see Figure 8.3). Personal contact with colleagues, and committee membership were also viewed as extremely important for more than a third of all firms. Paper-based methods of communicating, such as mailing lists and publications were given less importance than face-to-face communication for obtaining information. Sixteen per cent of firms are involved in joint research programmes or data sharing, suggesting a high level of external involvement.

When the influence of the reinsurers is removed, the share of companies ranking 'Joint research programmes' as 'extremely important' falls to just 6 per cent. Table 8.9 shows that 67 per cent of reinsurers rank as extremely important the role of joint research programmes. This suggests that both

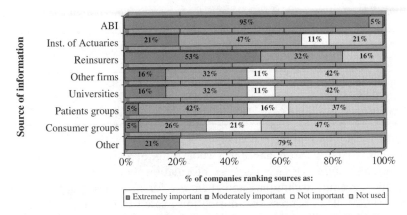

Figure 8.2   How important is each of the external sources of information to companies?

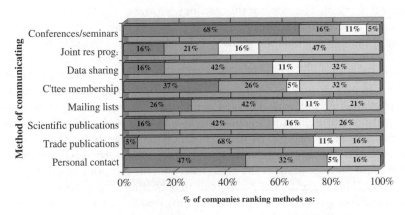

Figure 8.3   How important are each of the methods of communicating with external bodies to companies?

in-house (see Table 8.6) and external research may be more important for reinsurers than insurers. It has already been shown that reinsurers felt that both internal and external sources of information had increased in importance in the last three years, and were of equal importance now.

4.2   Discussion

Insurance companies are involved in competence-building activities in response to genetic testing. Both the interviews and the survey data

Table 8.9   Importance of joint research programmes by company type

Company Type	Extremely important	Moderately important	Not important	Not used
Reinsurers	67%	0%	0%	33%
Large firms	13%	25%	25%	38%
Small firms	0%	25%	13%	63%

highlight that firms are engaged in a whole range of activities aimed at gathering, analysing and even generating information about the development of genetic screening techniques and their use for insurance purposes. The importance of all sources of information related to genetics had not, in general, decreased over the last three years. Overall, all firms draw on both external sources of information and on in-house learning activities. However, the respondents to the survey considered external sources to be more important. This was not the case for reinsurers, who make use of both internal and external sources of information and have a different role to play in the way in which the industry is learning about genetic testing.

Within each organization, the main gate-keeping role is that of the Nominated Genetics Underwriter (NGU), who has the responsibility of collating and disseminating relevant information received from external sources and internal work. Very few companies (that is, only reinsurers and large firms) are attempting to carry out in-house research as a way to learn about genetic screening. The distinction between large and small firms is also emphasized by the fact that 'all sources of information' are more important for large than for the small ones, with the exception of trade publications. This suggests that small firms may be interested in what is going on research-wise, but do not get actively involved in it. This passive role is also illustrated in Table 4.6, with none of the small firms carrying out in-house research. Similarly, Table 4.7 suggests that few small firms see external activities requiring active involvement as important to their learning about genetic testing.

The evidence from the interviews, backed up by the questionnaire analysis, showed that there were two main external sources of information for insurance companies, the ABI and the reinsurers. This applies to both large and small companies, though large companies appeared to make use of other external sources of information as well. The ABI and reinsurers provide different services to the firms. The ABI appears to act as a filter for information from government sources such as the HGC and the GAIC, and represents the interests of the companies to those bodies. The reinsurers have a role in translating published scientific and technical literature into an actuarial form that can be used by companies in underwriting. Such a pivotal role is enabled by maintaining a position of centrality in the emerging network, but it also requires internal development of research that allows reinsurers to both absorb and to generate new information and knowledge that builds on published literature and external research activities.

Figure 8.4 visualizes the network of relations identified through the fieldwork. The arrows indicate the direction of the flow of information related to genetic testing. Reinsurers provide insurers with an informed assessment

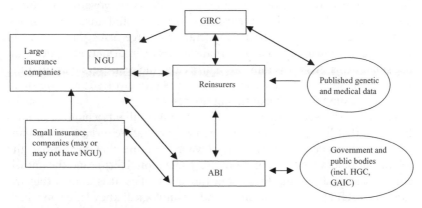

Figure 8.4   The learning network

about the risks attached to claims or applications containing genetic information (large firms do the same for small firms, which often do not have an in-house NGU). Besides the information embodied into proposed premiums, interaction with reinsurers also helps large insurers to understand the developments in and relevance of genetic tests, as it allows them some access to the research undertaken by reinsurers. Reinsurers also learn from large insurers in the sense that through them they can aggregate dispersed information about an issue with which they do not have direct contact, as they do not, in the main, issue policies to private individuals (that is, the carriers of genetic information), but only other companies. With regard to the ABI, as the industry representative, it is both the receiver of and provider to government bodies of information, for example, in the form of submissions to the GAIC. Companies also receive information from the ABI and are providers of data on genetic tests (e.g. how many applications containing genetic information they process over a period of time). As members of the ABI, reinsurers also receive from and provide information to insurers. Figure 8.4 is a general structure that does not address the relative strengths of the various links in the network, but focuses only on the kind of information that is exchanged among different types of organizations.

In a way, a gate-keeping role, similar to that discussed by Cohen and Levinthal (1989) with respect to firm-level activities, is performed for the industry by reinsurers. However, this role is somewhat different to that discussed by Cohen and Levinthal. The gatekeepers identified by the survey are a specific class of organizations (i.e. reinsurers and, to some extent, the ABI), which connect the industry with the scientific or policy-making community. It was not possible to identify a 'technical' role for the ABI, whose key function is to provide a forum for the industry, government and a number of public bodies interested in genetics and insurance issues. Networking activities are more important for large firms than for small firms and large firms rated all external sources of information higher than did small firms. Also it is only large firms and the reinsurers that fund the GIRC. This suggests that the large firms are more involved than are small firms in learning through external networks.

The network structure depicted in Figure 8.4 highlights a number of relationships that link this analysis to the wider literature on network organizations. In particular, there is a growing literature that focuses on the relationship between the evolution of organizational forms and the knowledge bases underpinning firms' innovative activities. It is argued that the range of useful bodies of scientific and technological knowledge on which firms rely is also expanding, with the continuous development of new disciplinary specializations, or even completely new disciplines (see, e.g. Pavitt 1998). Wang and von Tunzelmann (2000) stressed that the range of disciplines relevant to firms' innovative processes is expanding in both breadth, that is, the number of relevant disciplines is increasing, and depth, that is, their sophistication and specialization is increasing. Such increasing complexity at product and technological level in turn challenges firms' in-house learning and innovative processes. Hence, the increasing attention being devoted to the analysis of networks, hybrid organizations and, generally speaking, distributed innovation processes (e.g. Coombs et al. 2001).

One line of research within this recent literature examines the changing relationship between what firms do and what they know. For example, Brusoni et al. (2001) argue that, under certain conditions, firms that rely on wide networks of specialized suppliers for the design and manufacture of specialized components need to maintain in-house competences about the components they outsource. Thus, their knowledge boundaries are wider than their production boundaries. Brusoni et al. (2004) have developed a simple indicator to capture the extent of 'knowledge integration' at firm level. The case of insurance, and the development of genetic screening capabilities, also touches on the relationship between who does what in this industry and who is developing relevant capabilities. This specific service industry is structured around a tight hierarchy, which reflects the traditional division of labour between small and large firms: a small set of large, global reinsurance firms lie at the root of a network of increasingly specialized (in terms of the type of risk they underwrite) and local (in terms of geographic distribution) insurance companies. What is interesting is that this specific pattern of division of labour is reflected in the division of (genetic) knowledge. That is to say, reinsurers are doing in-house research to a greater extent than any other company type in the industry. Also, the vast majority of the insurance companies interviewed confirmed their reliance on reinsurers for assessment of genetic information.

This study also suggests that firms (whether insurers or reinsurers) that perform in-house R&D are more likely to engage in joint R&D programmes with external organizations. This result is similar to the results of a number of case studies in various manufacturing sectors, which show that large firms lie at the root of wide networks of suppliers that are specialized in individual pieces of equipment or bodies of knowledge, but need the coordinating efforts of larger organizations to assemble the final product (Granstrand et al. 1997; Prencipe 1997; Brusoni et al. 2001). Also, it is well known that firms that carry out in-house R&D also rely on contract R&D (Mowery 1983; Cohen and Levinthal 1989).

5. CONCLUSIONS

This chapter has focused on the competence-building activities of insurance companies in response to genetic testing. The results of the empirical research show that companies are making efforts to learn about genetic testing, and that external sources of information are more important to insurers than internal sources. It is therefore reasonable to suggest that companies are making use of networks in order to facilitate their learning. Reinsurers and, more particularly, the ABI, are both very important sources of information for companies and in a network structure reinsurers can be seen as the 'gatekeepers', playing the role of translators of medical and genetic data into actuarial data.

This study has focused on a specific point in the evolution of genetic screening technologies. The sheer novelty of the technology implies that its application for insurance purposes is still limited. The range of tests technically reliable (besides issues of national bans on their use) is so limited that this type of technique has not had an impact on the models insurance firms use to assess individual risk. However, some insurance firms have started to develop in-house and relational capabilities to assess the opportunities and

threats arising from genetic screening. The possibility of using genetic information fired public debate, such that government action was provoked to
quickly establish forums for debate. Developing a working knowledge about
these issues has suddenly become of fundamental importance to an industry that needs and wants to be involved in, and possible even lead, future
debates in the policy-making process.

On a more speculative basis, this chapter raises a number of issues related
to broader questions. First, what is the right notion of productivity to be
applied to this industry? Traditional studies that focus on the impact of
ICTs focus on the role played by scale economies in retailing, that is, productivity is about pooling the risks of as many customers as possible.
Pooling in the insurance industry has been made possible by the availability of new, sophisticated database systems. However, pooling risk is not the
same as 'assessing' risk. Genetics potentially allow firms to assess more and
more precisely individual (as opposed to pooled) risks.

In terms of competition analysis, traditional indicators of 'competitiveness' at industry level rely on numbers of firms, rates of entry and exit,
market shares and concentration. While all these indicators are useful in
stable contexts, they are not necessarily so in fast-changing environments
characterized by the emergence of new bodies of knowledge. This study has
revealed that, while the UK-based industry is highly competitive in terms
of the number of firms active in it, the division of knowledge (related to
genetic testing) is concentrated in the hands of a few global firms (that is,
the reinsurers). This might be interesting from a 'dynamic' competition
analysis point of view. Related to this point is the finding that firms in the
UK market are outsourcing their underwriting function (on the grounds of
the operating costs associated with it). While this study did not focus on this
specific issue, it opens up the question of at what level of analysis should
competitiveness in the insurance industry be assessed. Does 'competition'
involve the number of firms that offer end products to clients through independent retailing networks? Or should it be based on the number of firms
that maintain independent underwriting (and thus risk assessment) capabilities? So far, information has been based on the number of firms that
exist in the market as independent retailers.

In terms of the policy implications of this study, we have seen that EU
member countries have so far adopted their own regulatory frameworks for
use of and access to genetic information. Reinsurers have been shown by
this research to be the key actors in the network and reinsurance is a tightly
knit business managed by a few global firms. Thus, it is not clear to us
whether 'member country' is the right unit of regulation: it is possible that
the right counterpart to global reinsurers is the EU and not individual
member countries.

NOTES

* Participants in the NewKInd Workshop (Maastricht, 11–12 March 2002), the International Schumpeter Society Meeting (Gainesville, Florida, 27–30 March 2002), and 'The economics and business of bio-sciences & bio-technologies: what can be learnt from the Nordic Countries and the UK?' Workshop (Gothenburg, 25–27 September 2002) provided very useful comments. Also, we are grateful to three anonymous reviewers for their very insightful comments. The usual disclaimers apply. Financial support from EUROSTAT, NewKInd Project – IST 1999-20728 is gratefully acknowledged.

1. Underwriting determines the risk that each individual will add to the pool of those insured, and charges a premium to enter the pool based on the risk that is added at the time of application. Insurers calculate premiums so that the pool is sufficiently large to cover the payouts which may have to be made. Because the risk is pooled, the payouts will be matched by the premiums paid into the pool by those who do not need to make a claim.

2. The impact of multifactorial disorders on the insurance industry may be small, because the increased risk associated with a positive test result is small, and may be counteracted (Macdonald 2001). In the case of these diseases a positive test result is an indication of increased susceptibility. Individuals may be able to reduce their susceptibility by making appropriate medical or lifestyle choices, such as opting for preventative surgery or stopping smoking.

3. Reinsurers are the insurance companies' insurers. Insurance companies have an agreement with their reinsurer allowing them to underwrite standard policies up to a certain value without consulting the reinsurer. Unusual policies, such as those containing a genetic test result or those of high value, will be underwritten separately by the reinsurer. The reinsurer will collect a certain proportion of the premiums paid by policy holders in return for taking a portion of the risk associated with each policy.

4. Copies of the questionnaire, semi-structured interview and list of companies to whom the questionnaire was sent are available on request, or can be downloaded with the report that provides the background for this chapter at website www.researchineurope.org/newkind /index.htm, viewed 2 October 2003.

5. Companies which appeared in the ABI 'Current largest 20 based on UK premiums in 1999' (ABI 1999b, p. 85) were considered for the purposes of this analysis as being 'large' companies. Those that did not appear in the list were classed as 'small' companies. This system was used as it was not possible from the data provided by companies or industry literature to determine the size of a company according to the 'gross premiums written', 'turnover' or 'number of policies sold'.

REFERENCES

Association of British Insurers (ABI) (1999a), 'Genetic testing code of practice', website www.abi.org.uk/members/industrybrief/abikey/genetics/gentest99/gentest 99.asp, viewed 20 March 2003.

Association of British Insurers (1999b), *Insurance Statistics Yearbook 1989–1999*, London: ABI.

Association of British Insurers (2001a), 'Insurers confirm decision to extend moratorium on use of genetic test results', website /www.abi.org.uk/Hottopic/nr 415 .asp, viewed 4 September 2001.

Association of British Insurers (2001b), 'Genetic tests approved by GAIC which may need to be disclosed to insurance companies', website www.abi.org.uk /Public/Consumer/Codes/disclosure.asp, viewed 20 March 2003.

Barras, R. (1990), 'Interactive innovation in financial and business services: the vanguard of the service revolution', *Research Policy*, **19**(3), 215–37.

Brusoni S., A. Prencipe and K. Pavitt (2001), 'Knowledge specialisation, organizational coupling and the boundaries of the firm: why firms know more than they make', *Administrative Science Quarterly*, **46**, 597–621.

Brusoni, S., P. Criscuolo and A. Geuna (2004) 'The knowledge bases of the world's largest pharmaceuticals groups: what do patent citations to non-patent literature reveal?' (forthcoming in *Economics of Innovation and New Technology*).

Chandler, A.D. (1990), *Scale and Scope: The Dynamics of Industrial Capitalism*, Cambridge, MA: Harvard University Press.

Cohen, W. and D. Levinthal (1989), 'Innovation and learning: the two faces of R&D', *Economic Journal*, (September), 569–96.

Cohen, W.M. and D.A. Levinthal (1990), 'Absorptive capacity: a new perspective on learning and innovation', *Administrative Science Quarterly*, **35**, 128–52.

Coombs, R., M. Harvey and B.S. Tether (2001), 'Analysing distributed innovation processes', CRIC discussion paper no. 43, University of Manchester, UK.

Department of Health (2001), Government response to the report from the House of Commons Select Committee on Science and Technology: 'Genetics and insurance', Cm5286, October, London: HMSO.

Eurostat (1999), *Insurance in Europe*, Eurostat-DGXV, Brussels: European Commission.

Freeman, C. (1991), 'Networks of innovators: a synthesis of research issues', *Research Policy*, **20**: 499–514.

Genetics and Insurance Committee (2000), 'Notes to accompany applications to GAIC for approval to use genetic test results for insurance risk assessment', GAIC, June, website www.doh.gov.uk/genetics/notes.pdf, viewed March 2003.

Genetics and Insurance Forum (GIF) (2001), 'Genetics and insurance forum research proposals', website www.geneticsinsuranceforum.org.uk/Research/menu.asp, viewed 20 March 2003.

Granstrand, O., P. Patel and K. Pavitt (1997), 'Multitechnology corporations: why they have "distributed" rather than "distinctive core" capabilities', *California Management Review*, **39**, 8–25.

Hecht, J. (2001), 'Classical labour-displacing technological change: the case of the US insurance industry', *Cambridge Journal of Economics*, **25**, 517–37.

Human Genetics Advisory Commission (HGAC) (1997), 'The implications of genetic testing for insurance', HGAC papers, December, website www.doh.gov.uk/hgac/papers/ papers_b.htm, viewed 20 March 2003.

Human Genetics Commission (HGC) (2001), 'The use of genetic information in insurance: interim recommendations of the Human Genetics Commission', London: HGC, website www.hgc.gov.uk/business_publications_statement_01may.htm, viewed 20 March 2003.

Howitt, S. (ed.) (1999), *Market Review: UK Insurance Market*, Hampton: Key Note Publications.

Macdonald, A.S. (2001), 'Genetics and health costs: some actuarial models', website www.ma.hw.ac.uk/~angus/papers/vienna.pdf, viewed 20 March 2003.

Meek, J. (2001), 'Insurers "broke code on gene information"', *The Guardian*, 2 May, Website www.guardian.co.uk/Print/0,3858,4179200,00.html, viewed 20 March 2003.

Mowery, D. (1983), 'The relationship between intrafirm and contractual firms of industrial research in American manufacturing, 1900–1940', *Explorations in Economic History*, **20**, 351–74.

O'Neill, O. (1997), 'Genetic information and insurance: some ethical issues', *Philosophical Transactions of the Royal Society of London, Series B*, 352, 1087–93.

Pavitt, K. (1998), 'Technologies, products and organization in the innovating firm: what Adam Smith tells us that Schumpeter doesn't', *Industrial and Corporate Change*, **7**(3), 433–52.

Powell, W.W. (1990), 'Neither market nor hierarchy: network forms of organization', *Research In Organizational Behavior*, **12**, 295–336.

Prencipe, A. (1997), 'Technological competencies and a product's evolutionary dynamics: a case study from the aero-engine industry', *Research Policy*, **25**, 1261–76.

Sigma (1999), 'No. 6 life insurance: will the urge to merge continue?' Swiss, Re Life & Health, Website www.swissre.com/.

Sigma (2000), 'No.9 world insurance in 1999: soaring life insurance business', Swiss Re Life & Health, website www.swissre.com/.

Wang, Q. and G.N. von Tunzelmann (2000), 'Complexity and the functions of the firm: breadth and depth', *Research Policy*, **29**, 805–18.

APPENDIX

Table 8.A.1   Interview codes

Interview Code	Interviewee	Date
A	NGU, large insurance company	8 June 2001
B	Technical manager, small insurance company	15 June 2001
C	Communications manager, large insurance company	20 June 2001
D	NGU, large insurance company	6 July 2001
E	New business manager, large insurance company	6 July 2001
F	Service manager, large insurance company	6 July 2001
G	Chief underwriter*, reinsurance company	15 August 2001
H	Professor Sandy Raeburn, Genetics Advisor to the ABI	12 June 2001
I	Patrick Mahon, Assistant Manager Life, ABI	3 July 2001
J	Richard Walsh, Head of Health, ABI	3 July 2001
K	Professor Angus Macdonald, Director, Genetics and Insurance Research Centre, Heriot-Watt University	21 August 2001

Notes:
NGU: nominated genetics underwriter.
* Chief underwriter also nominated genetics underwriter.

9. Emergent bioinformatics and newly distributed innovation processes

Andrew McMeekin, Mark Harvey and Sally Gee

1. INTRODUCTION

Although bioinformatics was recognized as a potentially revolutionary approach towards biology as early as 1968 (Ryback 1968, 1978), the application of computational methods to digitalized biological data has only begun to have a major impact across the whole area of bioscience and biotechnology in the last decade or so. With the genomics and now postgenomics revolutions in biology, attention has been increasingly focused on how the newly generated data can be stored, managed, analysed and ultimately used to improve innovation in the life science industries. It has been this proliferation of data, in quantity and of type, that has brought bioinformatics to prominence.

Bioinformatics is a combination of **BIO**logy, **INFOR**mation technology and mathe**MATICS** (Hodgman 2002), but there is some debate concerning its specific focus. Straightforward data management technologies and computational biology (involving modelling and simulation of biological problems, in the absence of data) could define the extremes, but there is a myriad of different activities between. It is the proliferation of new techniques and technologies within this space that forms the focus of this chapter.

Bioinformatics is at the core of the genomic revolution and therefore centrally involved in the significant uncertainties that are emerging in the foundations of biological understanding. New biological units of analysis have been developed and studied, including single nucleotide polymorphisms (SNPs), expressed sequence tags (ESTs) and protein complexes. The precise definition of the gene has been rewritten many times, and is still the centre of controversy. Sequencing the human genome and the subsequent bioinformatics analysis has seen the number of human genes drop from a predicted 175000 to around 36000; and new human genes are still being identified through comparisons with the genomes of other organisms.

Oliver (2002), who was involved with the yeast genome sequencing project, has recently written about gene identification as a protracted process that continues long after a draft sequence has been completed. He suggests that the process will become more efficient with cross-organism comparison and expression studies. In a recent paper, he paraphrases a poem called 'Naming of Parts'[1] applying it to current genomic biology to express the uncertainties that still surround gene expression and gene function:

> To-day we have the naming of parts. Yesterday,
> We had daily sequencing. And to-morrow morning,
> We shall have what to do after arraying. But to-day,
> To-day we have the naming of parts . . .

The nature of biological pathways and mechanisms linking genes, proteins, protein complexes and cell dynamics continues to be studied and emphasis in some quarters has shifted towards 'systems biology' where the ambition is to develop more integrated science at the cellular and organism level. There are ongoing debates concerning which units and levels of biological analysis will reap the greatest scientific yield.

Accompanying this scientific uncertainty, there has been significant fluidity in the organizational basis for genomics and bioinformatics science and innovation. Certainly those involved from the commercial side are as likely to discuss the types of deal that can be struck between firms and public science institutions of different types and sizes, as they are to talk of 'scientific visions'.[2] There has been considerable institutional flux at the leading edge of bioinformatics with the formation of new types of firm and the development of novel mergers, alliances, joint ventures and collaborative agreements. The structure and distribution of innovation within the life science industries have been shifting and there are competing views concerning the most economically viable modes of interaction. There are evidently different strategies that reflect alternative predictions about how and where within this distributed arrangement economic value can be extracted. The strategies are also about finding a valuable position within the whole 'biological picture'; whether to develop expertise at one biological level of analysis or to move towards the 'systems biology' approaches. Similarly, strategies differ according to how far down the innovation pipeline they extend and the extent to which they include the development of marketable consumer products.

One consequence of this organizational fluidity has been the emergence of new types of firms and new inter-firm relationships, some between previously unconnected types of firms. Thus, the life science universe is now populated by transnational life science and computing companies, dedicated biotech firms, dedicated bioinformatics firms, drug discovery firms, pre-competitive consortia, bioinformatics tool providers, universities, public science institutions and more. To understand how activities are coordinated between these different types of economic agent is to understand how innovation is distributed within the life sciences. But the life science universe is not static: first, the universe is expanding to include a wider range of innovation activities and new types of economic agent; second, coordination within the universe is changing to accommodate the new economic agents and the shifting orientations of existing agents. The study of these shifts amounts to an analysis of the dynamics of distributedness of life science innovation processes. So, just as there is a need in biological science 'to rename the parts', there is a parallel renaming of institutional parts within the newly distributed bioinformatics-based innovation processes. Given this fluidity both in scientific/technological activity and organizational structure, the problematic we address in this chapter concerns the changing relationships between the two.

In Section 2, we develop an approach to address this question, which we call 'distributed innovation processes', indicating how this adds to the systems and networks of innovation perspectives. We then briefly outline in Section 3 some methodological consequences of this approach, before presenting the case study in Section 4 which analyses the formation of a major bioinformatics firm, LION Bioscience, its developing science and technology capabilities and its changing interactions. The case study is divided into three parts: its initial formation, its subsequent development of internal capabilities and its emergent external interdependencies. The chapter concludes in Section 5 with an assessment of the analytical framework adopted in relation to the empirical material and the significance of emergent bioinformatics in redrawing the landscape of life science innovation.

2.   FROM NETWORKS AND SYSTEMS OF INNOVATION TO DISTRIBUTED INNOVATION PROCESSES

It is now well known that bioscience innovation is distributed beyond the boundary of the single firm. The concepts of regional cluster (see, e.g., Saxenian 1994; Dohse 2000; Storper and Harrison 1991), network (see, e.g., Orsenigo et al. 2001) and of technological system (see, e.g., Carlsson and Stankiewicz 1995; Carlsson 1997; Stankiewicz 2001; Carlsson et al. 2002), national system (Nelson 1993; and for a retrospective survey of the national systems of innovation approach see Lundvall et al. 2002) and sectoral system (see Malerba 2002) have been used to study the management and coordination of distributed innovation and to account for differences in performance. These approaches all share an interest in understanding how knowledge and resources flow between different agents, and have proposed a variety of processes and conditions through which this occurs. In each case, a particular and limiting parameter is used to explain the characteristics of interactions and interdependencies typically as defining the properties of the system or network (Coombs et al. 2003).

Another key feature of these perspectives is the focus on firms as the key innovating agents. However, this has proved particularly inadequate for understanding innovation processes in the biosciences, where it is widely recognized that interactions between public and private organizations are critical. A variety of different types of knowledge flow (Faulkner and Senker 1994) has been identified and these have been attributed to a variety of different institutional settings across public and private boundaries (Rappert et al. 1999; Fransman 2001). In aggregate, these studies confirm that the traditional public–private divide has been blurred. In addition, it has been argued that the nature of public–private interaction is a determinant of innovation performance. It has been argued (Powell and Owen-Smith 1998; Owen-Smith et al. 2002), that the more developed public–private innovation interactions in the USA compared to Europe are one primary source of comparative advantage in the life sciences. For example, in an earlier stage of our current research on bioinformatics, we identified extreme fluidity at the interface between public, private and hybrid organizations regarding whether bioinformatic knowledge is traded (and tradable) or not (McMeekin and Harvey 2002).

Taken together, these approaches all share an interest in analysing interdependence between innovating agents, whether interdependence is considered geographically, sectorally, technologically or across the public–private divide. Yet, what we observe with the development of bioinformatics is that emergent interactions and interdependencies transcend and transect established 'boundaries' or 'systems'. In this chapter our objective is to develop the analysis of interdependency and asymmetry of interactive relations involved in innovation by asking how innovation processes become organizationally and economically instituted, and how the configurations of such processes change over time. To do so we will develop a framework for the analysis of emergent institutionalization of bioinformatics from the 'distributed innovation process' (DIP) approach (Coombs et al. 2003).

The focus of a DIP approach adds two particular emphases to the networks, clusters and systems literature by taking innovation processes as the unit of analysis. First, it does not assume any particular scale or dimension of a system or population of organizations, but argues that innovation processes by nature potentially disrupt and transcend established institutional configurations. Second, by analysing innovation in terms of the emergences of new classes of economic agent, the reconfiguration of existing agents, and the resultant new patterns of interaction between them, the approach focuses on the dynamics of changing distributedness.

As with the systems and network approaches, the DIP approach is about understanding how different parts of the innovation jigsaw fit together, although in this case the jigsaw itself is changing. The particular focus is on identifying the different classes of economic agent that are connected within a particular innovation process and on the nature of relationships between them. Previous studies have pointed to the dichotomous classification of partners versus rivals or collaborators versus competitors. The DIP approach proposes that rivalry occurs within a class of economic agent; it also suggests that 'partnerships' are conditioned by the distribution of power and interdependencies between agents. These relationships occur between firms and other organizations belonging to different economic classes, and can be manifested through relationships of mutual dependency or by asymmetries of power.

One of the outcomes of these competitive processes is the creation of new classes of economic agent as an instance of a continuing division of innovation labour through increasing specialization. Accompanying the creation of new classes of economic agent is the rearrangement of innovation across the existing classes. This means that we observe changes to the patterns of distributedness of innovation over time and this is applicable across the public–private divide and within and between firms.

So, the nature of coordination (or mode of distributedness) between different innovating agents plays an important part in determining the rate and direction of innovation. Simultaneously, innovations can provide new opportunities or constraints in different ways to different agents, thereby causing a redistribution of innovative effort between them and changes to the extent of co-dependencies and distribution of power.

By following the changes to the distributedness of innovation, we are also analysing the negotiation of scale. The extent to which a given innovation process can transcend national boundaries, or radicalize sectoral boundaries is an outcome of the innovation process, and in the DIP approach, a matter for empirical investigation. This does not mean that geographical or sectoral matters are irrelevant. We will look at what is local, national and international as the innovation process unfolds. Similarly, whether the process of distributing innovation is understood through sectoral or technological logics will be questioned. In the case of this current study, science and technology are drawn together from mathematics, computation, information technology (IT) and biology. Applications for the innovative activity can be found in pharmaceuticals, agriculture, food processing and it is most likely that the range of domains will expand.

Our analysis is based on the proposition that bioinformatics, as a new scientific and technological activity, represents a change to the population of functions (through addition and substitution) that constitute the life science innovation 'universe', and has the additional potential to change the size, shape and orientation of that universe. Accompanying the introduction of a new function to the innovation universe is the question of where organizationally such activities take place. There are several possibilities for this. Bioinformatics could be assimilated into already existing economic agents or be associated with a new class of economic agent. In both cases, a change in the way that life science innovation is coordinated will ensue, with the creation of new classes of interdependence and interaction as well. Furthermore, existing economic agents not traditionally associated with life science innovation are drawn into the 'universe'. In this chapter, we are interested in the nature of interdependencies between different organizations involved in life science innovation and how these have changed with the emergence of bioinformatics.

The early stages of the development of bioinformatics have already witnessed the establishment of new classes of innovating agent. New dedicated bioinformatic firms such as Incyte and Celera were formed to produce and sell genomic databases based on proprietary bioinformatic capabilities. Similarly, new specialist public sector organizations, such as the European Bioinformatics Institute and Genbank, were instituted to provide bioinformatic research and services, and to host the public domain databases such as those emanating from the human genome project. In addition to these new organizations, life science innovation has developed engagements with hitherto more distant organizations including high-end computer, inkjet printing and silicon chip manufacturers and software houses. The combined effect of these shifts has been a change to the distributedness of innovation processes between previously unrelated classes of economic agent.

By taking innovation processes as the unit of analysis, the distinctive characteristic of this approach is to see instituted interdependencies as the outcome of processes of differentiation and integration, rather than being subject to any particular limiting parameters, such as technology, nation, sector or location. Applying our approach to the development of LION Bioscience leads us to examine the dynamic relationship between a changing internal and external capability differentiation.

3. METHODOLOGICAL CHALLENGES FOR RESEARCHING DISTRIBUTED INNOVATION PROCESSES

The case study reported in this chapter is part of a broader project researching the development of bioinformatics in Europe and the USA, comparing different types of new economic agent and distributed processes of innovation. LION Bioscience (hereafter abbreviated to LION) has emerged as one of a very limited number of integrated informatics companies worldwide and therefore offers a particularly interesting empirical probe. It is a key example of a new class of economic agent associated with bioinformatics, and demonstrates how old interdependencies have been disrupted and new ones created in the process. While LION is at the centre of the account, the story cannot be told without 'refracting in' developments within other organizations, some linked to LION and belonging to complementary or mutually dependent economic classes and others that could be competitors within the same economic class. That is to say, the boundary around any given DIP is negotiated in reality and relates to the nature of dependencies between different economic agents which change over time. It is important to note that there are many different ways that DIPs can be formed, constituted by a variety of different types of interdependence.

In terms of methodology too, which we call 'interviewing a DIP', LION is taken as an empirical 'probe' that we use to examine the changes to life science innovation more generally. In taking one firm as our initial probe, we recognize that we have taken decisions regarding where boundaries are drawn and where analysis is focused. For example, in the course of exploring the distributedness of innovation in and around LION, the discussion will cover Bayer, Nestlé, the European Bioinformatics Institute, the European Molecular Biology Laboratory, IBM and Celera and Netgenics. We could have chosen to pursue the distributedness of innovation with any of these organizations as our probe and in each case the story would have covered both similar and different ground to the LION-centred story. The 'LION DIP' that we will unravel has a range of interdependencies facing outwards, and these relate to the myriad connections that each of the organizations that are related to LION has with yet other organizations. In other words, DIPs can overlap, be complementary to each other and can be in competition; this reflects the variety of interdependence that we are interested in exploring. This chapter considers the shifting dimensions of the LION DIP as a necessary prerequisite to the next stage of our analysis, which will be to compare the dynamics of different DIPs. By exploring the changing interdependencies that characterize the LION DIP, we are investigating the dynamic processes common to many innovating firms.

By 'interviewing a DIP', we refer to a programme of ongoing discussions with a number of key informants who represent the key agents in the DIP. There are particular challenges to assembling a coherent suite of interviews where the unit of analysis is the innovation process, rather than one firm, a technological system, a sector or a nation. Thus, we have chosen contrasting types of interdependency (organizational or scientific-technical) of which LION is one. A further challenge is the researching of innovation processes in a context of extreme institutional fluidity.

We have conducted 30 semi-structured interviews with key informants representing interlinked organizations in Europe and the USA (see Appendix, Table 9.A.1). We have also collected and analysed information from the annual reports and accounts from key organizations, and a range of investor information, particularly regarding merger, acquisition and licensing activity. In analysing the data we have attempted to characterize different epistemic and technological activities and different organizational forms in order to address our research question concerning the relationship between the two.

4. LION BIOSCIENCE: NEW KING OF THE BIOINFORMATICS JUNGLE?

LION is at the centre of our case study that covers five years of bioinformatics-based, and ultimately other life science informatics-based, innovation across Europe and the USA. The study involves coordination between public science institutions, dedicated biotech firms, dedicated bioinformatic firms and transnational life sciences firms, including the big pharmaceuticals companies.

Founded in 1997, LION now has more than 500 employees, with headquarters in Heidelberg, Germany and subsidiaries in Cambridge, UK, Cambridge, MA, Cleveland, OH, Columbus, OH, and San Diego, CA, USA. To date, LION has established partnerships with some of the major life science companies, including Aventis, Bayer, Boehringer Ingelheim, Celera, DuPont, GlaxoSmithKline, IBM, Janssen, Merck Inc., Nestlé, Novartis, Paradigm Genetics, Pharmacia Corporation, Roche, Schering AG and Sumitomo Pharmaceuticals. Thus, it is already evident that LION has enjoyed some success in forging a space in the life sciences universe. The expansion of LION in terms of personnel and geographical location represents a period of rapid growth, achieved largely through merger and acquisition of other firms, some of which were rivals, others offering complementary products and services. The range of partners that LION currently interacts with illustrates the variety of interdependencies that have formed with diverse classes of economic agent, including large pharmaceuticals firms, large agri-food firms, large computing firms, dedicated biotech firms and other dedicated bioinformatic firms.

We present the case in three stages. First, we describe the formation of LION, from its roots in public science, as an instance of the creation of a new class of economic agent. The principle focus of this stage will be on the types of knowledge and technology that formed the basis for LION and the process of spinning out from a major European public science institute. We will consider how this created new dependencies within the life science community, concerning the particular technological functions offered. The second stage explains how LION expanded its informatics capability through several phases of integration (merger, acquisition and cross-licensing) with other companies. So, having established that LION exemplifies the formation of a new class of economic agent in the first stage of the case study, the second stage looks at the emergence of a new division of innovative labour within this new class of economic agent. In a sense, by analytically zooming in, we are able to analyse a process of changing distributedness with something like fractal self-similarity: the dynamic process looks the same at which ever level of analysis. The third stage moves the fractal lens back out to reconsider the changing life sciences universe. It considers the evolving business model of LION in relation to the broader picture of life sciences innovation and in doing so evaluates the changing position of informatics within the life science industry and within academic biological sciences.

The Formation of LION Bioscience – and a New Class of Economic Agent

While the formation of LION was based around personnel and technologies from the European Molecular Biology Laboratory (EMBL), it was not in any sense a standard spin-out especially since, in 1997, technology transfer at EMBL was in an early stage.[3] Previous commercialization attempts by two EMBL-based research groups, one based around sequencing technologies, the other on sequence analysis software, had been unsuccessful. The German government's BioRegio[4] competition was announced in 1996 and both groups independently developed proposals for seed funding. Heidelberg Innovation, a local venture capital fund suggested that amalgamating the two groups could provide a stronger bid. LION was formed as a result, and received start-up funding as part of the region's successful application. In addition to the DM5 million BioRegio grant, a further DM3 million came from entrepreneur von Bohlen, who also joined the company as chief executive officer (CEO).

There are several important features regarding the start-up process and the 'transfer' of technologies from EMBL to a new firm. The first aspect concerns the financial arrangements between LION and EMBL, involving a combination of licence agreements (over the analysis software) and an equity stake (approximately 10 per cent). There are perhaps two principal forces underlying the desire or at least acceptance of EMBL to commercializing the technologies. The remit and culture of EMBL is strongly associated with leading-edge biological research. By 1997, sequencing activities were becoming factory-based and routine; a service required for leading-edge science, but not an activity in that category itself. During the same period, there had been a general shift at the European Commission (EC) level towards promoting opportunities to commercialize outputs from the science base (arguably in response to the considerably more entrepreneurial US academic system).

So LION was established as a dedicated bioinformatic firm (DBIF) essentially offering sequence analysis software.[5] The software, while being developed in EMBL, was called GeneQuiz, and was renamed BioScout[6] in its LION incarnation. The 'transfer' itself involved the source code, with a licence agreement and two key members of the EMBL GeneQuiz group, Reinhard Schneider and Georg Casari. During the following years the BioScout source code has been continually rewritten and incrementally became increasingly distinctive from the GeneQuiz version. LION and EMBL have held meetings every six months to discuss the licence agreement surrounding the GeneQuiz source code. In 2002, both parties agreed that the BioScout package was sufficiently distinct from the original that the licence should be terminated. Prior to this agreement, the meetings were used to determine the royalty payment due to EMBL, based on the similarity of source code between GeneQuiz and BioScout.

The other major technology platform of LION was also initially developed within EMBL (and subsequently the European Bioinformatics Institute – EBI). The sequence retrieval system (SRS) has been a key technology within the field of bioinformatics as 'a query and navigation system that brings the world of biological databases together into a single interface and environment' (LION Bioscience 1998). It was developed in response to the proliferation of new biological databases with different structures, fields and data types. By providing the means to conduct multi-database searches and analyses, SRS became a dominant technological standard. The development of SRS within the EMBL–EBI infrastructure faced similar constraints as the other LION technologies. Not considered as state-of-the-art scientific research, and requiring significant resource for its maintenance and development, Thure Etzold, the inventor, had been interested in finding commercial backing for some time.[7] Social connections formed through the EMBL system favoured the approach made by LION, who were particularly keen to own the SRS technology.

Developing the BioScout software as the primary business for LION potentially exposed the company to excessive dependency on whoever owned SRS. This was tolerable where the source code remained fully in the public realm, but the prospect of SRS becoming a proprietary technology of one of their competitors was of considerable concern. The reason for this vulnerability was that the BioScout technology was developed within the SRS environment and for it to operate most effectively, ongoing development required access to the SRS source code. Owning SRS would enable LION to develop the integration technologies and application technologies with a greater degree of synergy. This points to a second and perhaps more significant motivation for LION. Owning SRS and alleviating their own vulnerabilities in application design would place them in a position where their competitors would become increasingly dependent on them. Equally, it would allow them to commercialize the SRS technology and to create a market for it within the life science industries. With this agreement, LION was in a position to modify its business model, with SRS as the cornerstone.

The EMBL were reluctant to lose the resource of SRS, but without the capacity to support its widespread use, maintenance and further development, it had little choice. The particular licence agreement provided a framework whereby LION would have all commercial rights to SRS, but that it would always be freely available to the academic community through the EBI. An outcome of this deal was that LION immediately opened an office in Cambridge, UK, within easy access of the Hinxton Campus. Thure Etzold was made Managing Director of the new LION UK, but retained a position (25 per cent of his time) at the EBI. This location, within the Cambridge cluster, was also of strategic importance to LION, and there is an ongoing relationship between the two organizations concerning the development and use of SRS. From the LION perspective, it has been critical that both EMBL and EBI use SRS, as this has helped to create SRS as an industrial standard. Academic bioinformaticians moving to industry have, in almost all cases, been trained in using SRS and the transfer of their skills is typically tied to the SRS interface. However, as SRS has evolved through different versions, the academic community has become more removed from the source code. The latest version is still free to academics, but access to the source code is limited and therefore restricts the academic's opportunity to develop customized applications.

The particular mode through which the technologies were transferred from public science to form LION, which we can term a 'slow appropriation of knowledge' demonstrates well the manner in which new interdependencies are formed and how they are negotiated in a variety of ways. By relinquishing SRS, EMBL effectively placed itself in a relationship of interdependency with LION and this explains the attention to providing an agreement whereby academics would be able to access the service for no cost. Equally, the business model for LION was based on the assumption that the commercial life science firms would also be dependent on SRS, but the interdependencies formed here would be constituted by market exchanges. From a DIP perspective, the formation of LION was both a process of separation and differentiation of functionality, and the formation of a new frontier between market and non-market exchanges, with new relations of dependency and asymmetry of power.

At this stage, LION had formed as one of a new class of economic agent broadly defined as firms offering bioinformatics tools and services, (DBIFs). Even at this stage, however, there was internal differentiation within this class between those firms more oriented towards tools and those more oriented towards providing content for databases. Although the distinction between the two was blurred in the case of Incyte for example, LION was certainly more oriented towards tool and platform provision. In the next section we will see how this orientation was extended, using the LION story as a probe again to investigate the changing distributedness of innovation within life science informatics.

Creating a New Internal Division of Innovative Labour Within LION Bioscience

The issue of analytical importance here concerns the nature of a division of labour within the class of innovating agent associated with life science informatics, accompanied by a simultaneous redefinition of the class itself. To conduct this analysis we describe the process through which LION has extended its own internal informatic capability to cover a progressively greater range of life science activities. This is the story of the changing distributedness of innovation within LION, which we will place in an external context in the subsequent third section.

Figure 9.1 illustrates one element of LION's business model and its stance towards life science informatics innovation.[8] This matrix is one possible depiction of the range of life science informatic functions. It is a representation provided by LION itself to explain retrospectively how the company has developed. In fact, it is only with the benefit of hindsight that LION are able to portray their activities in this way since they did not know *ex ante* the path that they were to pursue. For our purposes, it is a useful device for showing dynamics within the life science informatics domain between 1997 and 2002. Moreover, although in retrospect it looks like a space to be progressively filled, it was more like the expanding jigsaw model, with new functionality pieces to be fitted on to continuously redesigned old

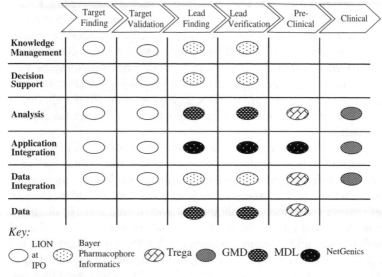

Key:

LION at IPO   Bayer Pharmacophore Informatics   Trega   GMD   MDL   NetGenics

Source: LION Bioscience

Figure 9.1   Distributing bioinformatic innovation within LION Bioscience across the drug discovery knowledge space

ones, suggesting also the emergence of new constraints precisely because it is an interlocking puzzle.

Along the top of the scheme are the familiar steps in the pharmaceutical innovation pipeline. At the left-hand side are two functions concerned with the detection of potential (biological) targets and this was the initial focus of bioinformatics. Then, moving towards the right are the successive steps in the pharmaceuticals innovation process, as conceived in a linear fashion. In turn, these steps involve the identification of a potentially effective chemical entity (lead finding) and subsequent testing (lead validation) followed by initial toxicology testing (pre-clinical) and finally the clinical trials.

The vertical axis of the matrix represents different aspects of informatic technologies. The bottom level depicts data content, which towards the left-hand side pertains to biological data such as SNPs or protein structures and moving right refers to chemical databases (traditionally the domain of chemo-informatic firms). Moving upwards, the next two levels refer to software solutions for integrating data from diverse sources and providing a computational environment for integrating different software applications. The next level refers to analysis of data itself, and includes genomic sequence analysis software. The final two levels refer to higher-level information and knowledge management systems.

The markers represent the activities progressively covered by LION or a 'partner' and as the key shows, the different shades correspond to other firms that LION has formed a connection with to extend their activities throughout the matrix. The shading of the markers also corresponds to a chronological ordering, going from light to dark, providing us with the opportunity to emphasize the process of creating a new internal distributedness of innovation within LION, as it progressed from a specific bioinformatics focus to a more general 'one-stop-shop' life science informatics orientation.

The clear markers represent the technology platform at its initial public offering (IPO). It is essentially a combination of BioScout (covering target finding and validation) and SRS (data and application integration). At this stage LION was the exemplar bioinformatic tool provider. Its activities were entirely oriented towards biology and offered a range of informatic solutions to that end; it did not however have any proprietary data. Thus, the functional and economic existence of LION was only possible if other organizations were producing biological data, thereby constituting a fundamental dependency on the data producers. From this starting-point, and with SRS as the enabling technology, LION has subsequently extended its capability throughout the matrix (and in fact the axes of the matrix were developed in the process). The following five 'deals' incrementally and sequentially allowed LION to extend its functional scope through the pharmaceutical innovation pipeline/informatic technologies space. These are now described in chronological sequence.

In a DIP perspective, the first and major deal involved the formation of a specific interface between a bioinformatic functionality and a major pharmaceuticals company, involved in drug development through to drug marketing. The Bayer deal was agreed in June 1999 and involved the establishment of an alliance worth $100 million over five years (LION Bioscience 1999). The agreement was based around several intriguing features. The estimated $100 million included an initial equity stake in LION AG as well as fees for use of LION's existing IT systems, research and set up costs for a new subsidiary. This new subsidiary was a contractual obligation of the agreement and led to the formation of LION Bioscience Research Inc. (LIBR) in Cambridge, MA. The agreement stated that LIBR would deliver to Bayer, 500 new target genes, 70 new annotations on existing Bayer-owned gene targets and an undisclosed number of gene expression markers and SNPs.[9] A Bayer-wide intranet system was also developed to ensure that the resulting information was globally accessible to Bayer's scientists. The LION–Bayer agreement has been subsequently extended into pharmacophore informatics (LION Bioscience 2000), at the interface of biology and chemistry, and this is represented in Figure 9.1 by the relevant shaded markers. It appears that the partnership with Bayer required LION to shift away from its traditional biological orientation, triggering the start of a process of further functional extension.

Considering the proximity that LIBR's activities had to commercially sensitive activities of Bayer, it is interesting to look at the intellectual property aspects of the contract. Essentially, there is a relatively straightforward informatics–biology split between LION and Bayer. LION ultimately retain all rights over any IT developments, to which Bayer have unlimited and non-exclusive access. In addition, Bayer have the option, after five years of the alliance, to acquire all the shares of LIBR, thereby providing the opportunity to wholly own an informatics subsidiary. To retain control over the IT component of LIBR, this part of the arrangement includes a provision for LION to have pre-emptive rights to the commercial exploitation of any new informatic technologies in the event that these are in competition with LION's own activities and Bayer has decided to market them. It is evident that where interdependencies occur within distributed innovation processes, close attention is paid to questions concerning the ownership of knowledge and technology. This step offers an early indication of how LION and Bayer, at that time, perceived the most appropriate division of innovative labour, and created a specific dependency and asymmetry of power. It will become apparent later that this model of the life science universe is not shared by all firms and public science institutes (PSIs).

In step two, December 2000, LION extended its informatic functionality into toxicology-informatics and chemo-informatics by acquiring 100 per cent of the equity of Trega valued through the agreement at approximately $35 million (LION Bioscience and Trega 2000). Upon completion of the transaction, Trega became a wholly owned subsidiary of LION. From the DIP's perspective, as Figure 9.1 shows, the acquisition changed the shape of LION by adding further chemistry-based capabilities to their own biology-based strengths. In particular, this included Trega's iDEA™ Predictive ADME (absorption, distribution, metabolism and excretion) Simulation System, which simulates, *in silico*, how drug candidates will be processed in humans.

The next addition to LION's functional repertoire expanded their universe into medical informatics and e-health solutions, as well as gaining entry into new geographical spaces. In February 2001, LION took a minority stake in Gesellschaft für Medizinische Datenverarbeitung (GMD) by acquiring 16.1 per cent of GMD's shares for €8.75 million (LION Bioscience and GMD 2001). GMD offers medical information management technology currently in use in hospitals throughout Germany, France and the USA. The Web-based electronic patient file of 'e-health.solutions' assists medical professionals with gathering and managing documentation relating to diagnosis, therapy and accounting. Having integrated data within the hospital will allow doctors to identify appropriate candidates for the testing of novel diagnostics and therapies, allowing clinical trials to be conducted more efficiently and more rapidly. LION and GMD are also collaborating on a pilot project to develop software for discovering the causes of diseases through clinical trials and to further explore opportunities for individualized diagnostics and therapy. Technologically, the idea here is to combine the health informatics capability of GMD with the biology-oriented expertise of LION using the integrative technologies of LION. The contract included the possibility for further acquisition by LION and for further strategic alliances in the future.

The process of extension and redesign of functionalities increased the need for higher levels of informatics integration within LION, progressing towards a conception of integrated life science informatics. In July 2001, LION announced that it would licence a wide range of MDL's industry-standard informatics applications and databases in order to develop and commercialize software applications that can interface with MDL's widely deployed products. Together with MDL's applications, LION's integrated life science informatics offering will further accommodate the entire research and development (R&D) chain of the life science industry. At the same time MDL will be the exclusive, global reseller for LION's iDEA™ product line for the prediction and analysis of ADME (absorption, distribution, metabolism and excretion). Dr Friedrich von Bohlen, CEO of LION said:

> This arrangement with MDL is another important step in facilitating the creation of knowledge management solutions that can integrate all disciplines in discovery that are required by the life science industry. The distribution of our unique iDEA products through MDL's sales force will enable us at the same time to introduce our software to an even larger customer base (LION Bioscience and MDL 2001).

MDL has been established in the field of chemo-informatics for some time, and therefore has very well-established and quite entrenched relationships with large life science companies. This offers LION a marketing and distribution capability, particularly in the USA, in addition to the chemo-informatics. From the standpoint of MDL, bioinformatics has emerged as a potential new business opportunity, and a means of extending their chemistry-based informatics technologies into the biological realm. Entering into the cross-licensing deal with LION allowed MDL the opportunity for relatively low-risk experimentation in this new domain.[10]

A further move towards a higher-level informatics platform was then targeted with the acquisition of NetGenics, based in Cleveland, OH, USA, and at the time a direct competitor. So, as well as constituting an expansion of technological capability, this deal removed one of LION's major competitors. NetGenics' DiscoveryCenter technology was a rival of SRS, and had the advantage of being able to integrate relational and flat file databases. The inability of SRS to cope with relational databases had been seen as one of its major shortcomings. DiscoveryCenter had been developed in cooperation with IBM (building on the latter's DiscoveryLink middle-ware technology). From an initial US government grant of $60 million, LION valued NetGenics at $17 million on acquisition: with dwindling financial assets, the competitive rival might not have survived for many more months.[11] Completion of the NetGenics acquisition and gaining ownership of their integration software positioned LION, at that time, as the most comprehensive life science informatics provider globally.

From its formation, through IPO and a succession of mergers, acquisitions and alliances with other informatics providers, LION had extended its capabilities across the matrix represented, from a bioinformatics company to a life science informatics company. The preceding discussion suggests that LION grew through an aggressive acquisition strategy oriented towards creating a more extensive internal distributedness of life science informatics better able to articulate with the changing distributedness in the wider life sciences. Before we return our analytical lens to that level, it is worth remarking on the fact that LION stopped short of assimilating biological content and therefore maintained a distinctive position to the database-oriented DBIFs, such as Incyte and Celera.

This study has shown how, over a period of five years, there has been considerable integration and consolidation within the global life science informatics community and how bioinformatics has extended into life science/x-informatics. The nature of the technologies involved and the push for ever greater interoperability combined with competitive pressures among firms offering services across the informatics domains have led to this concentration. LION has been a major player in this process. At the same time, LION has pulled back from in-house drug-discovery activities and has a policy for seeking partnerships to exploit any potential drug targets revealed through its I-biology operations. This process of realignment at LION has taken place in a broader context of shifting distributedness of life sciences innovation in general. The next section examines some of these wider shifts.

The Shifting Position of LION in the Distributed Life Science Innovation Process

In turning to the interdependencies and asymmetric relations external to LION as a firm, LION's distributed internal expanding informatic capacity is now located within an expanding universe of constantly reshaping entities. For example, we know that many previously integrated large life science companies have now divided into agri-business (e.g. Syngenta) and pharmaceuticals (e.g. AstraZeneca). In this final stage of our case study, we explore some dimensions of the external distributed innovation process. We have already pointed to the creation of an interface between bioinformatics and large pharmaceuticals manufacture and marketing. Figure 9.2 represents a by-no-means exhaustive universe of different economic agents with which LION's Life Science Informatics is engaged in innovation processes essential for its economic survival.

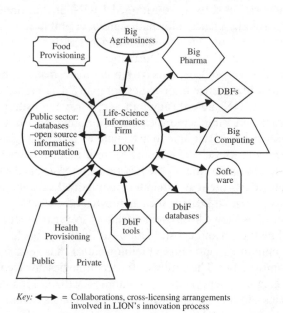

Key: ◆━━▶ = Collaborations, cross-licensing arrangements
                 involved in LION's innovation process

*Figure 9.2   The life science informatics-centric view of distributed
            innovation processes*

We should stress that this is the view as seen from our choice of LION as a probe, and the universe of distributed innovation processes would look quite different if centred on any of the other classes of economic agent contained within it. Moreover, the LION business model, as an integrated life science informatics company, represents one particular view of how and where value can be created within the overall life sciences industries. We now examine the changing business model of LION in the context of shifting forms of distributed innovation within the life sciences and across the public–private divide.

At the centre of Figure 9.2, there is the key and continuing innovation interaction between public and private that we have explored in more depth elsewhere (McMeekin and Harvey 2002). What is significant from a DIP's perspective here is the way that LION emerged and grew at a time when the nature and functionality of the interface between public and private, and the flows of knowledge between them, was changing. The distributedness of innovation between public and private was altering in ways that opened up some possibilities for firms, while closing down others. Nonetheless, the interdependence between public and private sector informatics and knowledge flows has been critical to private sector firms' continuing market functionality. LION however has been quite distinctive in the extent of the interpenetration between public and private from its origins through to the present even while developing purely private relationships.

The number of biological databases has continued to grow over the past five years, and in parallel the types of arrangement have changed for their ownership, intellectual property status, ongoing maintenance and development, accessibility and interoperability. Broadly speaking, databases are owned by PSIs (e.g. EBI), pre-competitive (hybrid public–private) consortia (SNP consortium), dedicated commercial content providers (Celera), life science companies (e.g. GlaxoSmithKline and Syngenta) and dedicated biotech companies (e.g. Oxford Glycoscience). The balance of ownership across these different classes of economic agent has shifted, as have the flows of knowledge between them. The most dramatic shift has been with the major content providers. Both Incyte and Celera have been forced to extend their activities into drug discovery, as the economic viability of major proprietary databases has been increasingly difficult to sustain. There are several reasons why a database content dominated strategy is no longer deemed sustainable. First, there is continued rivalry from public sector activities usually providing equivalent resources with a time lag that is now perceived as less critical. Second, the content providers have been 'squeezed' from the other direction, as life science transnational corporations (TNCs) have started to invest more in their own in-house databases. It was notable that while the race over the human genome pitted the public effort against Celera, the equivalent and later race over the rice genome involved Syngenta, the fully integrated agro-food TNC.

The specific functionality of dedicated content providers has also been challenged from another source, namely pre-competitive consortia spanning the public–private divide. The SNP consortium is one such example that included a range of transnational life science companies and public science institutions and was supported by a non-profit organization, The Wellcome Trust. It would appear that the life of major data content provider as a discrete class of economic agent has been short, with the specific functionality migrating into more long-standing institutions. Thus these new economic agents forged a position within life science innovation processes for only a short time, ultimately to be squeezed back out and forced to shift their activities in different directions.

This jostling of position has influenced the activities of LION, despite being one step removed, demonstrating the nature of interdependencies within DIPs. The basis of the partnership between Celera and LION was an SRS-based Internet interface for Celera customers to access the proprietary databases. At the outset, there was a significant cross-over with respect to analytical software and general bioinformatic expertise, despite the fact that the core product orientations of the two firms were different.[12] This confusion most likely stemmed from the extreme uncertainty concerning what could be a tradable bioinformatic-based product and what form any markets would take. Since then, Celera has transformed into a drug discovery firm and it is now considerably clearer that LION and Celera are complementary partners rather than rivals. This is an indication of the ongoing specialization and division of innovative labour in the sense that LION and Celera now represent different classes of economic agent and for the time being the nature of interdependencies are clearer and more stable.

By withdrawing from internal drug discovery operations and seeking large life science firms as partners, LION is attempting to form a new type of mutual dependency between the institutionalization of life science informatics and the development, marketing and distribution of life science-based product innovation. As we have already seen, LION has a close relationship with Bayer; it also has a similar relationship with Nestlé, the food company. Both these relationships involve LION working behind these companies' firewalls and their activities are closely intertwined with their respective life science innovation process. As a consequence of this high level of integration, these partnerships are long running and involve considerable resources.

LION also has a customer base of considerably smaller DBFs, and these necessarily operate at a smaller scale, though there is a wide spectrum of possible arrangements owing to the variety of strategies amongst firms of this size and orientation. Many of these firms are generating very specific and niche proprietary databases based on new types of data. With ACE Biosciences, for example, the focus is on parasite proteomics and ACE has developed a database for storing this type of data.[13] The intention is to discover and develop therapies based on their own knowledge rather than form alliances with the big life science firms. The strategy is towards integration along the life science innovation pipeline in a manner similar to the major content providers, Incyte and Celera. However, the principal difference is that while Celera and Incyte have competed with the major public science institutions (PSIs) and large life science firms over the genomes of prominent organisms such as human and rice, ACE has a niche organism strategy, which could prove more sustainable in life science distributed innovation processes.

The DIP's perspective stresses that sectoral boundaries are radicalized by innovation and this is well illustrated when we look at the orientation of LION towards pharmaceuticals or agri-food innovation. LION's own market, and indeed the final consumer markets are highly segmented with respect to these end-product domains. However, in terms of LION's core technologies, in particular the integration platforms, both markets can be accessed. In fact developments at the leading edge of post-genomic biology and bioinformatics are revealing new opportunities for integrated activities across agriculture, food and health. Indeed, the work that ACE and LION are involved with is connected to new research into general models of pathogenecity, which could radically transform the drug discovery process by providing plants as test organisms. This is good evidence of how new types of distributed innovation processes, creating novel interactions across previously clear boundaries, have been instituted within a constantly fluid life sciences universe.

There have also been shifts regarding the development and use of informatic applications (e.g. sequence analysis technologies) and the formation of markets for these products. Again, there has been a tension between the development of tools within the public sphere, based on an open source arrangement, and those developed in commercial institutions, where some or all of the source code is hidden and protected. The culture of open source software development for bioinformatics has been disrupted over the previous few years. There has been a progressive 'privatization' of software as evidenced by both the development of LION's BioScout from the EMBLs GeneQuiz and the development of proprietary software at GCG (now Accelrys, which is the major rival of LION) from the public domain Wisconsin Package.[14] In both cases, public domain versions remain available and are still being developed. The capacity of public science institutions to develop their own analysis applications has become increasingly restricted as access to the SRS code becomes more elusive. The over-dependency of the EBI and other PSIs on proprietary integration technologies in the future could, in the view of EBI, compromise the ability of public science to conduct leading-edge science.

To conclude this stage on the life science industries, we turn to the arrival of already established firms that are becoming part of distributed life science innovation processes. These firms had not previously had significant involvement in life sciences, but given the widespread perception of market expansion in this area, they have sought to find a position within the new division of innovative labour. For example, IBM has recently started to invest significant and 'ring-fenced' resources into this field by establishing a dedicated life science division.[15] LION and IBM have been exploring possibilities for collaborative research, but to some extent they are rivals as well as collaborators. The types of service that the two firms offer are gradually becoming closer. LION has been shifting from biology-based informatics towards offering complete IT-based knowledge management systems. IBM has essentially been moving in the other direction. Moreover, IBM has been joined by all other major computer hardware companies in reaching agreements for co-development and collaboration in innovation processes in the expanding informatics universe. For computer firms, this shift into life science signals a major shift in the distributedness of their innovation processes, as companies like LION develop new relations of dependency and asymmetric power with them for life science informatics innovation.

This section has shown how there has been considerable flux in the organization of life science innovation. The patterns of interdependency between different classes of economic agent have shifted, as have the definitions of those classes. These shifts have had a marked effect on which organizations are rivals and which have relationships of mutual dependency. It seems that informatics will play a critical role in future life science innovation processes, but it is not yet clear whether the large life science firms will have these activities in-house or whether a dedicated life science informatics firm is economically sustainable.

CONCLUSION

The fundamental question underlying this chapter has concerned the dynamic relationship between knowledge and organization within innovation processes. The approach that we are developing focuses on processes of differentiation and integration, interdependency and asymmetry of interactive relations involved in innovation. One of the most interesting features of emergent bioinformatics has been the way that it has transgressed and transected existing boundaries, so proving particularly fruitful for a distributed innovation process approach.

In this chapter, we have chosen to describe the distributedness of innovation within and around one company, LION. This company exemplifies how new or previously distinct areas of scientific and technological activity have emerged and become interdependent in new and unanticipated ways within and across organizations, both public and private. The successive stages of instituting an innovation process around LION have demonstrated the propensity for innovation to disrupt given classifications, since it would be difficult to draw geographical, sectoral, scientific or technological boundaries.

Stage one of the story involved the formation of LION as an example of a new class of economic agent. In this case the new firm was borne out of a public science institute and received pump-priming state support. Two technologies that were initially developed in the public science domain were translated into tradable products, and this was achieved through particular arrangements with the academic community to ensure that they still had access rights.

Stage two involved the creation of an internal division of innovative labour within LION. Successive acquisitions and other 'deals' aimed at integrating bioinformatics with other informatics within the life sciences field. Through this process, LION has been able to combine expertise in biology, chemistry, toxicology and medicine-based informatics to create a 'one-stop-shop' for the life sciences industry. This strategy is about trying to create dependencies within the life sciences community for LION's services.

Stage three of the story placed LION within the changing configurations of life science innovation. The current LION strategy is to consolidate the position of life science informatics provider as a stable new class of economic agent within life science DIPs. There has been a decision at LION not to extend their activities into drug discovery on the one hand, or to the provision of biological database content on the other. For LION to exist in this form, they are dependent on other economic agents to undertake these activities and on these other economic agents to not encroach on the particular services that LION offers.

The particular pattern of interdependencies at the current time suggests that LION are at a make or break point. If they are able to institute life science informatics as a discrete economic function, then LION could be in a position of developing a standard integration platform within the associated industries to which other economic agents would become dependent. Conversely, if public science institutions or pre-competitive consortia were to develop rival platforms, the LION business model would be highly vulnerable. There is little doubt that genomics, post-genomics and life science informatics will have a profound impact on the life science innovation pipeline in the future. However, predicting the position of LION within future configurations is impossible.

In presenting the story of the distributedness of innovation within and around LION, we have made an attempt at 'naming the parts' within the bioinformatics innovation processes. At the same time, we have emphasized the importance of connections between the parts in shaping the ways the configurations have changed over time. The emergent characteristics of LION have demonstrated the highly contingent relationship between knowledge and organization, where there is no inevitable one-to-one mapping. It is clear that LION represented one particular historical outcome of combinations between changing bioinformatic knowledge and innovation processes and changing organizational configurations. As an innovation process account, an extreme fluidity of organizational and knowledge boundaries over time was demonstrated, as was the dynamic relationship between internal and external differentiation, integration and interactions. There was a distinctive starting point and a singular trajectory, resulting in a particular internal combination of capabilities in a newly configured life science universe. The distributed innovation process framework provided the conceptual tools for an analysis that demonstrated the basis for this variation and singularity.

NOTES

1. The original poem is by Henry Reed.
2. Interview with Walter Blackstock, Technology Director, Cellzome and former Glaxo Wellcome.
3. EMBL now has a considerably more developed commercialization body, called EMBLEM. As a result, the formation of Cellzome, the company referred to in the introduction, was a far more typical spin-off. From an interview with Iain Mattaj, Director of Research, EMBL.
4. The BioRegio Competition was set up by the German Government to compensate for the slow start the country had in biotechnology compared to the USA and the UK (Dohse 2000).
5. The sequencing service, only low throughput, was secondary and offered mainly to the Heidelberg community.
6. GeneQuiz has continued as a public science resource and has been further developed at the European Bioinformatics Institute (EBI).
7. Interview with Thure Etzold.
8. In this matrix, and here, the focus is on activities within the pharmaceuticals sector. We discuss below the position of LION within the broader life science community.
9. LION subsequently reported that they had identified 250 of the targets considerably ahead of schedule (LION Bioscience 2001).
10. Interview with Phil McHale, MDL.
11. Interview with Reinhard Schneider, Director, LION Bioscience.
12. Interview with Reinhard Schneider, Director, LION Bioscience.
13. Interview with Angus King, ACE Biosciences.
14. The Wisconsin Package was one of the earliest and most important suites of sequence analysis programs, developed at the University of Wisconsin Genetics Computer Group (GCG). Interview with John Devereux, founder of GCG.
15. Interview with David Zirl, IBM Life Sciences Division.

REFERENCES

Carlsson, B. (ed.) (1997), *Technological Systems and Industrial Dynamics*, Boston and Dordrecht: Kluwer Academic Publishers.

Carlsson, B. and R. Stankiewicz (1995), 'On the nature, function and composition of technological systems', in B. Carlsson (ed.), *Technological Stystems and Economic Performance: the Case of Factory Automation*, Boston and Dordrecht: Kluwer Academic Publishers.

Carlsson, B., S. Jacobsson, M. Holmen and A. Rickne (2002), 'Innovation systems: analytical and methodological issues', *Research Policy*, **31**(2), 233–45.

Cooke, P. (2001), 'New economy innovation systems: biotechnology in Europe and the USA', *Industry & Innovation*, **8**, 267–89.

Cooke, P. (2002), *Knowledge Economies*, London: Routledge.

Coombs, R., M. Harvey and B. Tether (forthcoming), 'Analysing distributed processes of provision and innovation', *Industrial and Corporate Change*.

DTI (1999), *Biotechnology Clusters*, London: Department of Trade & Industry.

Dohse, D. (2000), 'Technology policy and the regions: the case of the BioRegio Contest', *Research Policy*, **29**, 1111–33.

Faulkner, W. and J. Senker (1994), 'Making sense of diversity: research linkage between industry and public sector research in three technologies', *Research Policy*, **23**(6), 673–95.

Fransman, M. (2001), 'Designing Dolly: interactions between economics, technology and science and the evolution of hybrid institutions', *Research Policy*, **30**(2), 263–73.

Harvey, M. and A. McMeekin (2002), 'UK bioinformatics: current landscapes and future horizons', DTI report, London, March.

Hodgman, H. (2002), Presentation to DTI, London.

Lemarié, S., V. Mangematin, and A. Torre (2001), 'Is the creation and development of biotech SMEs localised? Conclusions drawn from the French case', *Small Business Economics*, 17, 61–76.

LION Bioscience (1998), 'LION Bioscience acquires SRS technolgy from EMBL to accelerate genomics integration program', press release, Florida, USA, September 17.

LION Bioscience (1999), 'LION Bioscience and Bayer enter US$100M research alliance', press release, Heidelberg, Germany, 24 June.

LION Bioscience (2000), 'LION Bioscience and Bayer pioneer linkage between chemistry & genomics to speed life science discovery research', press release, Heidelberg, Germany, 16 October.

LION Bioscience (2001), 'LION Bioscience reaches major milestone in collaboration with Bayer AG for the identification and validation of drug targets', press release, Heidelberg, Germany, 5 April.

LION and MDL (2001), 'LION Bioscience AG and MDL Information Systems, Inc. sign software and reseller agreements', press release, Heidelberg, Germany and San Leandro, USA, 19 July.

LION Bioscience and GMD (2001), 'LION Bioscience supports expansion of Gesellschaft für Medizinische Datenverarbeitung', press release, Heidelberg, Germany and Munich, Germany, 21 February.

LION Bioscience and Trega (2000), 'LION Bioscience to acquire Trega Biosciences', press release, Heidelberg, Germany, and San Diego, USA, 27 December.

Lundvall, B.-A., B. Johnson, E.S. Andersen and B. Dalum (2002), 'National systems of production, innovation and competence building', *Research Policy*, **31**, 213–231.

Malerba, F. (2002), 'Sectoral systems of innovation and production', *Research Policy*, **31**(2), 247–64.

McMeekin, A. and M. Harvey (2002), 'The formation of bioinformatic knowledge markets: an "economies of knowledge" approach', *Revue d'Economie Industrielle*, **101**(4), 47–64.

Nelson, R. (ed.) (1993), *National Innovation Systems*, Oxford: Oxford University Press.

Oliver, S. (2002), 'To-day, we have naming of parts . . .', *Nature Biotechnology*, **20**(1), 27–8.

Orsenigo, L., F. Pammolli and M. Riccaboni (2001), 'Technological change and network dynamics: lessons from the pharmaceutical industry', *Research Policy*, **30**, 485–508.

Owen-Smith, J., M. Riccaboni, F. Pammolli and W.W. Powell. (2002), 'A comparison of US and European university-industry relations in the life sciences', *Management Science*.

Powell, W.W. and J. Owen-Smith (1998), 'Universities and the market for intellectual property in the life sciences', *Journal of Policy Analysis and Management*, **17**(2), 253–77.

Powell, W., K. Koput and L. Smith-Doerr (1996), 'Interorganizational collaboration and the locus of innovation: networks of learning in biotechnology', *Administrative Sciences Quarterly*, 41, 116–45.

Rappert, B., A. Webster and C. Charles (1999), 'Making sense of diversity and reluctance: academic – industrial relations and intellectual property', *Research Policy*, **28**, 873–90.

Rybak, B. (1968), *Psyché, Soma, Germen*, Paris: Gallimard.

Rybak, B. (1978), 'Bio-informatics and Bio-process in the physiology of communication', *Biosciences Communications*, **4**(3–4), 158–9.

Saxenian, A. (1994), *Regional Advantage: Culture and Competition in Silicon Valley and Route 128*, Cambridge, MA and London, UK: Harvard University Press.

Senker, J. and P. van Zwanenberg (2001), 'European biotechnology innovation system,' final report for EC TSER project (SOE1-CT98-1117), Brighton: SPRU, University of Sussex.

Stankiewicz, R. (2001), 'The cognitive dynamics of technology and the evolution of its technological system', in B. Carlsson (ed.), *New Technological Systems in the Bio-Industry – an International Comparison*, Dordrecht: Kluwer Academic Publishers.

Storper, M. and B. Harrison (1991), 'Flexibility, hierarchy and regional development: the changing structure of production systems and their governance in the 1990s', *Research Policy*, **20**, 407–22.

APPENDIX: INTERVIEWS

Table 9.A.1   Interview codes

University of Manchester, UK	Terry Attwood, Erich Bornberg-Bauer, Carol Goble, Andy Brass, David Broomhead, Olaf Wolkenhauer
GlaxoSmithKline, Stevenage, UK	Charlie Hodgman, Malcolm Skingle, Patrick Devitt
Human Genome Mapping Project, Hinxton, UK	Alan Bleasby
European Molecular Biology Laboratory, Heidelberg, Germany	Iain Mattaj, Halldor Stefansson
Eueopean Bioinformatics Institute, Hinxton, UK	Janet Thornton, Ewan Birney, Kim Henrick, Nick Goldman, Helen Parkinson, Paul Mathews
Cellzome, Heidelberg, Germany and London, UK	Georg Casari, Walter Blackstock
LION Bioscience, Heidelberg, Germany and Cambridge, UK	Reinhard Schneider, Thure Etzold, Peter Rice
IBM, San Jose, USA	David Zirl
Ace Biosciences, Odense, Denmark	Angus King
Accelrys, Cambridge, UK and San Diego, USA	Tom Flores, Scott Kahn, John Garrison
Genetics Computer Group, Madison, USA	John Devereux
MDL Informatics, San Leandra, USA	Phil McHale
Oxford GlycoScience, Abingdon, UK	Andrew Lyall
Confirmant, Abingdon, UK	Martin Gouldstone
Institute for Systems Biology, Seattle, USA	Leroy Hood
Dupont, Wilmington, USA	Jean-François Tomb
University of Washington Genome Centre, Seattle, USA	Rajinder Kaul

PART IV

Forming the new

10. The dynamics of regional specialization in modern biotechnology: comparing two regions in Sweden and two regions in Australia, 1977–2001

Johan Brink, Linus Dahlander, and Maureen McKelvey

1. INTRODUCTION

A crucial issue within evolutionary economics and innovation studies is the extent to which scientific and technological knowledge affect economic transformation. This chapter therefore addresses whether, and how, regions may transform and achieve global specialization in modern biotechnology over time. The broader issue is, to what extent, and why, does knowledge affect industrial dynamics and the relative competitiveness of specific regions? This chapter addresses more specific questions about specialization and about the long-term trajectories of specialization at the regional level over 25 years. It examines regional specialization in modern biotechnology in two regions in Sweden (Gothenburg and East Gothia) and two regions in Australia (Melbourne and Brisbane).

Specialization depends on the development and diffusion of new scientific and technological knowledge *per se* as well as on the market-related processes arising from the application of such knowledge within different manufacturing and service sectors. Theoretically and analytically, industrial sectors may be differentiated from fields of scientific and technological knowledge. This differentiation applies particularly to analyses involving modern biotechnology.

We analyse the long-term relative specialization of four regions into modern biotechnology within scientific, technological and business activities. The empirical focus is thus on one area of new knowledge argued to have major future impacts on society and industry. Our starting-point here is that scientific and technological knowledge should be differentiated from sectors and the broad category modern biotechnology is further divided into four distinct categories: (1) 'core biotechnology'; (2) 'drugs'; (3) 'medical technologies'; and (4) 'agriculture' (Brink, Dahlander and McKelvey 2003).[1] 'Core biotechnology' is used as a unique category and is here differentiated from biotechnology as one input – albeit important input – into goods and service products in other sectors.[2] Therefore, in addition to 'core biotechnology', three additional categories were chosen for the overlaps between modern biotechnology and the sector.

The patterns of long-term specialization are presented, based on extensive empirical material. The empirical analysis measures relative specialization at the regional level, based on global indicators for science and technology and local indicators for business. Thus, scientific activities are measured through scientific publications; technological activities are measured in terms of the United States Patent and Trademark Office (USPTO) patents; and business activities are measured through a new database of all relevant firm entries in the respective regions. While the indicators are relatively standard each in themselves, this chapter does demonstrate a methodological development. The contribution is that the empirical work (1) compares all three indicators over a long time period, and (2) compares the three types of activities at the regional level – as opposed to more common levels of the firm or nation.

This article proceeds as follows. Section 2 presents an overview of theories purporting to explain regional development. Section 3 first discusses how and why indicators are used to measure the three activities, for example scientific, technological and business ones, in regions around the world over time. This section then discusses how the broad category of modern biotechnology is divided into four distinct categories. Section 4 briefly presents the rationale for comparing two regions in these two countries as well as outlining some stylized facts about each region. Section 5 presents a correlation analysis of all three activities for modern biotechnology at the regional level while Section 6 compares and contrasts the trends within each activity, for each biotechnology category and region. Section 7 summarizes the arguments and draws conclusions.

2.   REGIONAL SPECIALIZATION AND CO-LOCATED INNOVATIVE SEARCH ACTIVITIES

This section presents an overview of theories purporting to explain regional development in terms of externalities, spillovers and knowledge flows. This literature – and critiques thereof – are used to motivate our focus on regions and long-term regional specialization. The section ends with explicit arguments about the co-location of 'innovative search activities' as well as two questions, which can be addressed through the data about modern biotechnology in two regions in Sweden and two regions in Australia.

2.1 Theoretical Foundations of Regional Specialization

Many innovation studies among economic geographers as well as economists focus on the importance of the region as the unit of analysis (Audretsch and Stephan 1996; Malecki 1985). Arguments about the importance of the local milieu do not, however, reflect only a recent phenomenon. Alfred Marshall was one of the first to emphasize the role of local connections in 1890 and in his work on industrial districts he argued that industries benefit when they were geographically close to other industries (Marshall 1947). In other words, concentration of firms in space creates a pooled labour market with unique resources that are less mobile than other resources that can be traded to other locations. The accumulated knowledge of the local workforce might be so unique that it is only demanded in the local settings. Production in industrial districts is carried on as a social process embedded in a social structure that encourages interactions among a lot of people (Brusco 1990) and results in a 'collective' competence base, consisting of specialized knowledge among the workforce. These competences can create 'non-tradables' of specialized inputs, which, in turn, support the development of specialized industrial sectors (or groups of related sectors).

A related theory to the industrial district is the work on industrial clusters, which highlights local interaction and specialization, and which has been influential in recent years (Porter 1990, 1998). Porter argues that these clusters are sets of industries, which are related by horizontal and vertical links of various kinds, although not only of input–output type. A cluster can consist of a city, region or even nations, which implies that the definition itself is rather fuzzy. Although the concept has been influential among scholars in general and policy-makers in particular, it has also been heavily criticized (Martin and Sunley 2001).

Another influential set of ideas recently is what is known as the 'new economic geography', which highlights the role of external economies and argues that these are more likely to emerge at the regional level as opposed to the national or global level (Krugman 1991, 1994). The main argument of 'new economic geography' is that externalities within a geographically bounded area result in increasing returns that, in turn, results in higher rates of economic growth (Krugman 1991). Other studies on similar issues have found that research and development (R&D) and other knowledge also generate externalities, and that the positive economic benefits of such externalities tend to be spatially close to where they were created. This 'new' approach is different *vis-à-vis* the 'old' economic geography approach in the sense that natural resources and factor endowments do not explain specialization and international trade. In fact, most of the international trade occurs between countries with rather similar structure – rather than between countries with quite different industrial structures, as would be predicted by Ricardian trade theory – and moreover, these countries exchange rather similar products (Martin and Sunley 1996).

Regional systems of innovation and evolutionary economics give further explanations relevant here, due to their focus on processes of learning and interaction, rather than static analysis (Cooke et al. 1997; Cooke 2001; Storper and Scott 1995). Various actors are involved in learning and search activities, and their actions, routines and norms are embedded in an institutional framework. Co-location of activities or individuals within a geographically bounded area generates opportunities for personal communication. Saxenian emphasized this in her studies of why Silicon Valley succeeded in electronics when Route 128 failed (Saxenian 1990, 1994). The relatively high turnover of employees in the Valley resulted in the diffusion of knowledge among the firms, and this, in turn, positively affected innovativeness and firm competitiveness. Saxenian's (1994) influential work states that one of the key success factors is the informal networks, which transcend organizational barriers and encourage interaction and learning.

It has been argued in a recent study, however, that despite all the theoretical attention and scattered empirical evidence, surprisingly few empirical studies have actually provided convincing and systematic, empirical evidence of the superiority of local over non-local interactions to explain regional development and specialization (Bathelt et al. 2002).

An often-used explanation for trajectories of regional development and competitiveness is the concept of localized knowledge spillovers. The concept of knowledge spillovers stems from the assumption that knowledge and related externalities can be seen as a non-rival production asset arising from both private and public R&D, and that it may be spread unintentionally as well as intentionally. Localized knowledge spillovers may include localized benefits arising from economic transactions as well as other aspects such as free sharing agreements and the firm's failure to fully appropriate the returns of innovative ideas. A further examination of the concept reveals both types of Marshallian externalities with both intra-industry economics of localization as well as urbanization externalities, that is inter-industry cross-fertilization of technologies. Marshallian externalities can be divided into pure knowledge spillover such as social bonds, trust and face-to-face interaction and 'rent externalities'. The latter type can further be divided into economics of specialization, for example a locally greater variety of inputs and services, and labour market economics, for example a larger pool of employers and employees which locally smoothes the effect of business cycles in both employment and wages. These rent externalities occur through market interactions whereby local firms may lower their costs.

Localized knowledge spillover is a too vague concept to fully explain geographically bounded knowledge transmissions. Recent developments in the economics of knowledge reveal a wider variety of transmission mechanisms. Evidence is not conclusive about which role that geographical distance does – or does not – play in the economics of knowledge transmission. However, Breschi and Lissoni (2001) question whether regional factors and pure localized knowledge spillovers really explain regional development and specialization. They argue that the underlying logic behind the pure local knowledge spillover is a three-step logical chain:

1. Knowledge generated can be transmitted to other actors.
2. Knowledge is a public good, non-excludable and non-rival.
3. Knowledge is mainly highly contextual and difficult to codify and hence more easily transmitted locally.

Pure knowledge spillovers is a concept that, on first glance, ought to apply particularly to knowledge-intensive fields and related sectors such as modern biotechnology. To further discuss pure local knowledge spillover, it is necessary to explore both the concept a bit more as well as the various criticisms, which have been raised in the literature.

The first critique against pure localized knowledge spillovers has been made for high-tech industries such as biotechnology by Zucker et al. (1998). They argue that the notion of spillovers does not apply to biotechnology since the knowledge generated at universities and by individual researchers is highly contextual and excludable. Cutting-edge scientific research is neither easy to transmit nor non-excludable for non-experts, but is codified within the communities. Hence, these 'star-scientists' tend to enter contractual relations or start-up firms on their own on a local basis to extract returns from their knowledge.

A related critique is that scientific and technological knowledge is often highly exclusionary (particularly to non-experts), and thereby prevents even local actors from understanding the content. Conversely, scientific and technological knowledge is often highly codified for a particular scientific community and this can be sent long geographical distances. The knowledge can therefore be communicated internally within a group of experts or a community over long distances but at the same time not fully being a free public good nor easy to transmit locally. Therefore, it is not the geographical distance *per se* which reduces the potential knowledge transmission.[3]

The third critique raises questions about the assumption of knowledge as a local public good since the nature of knowledge refers to embodied knowledge within individuals. This assumption emphasizes the importance of labour mobility and the local pool of employers. The local mobility of workers must however spread and further develop the knowledge, instead of merely shifting it from one firm to another. This positive effect for firms is not obtained if the knowledge is perfectly embodied in the individual. Neither is it achieved in the opposite view, where the organization itself is the place of knowledge. The mobility causes a loss at the departing firm, and may contribute to the organizational knowledge of the arrival firm through organizational learning.

The fourth critique regards the time lag between basic scientific discovery and industrial application. The indirect links between science and firms may (often) mean that this type of knowledge is transmitted far beyond the local region before it becomes useful in innovations. Moreover, the scientific activities often must be used in combinations with complementary knowledge and industrial specific knowledge, which may not necessarily be held by local actors.

Finally, the concept of localized knowledge spillover is not fully consistent with how R&D intensive firms and universities acquire new knowledge. Although local business partners may be important, the scientific and engineering communities acquire new knowledge through a variety of mechanisms, such as reverse engineering, patent disclosures, journals, conferences and fairs. The concept of absorptive capacity has been used to argue that the firm's ability to assimilate external knowledge is related to the extent of internal R&D (Cohen and Levinthal 1990). Thereby, the effects of local external knowledge may only affect a few R&D-intensive firms.[4]

2.2 Synthesis and Research Questions

The discussion regarding local knowledge spillover and theories of regional development links into theories within innovation studies and evolutionary economics about the reasons for the developments of particular trajectories. In this perspective, technology is cumulative and differentiated, that is industrial composition of innovative capacity in a given location reflects past technological accumulation (Nelson and Winter 1982). The underlying explanation is that technological advance is an incremental learning process in which co-location enables the creation and diffusion of new relevant knowledge (Antonelli 2001; Malmberg and Maskell 1997). We thereby suggest that the trajectory of specialization within a sector in a region depends on the co-location of multiple types of innovative search activities, for example scientific, technological and business ones that cumulatively affect the trajectory. Innovative search activities are defined as 'the investment that the actor makes into activities, routines, and behaviours, which influence the actor's capability to innovate' (McKelvey 2004).

In our analysis we use regions as one locus for the agglomeration of knowledge and innovative search activities. The issue of interest in this chapter is not the region *per se*, but rather using regions as the unit of analysis for analysing knowledge and economic transformation. Therefore, based on these views, two research questions can be formulated:

Research Question 1
Whether the three search activities of scientific, technological and business activities on an aggregated modern biotech level are correlated. In other words, do all activities within one region tend to move in approximately the same direction and at the same rate?

Research Question 2
The extent to which long-term trends show whether these regions tend to develop along specific trajectories of application. In other words, are the regions relatively specialized in 'biotechnology', 'drugs', 'medical technologies' and 'agriculture' across the indicators of scientific, technological and business activities?

3. CAPTURING MODERN BIOTECHNOLOGY IN TERMS OF THREE ACTIVITIES

This section first discusses how and why indicators are used to measure the three activities, for example scientific, technological and business ones. The criteria to choose indicators for these activities is that they must be comparable over countries, regions and time, and thereby allow long-term comparisons as well as comparisons of relative specialization. This section then discusses how and why the broad category modern biotechnology is divided into four distinct categories, useful for the indicators. The claim is that this choice represents a more correct understanding of how, when and why this type of new knowledge creates economic value.

We use specific indicators to understand long-term development in the search activities.[5] Twenty-five years were examined at the regional level, for each of the four regions. The indicators of these three activities are, respectively:

1. scientific activities through publication in a set of peer-reviewed and cited scientific journals;
2. technological activities through patents granted by USPTO;
3. business activities through databases of firms present in the regions.

Before presenting our method in more detail, it is necessary to mention that much discussion has focused on whether publications and patents are representative indicators of science and technology (Noyons et al. 1998; Schmookler 1971; Griliches 1990). One can say that science and technology indicators both represent protection of intellectual property rights, but in different ways (Noyons et al. 1998). Patents commonly protect the commercialization and the market-related aspect, while publications protect the intellectual ideas. Such differences partially reflect different incentives and 'selection environments' for actors engaged in different activities (Dasgupta and David 1987; McKelvey 1996).

The implication of these debates about what the indicators actually measure is relevant here. We interpret them to imply that science indicators represent the knowledge base and competence profile while technology indicators reflect technological activity and specialization related to possible commercialization. This interpretation thus means that the indicators are in line with the concept of co-located search activities for innovation, as discussed in Section 2.

Scientific activities are measured through publications of scientific articles, through gathering data on the number of articles in journals above a set 'journal impact factor'. The tradition of paper publication in the scientific community makes a bibliometric study a suitable measurement of the scientific activities. Clearly, a large part of the research and development activity inside private firms is not made public through journals, but nevertheless, a bibliometric study does reveal much of the basic research activity at universities, institutes and in university–industry collaborations.

This chapter uses bibliometrics, based on 'journal impact factor', for each of the categories of modern biotechnology.[6] A 'journal impact factor' is a method used to rank scientific journals according to their scientific prestige. The journal impact factor is a measure of the frequency with which the average article in a journal has been cited in a particular year.

This indicator was here applied at the regional level. This means that the indicator of scientific activities is measured in terms of those papers published in journals above a 'journal impact factor' and where the author can be identified to be active in that region. The data is based on the top cited journals within each modern biotech category, such that the 10 per cent highest ranked within that category are included. Thus, this indicator measures relative specialization of the region in what would be considered internationally as high-quality scientific activities.

Narin and Noma (1985) found that within modern biotechnology and biosciences, science and technology are more closely linked such that 'the division between leading edge biotechnology and modern bioscience has almost completely disappeared'. The argument is hence that scientific activities – or the relative performance of research output in a region – is relevant to reveal strengths within specific categories of modern biotechnology. Despite linkages, we take this to mean that both scientific and technological activities need to be considered.

Technological activities are measured through patents granted by the USPTO, based on a search criterion to pick up all activities where any of the actors identified with the patent can be identified to be active within the municipalities in the four regions.

As with publications and citations, patents are an accepted indicator of technological knowledge, or at least of technological activities.[7] Patents are a time-limited monopoly awarded to the owner. A patent has to be novel to the world and reflect technological activities. Granted patents have to be 'non-trivial' in the sense that they are not obvious for a skilled person within the technology and that they have a commercial value. Furthermore, the innovation has to be reproducible and industrially exploitable, that is the patent application is filed in order to create a business impact (Jacobsson et al. 1996). The USPTO database is often used as a type of global indicator of technological activities. Because, given the cost and resources for a non-US based inventor to apply for a US patent, the assumption is that such applications represent somewhat more major and/or more commercializable ideas for non-US inventors (Pavitt 1985).

There are a number of advantages of using patents as a technology indicator for international comparisons. Patents cover almost all field of technology and there is also a large amount of detailed information available about the innovation itself, the inventors, assignees, technological areas and so on. They provide information about the composition and direction of inventive and innovative activity at a very detailed level (Mogee 1991). Patents, therefore, provide information about what firms see as their competitive advantage. In this chapter, the indicator of USTPO patents are only used as a measure of technological activity – not as an indication of commercialization or firm competitive advantage.[8] Limitations with patents affect level comparisons across industries at some point, but they do not affect the analysis of trends and changes over time (Trajtenberg 1999). Hence, the trends and changes should anyway be valid for our comparison.

Business activities are here measured through a new database of all relevant firms in the regions. As with publications/citations and patents, company data are a traditionally used measure of industrial activities. Some analyses focus mainly on start-up firms while others analyse the overall activities within an industrial sector, based on existing firms, exits and entries (Klepper 2001; Suarez and Utterback 1995). Here, however, the databases do not (yet) have the type of comprehensive and internationally comparable data gathered by a third party, in the same way that exist for publications and patents. The inclusion criterion for our database has been at least one employee as well as some kind of biotech-related R&D within the firm.[9] Business activities are not compared with a global average since global data on firm entries could not be found. Thus, to measure business activities, the authors therefore built a new database, to include all firms in the four fields and four regions, which exist today and which entered the categories in the past 25 years.

Obviously, there are some problems with simply using the number of firms. The most important reservation is that firms included are of very differing sizes – ranging from zero full-time employees to many thousand employees and ranging from being in the red, zero results to millions of US dollars in profits. Indicators or indexes that take into consideration the size of companies and changes in size over time would therefore be useful – but resource consuming to collect, when reliable international statistics do not already exist. Thus, for this initial study, we use the number of firms – and the entry of firms – as a measure of changes in business activities. This was deemed as a very first step, to be developed in later work.

While the activities here defined as scientific, technological and business may overlap, the indicators used to measure each activity are, in fact, separate and distinct.[10] Scientific activities are classified according to publications within specific sets of journals above a 'journal impact factor' of 10 per cent for each category. Technological activities are classified according to the different application areas of each patent class. Similarly, business activities are used to categorize firms by their main products they aim to commercialize.

For each of these three indicators, this empirical study collectively calls them modern biotechnology, a term useful for aggregation of the categories, as found in Section 5. Sections 6 and 7 analyse the four categories. Much of the methodological work underlying this chapter developed detailed proxies for classifying papers, patents and firms into the four categories of (1) core 'biotechnology', (2) 'drugs and medicals', (3) 'medical technologies' and (4) 'agriculture'. In this way, this chapter also relies on four distinct and non-overlapping categories of modern biotechnology (see Table 10.1).[11]

In contrast to only 'core' knowledge fields of modern biotechnology, the empirical material presented here represents a testing of the ideas about the broader impacts of modern biotechnology on a variety of related sectors.

Table 10.1   Schematic overview of classification: categories of modern biotechnology, by type of activity

	Scientific activities	Technological	Business activities
Core biotechnology	Core biotechnology	Core biotechnology	Core biotechnology
Drugs	Drugs	Drugs	Drugs
Medical technologies	Medical technologies	Medical technologies	Medical technologies
Agriculture	Agriculture	Agriculture	Agriculture

Note: For publications, 'medicine' was also used in the descriptive statistics. Because 'medicine' is a more generic level than the others, the trends are included in the aggregate analysis of publications, but not included here as separate categories.

Hence, the more nuanced and different definition of modern biotechnology is tested within an expanding field of knowledge, which affects many industrial sectors.

4.   A BRIEF INTRODUCTION TO THE FOUR REGIONS

This section briefly presents the rationale for comparing two regions in these two countries as well as outlines some facts about each region. This overview gives background information, which is useful to contextualize the specific empirical results presented in later sections.

This chapter is based on a choice to compare and contrast two regions in two countries. While comparisons with leaders are common ways of benchmarking, the problem is that the 'exceptional leaders' often provide little insight into the dynamics of the followers. For the same reason, this chapter excludes a priori, the region in each country which would historically be seen as a leading biotechnology region – that is Stockholm/Uppsala and Sydney, respectively.

The two countries, Sweden and Australia, have some similarities, which thereby allow comparison. Each of them has a relatively small economy in a global perspective; each has been historically dependent on more traditional industries; each claims to be moving towards the 'knowledge economy', based on Organisation for Economic Co-operation and Development (OECD) figures; and each has identified biotechnology as one of the most important areas for future societal and economic development (OECD 2001a, 2001b).

By regions, we mean regions including a main metropolitan area as well as the hinterland of each city. For Sweden, Gothenburg and East Gothia were chosen, partly based on previous work which identified them as lagging regions (Gökbayrak and McKelvey 2002; McKelvey et al. 2003). For Australia, Brisbane and Melbourne were chosen, namely two large metropolitan centres in different states. Thus, the four regions chosen are Gothenburg, Sweden; East Gothia, Sweden; Melbourne, Australia; and Brisbane, Australia.

In Sweden, Gothenburg is on the western coast and has a population of approximately 600 000 people. Gothenburg is the second largest Swedish city. It has a history as a strong international trading centre, but from the mid-twentieth century, it has been perceived as more dominated by heavy manufacturing such as shipbuilding and automotives. In the early twenty-first century, the city hopes to profile itself as a city for wireless Internet and mobile data. Around 2000–2001, Gothenburg actors have been involved in major efforts to finance and promote research at Gothenburg University, Chalmers and Sahlgrenska Akademi, also as part of national and inter-national initiatives. Moreover, Gothenburg is also a major site for AstraZeneca – and the pre-merger Astra R&D units have been in this region for many decades. In terms of biotechnology, Gothenburg was per-ceived at the start of this study as relatively underdeveloped as compared to Stockholm. More research and more companies are thought to exist than in East Gothia – likely doing other activities like bioengineering and biomaterials more than drug discovery.

East Gothia is located in eastern Sweden and has approximately 300 000 inhabitants. The region is a major agriculture area, but also has a strong tra-dition in iron ore and steel making. Our focus has been on the two cities Norrköping and Linköping. Traditionally, Norrköping has relatively more specialization in heavy manufacturing, drawing on a long legacy. East Gothia is more dominated previously by Saab (aircraft) as well as the uni-versity after the 1970s. In the 1990s, small and larger 'high-tech' companies developed, often in fields such as telecommunication and sensors. Around 2001, however, Linköping University implemented a major push into biotech research, by announcing positions for 16 new professorships, with associated groups. In terms of biotechnology, East Gothia was perceived at the start of this study as relatively 'underdeveloped', with little going on either in terms of research or in terms of commercialization and companies.

In Australia, Melbourne is the capital of the state of Victoria and has approximately 3.3 million inhabitants. The city has traditionally been strong in the food, textile, metal and chemical industry. The production is to a large extent focused on the Australian market. There is a high concentration of universities and research institutes with research in biotechnology, for example University of Melbourne, Monash University, RMIT University, LaTrobe University and Victoria University of Technology. Australia's research environment is characterized by a diverse structure with a number of research institutes. Moreover, several well-known research institutes are situated there, for example Walter and Eliza Hall Institute of Medical Research, Howard Florey Institute, the Ludwig Institute for Cancer Research, the Baker Institute, the Austin Research Institute, the Victorian Institute of Animal Science and Royal Melbourne Hospital. In terms of biotechnology, Melbourne was perceived at the start of this study as not traditionally being active in these fields but more recently moving into 'biotech' research and possibly commercialization. Biotechnology in the region was seen as having had much political recognition and as having research activities in a variety of related medical, natural science and engineering research.

Brisbane, with its 1.5 million inhabitants, is the capital of Queensland in eastern Australia. A number of universities are situated in Brisbane and its surroundings, for example Central Queensland University, Griffith University, University of Queensland and Queensland University of Technology. Additionally, there are research institutes with efforts in biotechnology, for example Bureau of Sugar Experiment Stations, Mater Medical Research Institute, Master Paediatric Respiratory Research Centre, Queensland Institute of Medical Research and Sir Albert Sakzewski Virus Research Centre. In terms of biotechnology, Brisbane was perceived at the start of this study as similar to Melbourne, but at a lower overall level.

5.   CO-LOCATION OF INNOVATIVE SEARCH ACTIVITIES IN MODERN BIOTECHNOLOGY

This section answers the first question, namely whether or not activities in all three activities tend to change at the same rate and direction over time. It presents a correlation analysis of the overall trends within each of the four regions. The correlation analysis is made for modern biotechnology, that is the aggregation of all biotech categories described in Table 10.1.

Table 10.2 shows the results from the overall correlation analysis of publications, patents and new firms in all four regions. In Gothenburg there is a strong positive correlation between patents and firms (0.788), which is significant at the 0.01 level. The patenting activity, in addition, is also significant at the same level with the firm category with a correlation of 0.383. In East Gothia, however, the only significant (0.05 level) is the relation between publications and patents. This can partly be explained by the fact that the firm data on East Gothia only consists of nine firms. The number

Table 10.2   Correlation matrix of patents, publications and firms for
*   Gothenburg*

	Publications	Patents	Firms
Gothenburg			
Publications	1	0.788**	0.713**
Patents		1	0.383**
Firms			1
East Gothia			
Publications	1	0.504*	0.965
Patents		1	0.277
Firms			1
Melbourne			
Publications	1	0.844**	0.713**
Patents		1	0.616**
Firms			1
Brisbane			
Publications	1	0.781**	0.792**
Patents		1	0.534*
Firms			1

Notes: Correlation based on numbers between 1976–98: * correlation is significant at the
0.05 level; ** correlation is significant at the 0.01 level.

of data points is therefore too low to draw any conclusive conclusions. There are strong positive correlations between all variables in Melbourne. Publications and patents are correlated at the 0.01 level. Furthermore, firms and patents are also correlated (0.616) and firms and publications (0.713). In Brisbane patents and publications are correlated (0.718) and publications and firms (0.792) at the 0.01 level. Patents and firms are also correlated but with a lower significance level.

This method of describing the modern biotech field tells us that the three activities of science, technology and business are highly correlated, at least as shown by the aggregate data. The correlation analysis thereby provides some preliminary answers to the first question about regional development and the relations between the three types of search activities. Our data, however, has a number of shortcomings. First, using regions as the unit of analysis might also be complicated since in small regions, the data points might not be sufficient to draw conclusive conclusions. This was evident in the East Gothia case, for example, where the number of firm entries was low. Second, using the aggregate modern biotech category did not provide insight about the competitive advantage of the region. Hence, while more distinct categories than modern biotechnology would be necessary to answer the research question, the data provides too few data points for correlation analysis. Finally, and most importantly, the correlation analysis is a tool that provides no insight into addressing the relative trajectories of specialization in the regions.

6. TRENDS IN SPECIALIZATION ON A DISAGGREGATED LEVEL

This section presents the empirical data of the long-term trends in the specific categories of scientific, technological and business activities in the four regions. The section first describes how revealed scientific advantage (RSA) and revealed technological advantage (RTA) are measured for each region. As explained in Section 2, business activities are analysed in terms of firms, and this indicator is not as internationally comparable as publications and patents. In short, this section presents the relative comparative advantage of each region – in terms of trends over 25 years and in terms of the four categories of modern biotechnology.

The measures presented here allow an examination of overall activities as well as changes in the rate and direction of regional specialization for the categories of modern biotechnology. Obviously, because the regions differ in size, the results should be expected to differ in terms of total output. Indeed, the absolute output from each of the four regions is partially the result of different net resources as to their size, financial resources, inhabitants, propensity to patenting and so on. For scientific output of universities, other characteristics also matter, such as the degree of teaching and research. However, this critique is already addressed through this method. The analysis here does not aim to measure any 'absolute contribution' or the efficiency of these three types of activities. Instead, the analysis examines the overall activities in the region and relative changes in rate and direction of regional specialization.

Regional specialization in scientific and technological activities is measured through a modification of Balassa's revealed comparative advantage (RCA) that originally was used in trade theory. Soete and Wyatt (1983) modify the RCA index to measure relative specialization in technological activities. For clarity and for consistency with the rest of this chapter, we measure scientific activities through RSA, and technological activities through RTA. RSA and RTA are calculated in the same way, but based on different indicators of different activities.

RSA [RTA] is calculated by the ratio of the number of publications [patents] in a particular category divided by the total number of publications [patents] in the region, and the ratio of the number of world publications [patents] in the category divided by the total number of publications [patents] in the world. Consequently,

$$RSA/RTA = \cfrac{\left(\cfrac{P_{ij}}{\sum\limits_{k} P_{il}}\right)}{\left(\cfrac{\sum\limits_{k} P_{kj}}{\sum\limits_{k}\sum\limits_{l} P_{kl}}\right)}$$

where:

P_{ij} = number of publications [patents] of region i in category j

$\sum\limits_{l} P_{il}$ = total number of publications [patents] of region i

$\sum\limits_{k} P_{kj}$ = global number of publications [patents] in category j

$\sum\limits_{k}\sum\limits_{l} P_{kl}$ = global of publications [patents] of region i in all fields

RSA and RTA thus indicate how each region is performing in relation to a global average in scientific and technological activities. By comparing with a global average rather than with the other regions in the study or with the nation state, this research design allows comparisons with any given region in the world.

These measures enable us to indicate whether or not a region is specialized within a specific scientific and/or technological activity – as well as whether specialization has increased or decreased over the years. Note, however, that one limitation of this measure is that regions with a high overall average can be categorized as not being specialized even though the overall activity is high.[12] Or, vice versa, a region with a low overall average may appear overly specialized, even if the absolute level is low. The RSA and RTA do, however, allow comparisons of relative specialization in a meaningful way to a global average.

For business activities firm entries were used. Section 2 has already pointed out that firm entries are an extremely rough measure, which does not take into account the size of company in terms of employees or sales nor is it relative to the firm's global activities. Still, the indicator gives some idea of the relative changes in business activities over time – and thereby of relative activity within regions – in terms of business start-up or major changes to ownership.

Table 10.3 presents the RSA for each region and biotech category, for intervals in the period 1977–2001. Table 10.4 presents the RTA for each region and biotech category, for intervals in the period 1977–2001. Table 10.5 presents the firm entry for each region and biotech category, for intervals in the period 1977–2001.

6.1   Scientific Activities

The time period measured is from 1977 to 2001. The data consists of a total number of 8755 relevant articles for the four regions. To calculate the RSA in Table 10.3, the total number of publications in a particular category and the total world publications in natural science were also used.

There are significant differences in the overall number of articles within each region, over the total period studied. When the overall number of articles is assigned in descending order between the different regions, the results are as follows: Melbourne 4667, Gothenburg 2039, Brisbane 1529 and East Gothia 520.

In contrast to the other two activities, scientific activities measures five categories, that is the four biotech categories and the additional 'medicine'. The reason is that the objective was to identify scientific activities at a more generic level apart from both technological and business activities. Thus, the bibliometric result for the five different classes is measured as the number of articles per year under a time period ranging from 1977–2001.

The regions studied here start at different points – but also exhibit quantitative differences in scientific publications. In general, the results indicate that there is an upward trend in total output for all four regions in all five categories. This likely reflects not only a regional shift towards increasing specialization in bioscience research but also the overall increase in scientific activity. Since the calculated RSA takes the overall global increase into account, Table 10.3 shows the relative specialization per region and category.

The Swedish regions indicate upward trends in total scientific output for all five categories. In the East Gothia region, there is an upward trend for all five classes of scientific publications – but from a low level. Gothenburg similarly has a steady increase of published articles in all five categories of journals – but from a higher starting-point.

Both Swedish regions show a high and stable specialization in the category of 'medical technologies' and are to a less extent non-specialized in 'core biotechnology'. Overall, however, the results from the Gothenburg

Table 10.3   Results from science indicators as evidenced by RSA (by region and modern biotech category)

	Gothenburg					East Gothia					Melbourne					Brisbane				
	D	MT	BIO	AG	ME	D	MT	BIO	AG	ME	D	MT	BIO	AG	ME	D	MT	BIO	AG	ME
1977–81	2.09	1.19	1.04	1.53	1.07	5.52	3.93	0.00	0.80	1.35	1.76	0.26	0.15	0.57	1.31	1.32	0.53	0.27	0.53	0.32
1982–86	3.63	2.74	0.81	1.10	1.60	1.82	2.08	0.07	0.59	1.15	2.38	0.78	1.37	1.38	2.53	0.89	1.27	0.13	1.75	0.73
1987–91	4.25	1.77	0.71	1.19	1.39	3.20	1.90	0.44	0.93	1.45	1.64	0.93	1.70	1.10	2.01	1.16	1.84	0.42	1.57	0.68
1992–96	2.21	1.89	0.67	1.14	1.28	0.15	1.80	0.32	0.64	1.50	1.09	1.20	1.57	0.90	2.16	0.62	1.46	0.99	1.24	1.32
1997–2001	2.20	1.56	0.75	1.49	1.51	1.27	2.57	0.44	0.75	1.03	1.26	1.87	1.90	1.21	2.32	1.11	1.45	1.69	1.76	1.50

Note: Information is based on science citation index. D = drugs, BIO = core biotechnology, MT = medical technologies, AG = agriculture, ME = medicine.

Table 10.4 *Results from technology indicators as evidenced by RTA (by region and modern biotech category)*

	Gothenburg					East Gothia					Melbourne					Brisbane				
	D	MT	BIO	AG	ME	D	MT	BIO	AG	ME	D	MT	BIO	AG	ME	D	MT	BIO	AG	ME
1977–81	1.30	1.96	2.43	1.25	–	0.26	1.63	0.00	0.00	0.00	0.99	1.33	0.61	2.40	–	0.21	0.43	0.41	1.91	–
1982–86	1.27	3.69	0.52	0.60	–	0.00	1.14	0.00	0.00	0.00	0.95	1.02	1.50	4.27	–	0.80	1.69	0.76	4.34	–
1987–91	2.39	3.98	1.40	0.90	–	0.43	1.68	0.33	1.27	–	1.66	1.17	3.03	2.27	–	0.74	0.91	1.30	4.08	–
1992–96	2.65	4.36	0.30	0.90	–	0.16	1.42	0.55	0.00	–	2.54	0.85	3.74	2.84	–	1.21	0.82	2.18	2.65	–
1997–2001	1.92	3.65	0.54	0.89	–	0.12	0.79	0.45	0.54	–	3.39	1.31	3.11	2.04	–	1.92	1.44	2.85	2.94	–

Note: Information is based on USPTO. D = drugs, BIO = core biotechnology, MT = medical technologies, AG = agriculture, ME = medicine.

Table 10.5   Results from business indicators as evidenced by firm entry (by region and modern biotech category)

	Gothenburg					East Gothia					Melbourne					Brisbane				
	D	MT	BIO	AG	ME	D	MT	BIO	AG	ME	D	MT	BIO	AG	ME	D	MT	BIO	AG	ME
1977–81		2			–					–	2	2			–				1	–
1982–86	1	6	5		–					–	1	3	5	6	–	1		1	3	–
1987–91	1	9	4		–		1			–	1	6	9	3	–			3	3	–
1992–96	2	5			–		2			–	3	5	11	6	–	1	1	1	1	–
1997–2001	6	8	6		–		5			–	6	7	30	14	–	2	2	15	2	–

Note: Information is based on a new database. D = drugs, BIO = core biotechnology, MT = medical technologies, AG = agriculture, ME = medicine.

region indicate a strong and stable specialization in 'drug'-related research articles. This indicates a sustainable regional strength in this scientific area. For East Gothia, the regional specialization in 'drug'-related research articles is declining towards the world average. Thus, the general specialization in East Gothia and Gothenburg area are quite similar and stable, but the scientific output of East Gothia is around a quarter of Gothenburg's output. These overall differences may be expected, since much public research is funded nationally and hence reflects the size difference between the two Swedish regions.

In contrast, both Australian regions indicate a trend towards increasing specialization in 'medicine' and 'core biotechnology' classes. This is in addition to an overall upward trend in the absolute output of all five categories – which holds for both the areas of Melbourne and Brisbane

Melbourne has a significantly high specialization in 'medicine' and has increased towards regional specialization in both 'core biotechnology' and 'medical technologies'. The specialization in the Brisbane region in 'medicine' and 'core biotechnology' is prominently increasing.

As compared to the other three regions, Brisbane is the region with the highest relative growth in scientific activities from the early 1990s and on in areas of 'core biotechnology' and 'medicine'.

6.2 Technological Activities

The time period measured is 1977 to 2001. The data consists of a total number of 1866 relevant patents for the four regions, granted anytime between 1977 and 2001. To calculate the RTA in Table 10.4, the total number of patents in a particular category and the total world patents were also used.

When the overall number of patents are assigned in descending order between the different regions, the results are as follows: Gothenburg 838, Melbourne 766, Brisbane 207 and East Gothia 55.

Thus, there are significant differences in the overall number of patents granted within each region, over the total period studied. There is an increasing trend in technological activities, but not as striking as in the scientific activities. OECD (2001a, p. 11) argues that in OECD countries, the 'absolute number of USPTO and EPO biotechnology patents has grown substantially in comparison with the total number of patents' in the period 1990–2000.

The East Gothia area has a lower total number of patenting than the other regions. There is not much change or trends over the past 25 years, because all four categories remain at the same low level during the years. The most common type of patent, and hence the category where some regional activity is visible, is within 'medical technologies'. Hence, the region is specialized in this category (see Table 10.4).

The Gothenburg region is specialized in 'medical technologies' (see Table 10.4). This category has increased during 20 years, and it peaked in 1996 with 50 patents. For the Gothenburg area, the category of 'drugs' reveals a similar pattern, with a growth in number of patents that lies just below the trend for 'medical technologies'. Gothenburg is not specialized in patents in the other two areas of 'biotechnology' and 'agriculture'. For these latter two categories, there are too few samples to draw any conclusive interpretation about trends, that is the total output is on a low level.

In contrast to Gothenburg's specialization, Melbourne has a high specialization of patents in both categories of 'core biotechnology' and 'agriculture'. Additionally, the patent results for Melbourne indicate that there is an upward trend in 'drugs'. In contrast, 'medical technologies' remains on the same level, over the whole period. While a specialization in 'agriculture' patents are shown, it is on a low absolute level; nevertheless Melbourne has more activity there than the other three regions, if examined in absolute numbers.

Since the mid-1990s, the Brisbane region reveals an increasing trend in three of our four categories – namely in 'core biotechnology', 'agriculture' and 'drugs'. The specialization is most pronounced in 'agriculture'; however, each of the other three categories is larger than 'agriculture' in absolute numbers. Still, as compared with the two Swedish regions, Brisbane has a relatively large proportion of the total amount of patents in 'agriculture'.

6.3 Business Activities

The time period measured is from 1977 to 2001. The data consists of a total of 220 firms, by merging the database for each region. The criteria were that the firm exists in the years 2001–2002, and they were started any time between 1977 and 2001. Table 10.5 shows firm entry, rather than a global comparison.

When the companies are assigned in descending order among the different regions, then the results are: Melbourne 124, Gothenburg 55, Brisbane 33 and East Gothia 8, respectively. Thus, as with scientific publications and patenting activities, the sizes of the regions in terms of number of firms differ drastically. Moreover, the regions have different patterns in terms of firm entries over the analysed years.

In the case of East Gothia region, all eight companies are classified as 'medical technologies' group (see Table 10.5). There is a small number of companies, but a clear specialization in that direction. In contrast, biotechnology and biosciences, which are commercialized in firms within the categories of 'core biotechnology', 'drugs' and 'agriculture', do not seem to exist in this region.

In the case of the Gothenburg region, the data shows 55 firm entries (see Table 10.5). Of these, 55 per cent of firms fall into the 'medical technologies' category, thereby indicating a clear specialization. Moreover, and as compared with the other regions, Gothenburg has a large proportion of firms in the 'drugs' category. In contrast, the region is not specialized in 'agriculture'. For 'core biotechnology', some firm entries have occurred, but at a lower rate and lower absolute number than, for example, Melbourne. Most of the Gothenburg 'core biotechnology' companies have been established in the mid-1990s and thereafter.

In the case of the Melbourne region, the database indicates 124 firms (see Table 10.5). This region has obviously the most number of firm entries. The majority of these are in 'core biotechnology', with also a relative specialization in 'agriculture', as compared to the other regions. Twenty-five per cent of the firm entries have been in the agricultural classification, which is a relatively large share. With our definition of biotechnology and related firms, a total number of 124 companies was identified in our database. Between 1976 and 1985 10 firms were established to be compared with 82 in 1996–2002. Consequently, entries from firms in Melbourne boomed in the latter half of the period 1996–2001.

For the Brisbane region, the pattern is quite similar to Melbourne, as differentiated from the Swedish cases (see Table 10.5). The Brisbane companies also tend to have a larger proportion of firm entries in 'core biotechnology' and in 'agriculture'.

7.   SUMMARY AND CONCLUSIONS

This section compares and contrasts the long-term changes in the four regions, across the three types of search activities, thereby the section provides a summary of the specialization of each region indicating similarities and differences. This analysis thereby enables us to link the two questions and the empirical material to theoretical discussions and conclusions. Table 10.6 summarizes the empirical results found in Section 6, in terms of the relative specialization of these four regions and whether trends are increasing or decreasing over time.

It also identifies significant differences between the four regions when compared to a global average. The regions differ in terms of overall output as well as in terms of specialization. Finally, Table 10.6 indicates the rate and direction of changes over time. The specialization and trends in the four regions may be summarized as follows:

Table 10.6   Summary of the specialization from scientific, technological and business indicators

	Gothenburg	East Gothia	Melbourne	Brisbane
Core biotechnology	Increasing in firms		Increasing in science, technology and firms	Increasing in science, technology and firms
Drugs	Specialized in science	Specialized but declining in science	Increasing in technology	
Medical technologies	Specialized in science, technology, firms	Specialized in science and firms	Increasing in science	
Agriculture			Specialized in technology. Increasing in firms	Specialized in technology
Medicine				Increasing in science

Note: Medicine is a separate category than the other four, and only for scientific activities.

1. Gothenburg specializes in 'medical technologies', seen both in papers, patents and firms, but also has significant activities in 'drugs'.
2. East Gothia specializes in 'medical technologies'. However, for all indicators (papers, patents and firms), the total output is low, even if the relative share is high.
3. Melbourne specializes in 'core biotechnology' and has specialization in 'agriculture'.
4. Brisbane specializes in 'core biotechnology', a relatively high degree of specialization in 'agriculture'.

Moreover, the two Swedish regions exhibit similar characteristics and trends, just as the two Australian regions do. The two Australian regions, however, show an increasing trend in all activities, considerably more than the Swedish ones.

To conclude Sections 5 and 6, the two research questions set at the end of Section 2 can be answered:

1. Whether the three search activities of scientific, technological and business activities on an aggregated modern biotech level are correlated. In other words, do all activities within one region tend to move in approximately the same direction and same rate?

    The empirical result in Section 5 indicates that the change in output is reflected in all three types of activities – scientific, technological and business ones. They do tend to move in approximately the same direction and rate, as seen over a long time period. The knowledge and the development of the region tend to move in the same direction over time.

2. The extent to which long-term trends show whether these regions tend to develop along specific trajectories of application. In other words, are the regions relatively specialized in 'core biotechnology', 'drugs', 'medical technologies' and 'agriculture' across the indicators of scientific, technological and business activities?

Here, the patterns are not clear. On one hand, the long-term trends do indicate specific trajectories at the regional level in some cases, such as 'medical technologies' in the Gothenburg region and 'core biotechnology' in the Melbourne region. On the other hand, some categories reveal no clear pattern such as 'core biotechnology' in Gothenburg and the categories from East Gothia. The empirical results also indicate that while some regional trajectories are 'history-bound' – in the sense that the region remains specialized over the whole period, other regional trajectories change at a certain point. These points of change represent some discontinuity – after which new trajectories are developed through the cumulative activities.

This chapter has presented analytical and methodological reasons for understanding the dynamics of regional specialization in modern biotechnology. Based on extensive empirical material about three types of activities, each region can be compared and contrasted to each other – also relative to global averages. The previous two sections thereby provide extensive evidence of the type of specialization and long-term trends for each of the four regions – Gothenburg, East Gothia, Melbourne and Brisbane. This section concludes by reflecting on the theoretical and analytical implications for understanding the dynamics of regional specialization.

Clearly, one important conclusion is that this research supports the theoretical view that economic transformation in some cases is the result of long-term development along trajectories (path-dependency) – and not due to stochastic processes. This implies that time has emerged as an interesting factor and that the dynamics of regional specialization are related to historical processes, as well as contemporary ones.

Second, however, while historical development matters for trajectories, there are still visible changes, which occur in the rate and direction (category) of specialization, over time. Such changes occur, possibly due to various types of discontinuities.

Despite a generally positive answer to the question about whether relative regional specialization arises, the question remains of how to categorize regional specialization along a trajectory. The evidence still demonstrates differences across regions and time periods. For example, there are differences in terms of whether the trends are reflected in all activities – or only some. Such differences may reflect time lags; that some activities precede others; and/or that not all activities are necessary at all times. These differences require additional consideration – in relation to theoretical explanations of economic transformation – in order to answer the puzzle.

Moreover, other variables may need to be included in the analysis – especially government (national or state) policy to support scientific, technological and/or business activities. Such variables may help explain the 'discontinuities' and changes within trajectories – as, after all, their purpose is to aid regional development in a certain direction. Also, interestingly enough, the two regions from each country tend to be more like each other than would be expected from a random sample. This indicates that the national level ought to be more specifically considered in terms of some variable(s).

Third, in order to incorporate time and the development of historical trajectories, the communities of practitioners and researchers need more systematic research tools (Holmén and McKelvey 2002) as well as comprehensive methodology (Brink et al. 2003). Such tools and methodology enable more stringent comparisons as well as help communicate the results.

Fourth, the correlation analysis showed that at the modern biotech level, all three regions showed increasing trajectories within all three activities. In other words, significant correlation was visible between the activities in all regions – except East Gothia due to its smaller size.

Fifth, the division of modern biotechnology into four distinct categories works in practice, that is the categories of 'core biotechnology', 'drugs', 'medical technologies' and 'agriculture'. By deconstructing the modern biotech field, this enables a more detailed analysis of the competitive edge and underlying dynamics of each region.

Still, the underlying knowledge base within these categories might result in externalities across knowledge fields and/or across industrial sectors (see Section 2). Therefore, they must be analysed in relation to each other, for example it is incorrect to analyse only 'dedicated' or 'core' biotech firms in a regional perspective.

Sixth, the analysis within these four distinct categories does highlight trajectories of regional specialization. At this more disaggregate level, the similarities and differences among the four regions become visible.

Thus, finally, the chapter demonstrates that the conceptualization of the dynamics of regional specialization does make sense when regional trends are placed in relation to global trends and global competition (McKelvey 2003a and 2003b). In other words, the concepts of long-term trends in scientific, technological and business activities – as well as relative regional specialization – highlight the region relative to global averages and relative to international trends and international selection environments.

NOTES

1. The differences are manifested in that they have different departments at the universities and the technological applications serve different markets. They might be intertwined in practice, but they still reflect true differences.
2. The rationale is that this differentiation allows us to examine both the knowledge and market developments within the global biotechnology and biosciences (McKelvey 2003a).
3. At the same time, face-to-face interaction among the global community (or local scientific milieu) may be needed to create a common language, norms and rules.
4. Thereby, the flow may be reversed, such that the local universities and R&D intensive firms may play an important role for training and recruitment attracting brilliant students and researchers to the local region.
5. The method is fully explained in Brink et al. (2003).
6. Bibliometrics – for example indicators based on publications in scientific journals – is often used to analyse and examine the scientific advance in a given field. The number of publications is considered to be a quantitative indication of the scientific output in terms of activity and intensity of a group in a specific field. However, quantity is not necessarily the same as quality. To reflect the scientific quality, various bibliometric methods have been worked out, and these are usually based on the frequency of citations an article receives after the publication date. The most common measurement is to calculate the average citations under a certain time period and are often used to evaluate individual researchers, departments and universities.
7. Still, much debate has focused on what patents represent. Griliches (1990) raised the question of whether or not patents' statistics represent the level of innovation. Hall et al. (2001) argue that patents are a good indicator of technological activities but they do not measure the economic value of those technologies.
8. Even so, patents as an indicator of technological activity have several limitations. The patent application is a strategic decision that the firm takes in order to gain a competitive advantage. Firms might make a strategic decision to rely on other means of appropriability instead of patents (Levin et al. 1987). Patents are just one way of protection along with secrecy, lags and firm specific skills and knowledge (Pavitt 1985).
9. Note! This implies that our database does not include only 'dedicated biotech firms', but ones with activities in biotech-related R&D. Moreover, the criteria implies that firms with only sales offices in a region are excluded. Moreover, firms with zero full-time employees are excluded, which is an issue in Swedish biotech-pharmaceutical data (McKelvey et al. 2003). Much of the existing data on biotech-related firms – which purports to be internationally comparable – are highly questionable.

10. Naturally, classification of journals, patents and firms depends on judgement and evaluation, but Brink et al. (2003) argue why the methodology presented here develops non-overlapping indicators as well as reasonable judgement for each classification.
11. This categorization is novel as compared to much of the literature on 'biotechnology', with a few exceptions (see, e.g., some chapters in Senker, 1998 and Senker, Chapter 5 of this book).
12. International comparisons also imply that differences in patenting activity across countries occur, for example the Japanese propensity is higher than other countries in many industries. This implies that RSA and RTA need to be complemented with the absolute number of output in each region, respectively.

REFERENCES

Antonelli, C. (2001), *The Microeconomics of Technological Systems*, New York: Oxford University Press.

Audretsch, D.B. and P.E. Stephan (1996), 'Company-scientists locational links: the case of biotechnology', *The American Economic Review*, **86**(3), 641–652.

Bathelt, H., A. Malmberg and P. Maskell (2002), Cluster and knowledge: local buzz, global pipelines and the process of knowledge creation, DRUID working paper no. 02–12.

Breschi, S. and F. Lissoni (2001), 'Knowledge spillovers and local innovation systems: a critical survey', *Industrial and Corporate Change*, **10**(4), 975–1005.

Brink, J., L. Dahlander and M. McKelvey (2003), 'Analyzing emerging technological areas and emerging industries: test case of modern biotechnology in regions', paper read at Innovations and Entrepreneurship in Biotech/Pharmaceuticals and IT/Telecom at Chalmers University of Technology, Gothenburg.

Brusco, S. (1990), 'The idea of the industrial district: its genesis', in G. Becattini (ed.), *Industrial Districts and Inter-firm Co-operation in Italy*, Geneva: International Labour Office.

Cohen, W.M. and D.A. Levinthal (1990), 'Absorptive capacity: a new perspective on learning and innovation', *Administrative Science Quarterly*, **35**(1), special issue, 'Technology, Organizations, and Innovations', 128–52.

Cooke, P. (2001), 'Regional innovation systems, clusters, and the knowledge economy', *Industrial and Corporate Change*, **10**(4), 945–74.

Cooke, P., M.G. Uranga and G. Extebarria (1997), 'Regional innovation systems: institutional and organizational dimensions', *Research Policy*, **26**, 475–91.

Dasgupta, P. and P. David (1987), 'Information disclosure and the economics of science and technology', in G. Feiwel *Arrow and the Ascent of Modern Economic Theory*, New York: Macmillan Press.

Gökbayrak, E. and M. McKelvey (2002), 'An inquiry into the dynamics of the east Sweden life sciences cluster', unpublished manuscript.

Griliches, Z. (1990), 'Patent statistics as economic indicators: a survey', *Journal of Economic Literature*, **28**, 1661–1707.

Hall, B.H., A. Jaffe and M. Trajtenberg (2001), 'The NBER patent citations data file: lessons, insights and methodological tools', NBER working paper series, working paper no. 8498.

Holmén, M. and M. McKelvey (2002), 'Systematic studies of novelty, destruction and renewal in economic systems', paper presented at the International Joseph A. Schumpeter Conference, University of Florida, Gainesville.

Jacobsson, S., C. Oskarsson and J. Philipsson (1996), 'Indicators of technological activities: comparing educational, patent and R&D statistics in the case of Sweden', *Research Policy*, 25, 573–585.

Klepper, S. (2002), 'The capabilities of new firms and the evolution of the US automobile industry', *Industrial and Corporate Change*, **11**(4).

Krugman, P. (1991), 'Increasing returns and economic geography'. *Journal of Political Economy*, **99**(3), 483–99.

Krugman, P. (1994), 'Complex landscapes in economic geography', *The American Economic Review*, **84**(2), 412–16.

Levin, R.C., A.K. Klevoric, B. Nelson and S. Winter (1987), 'Appropriating the returns from industrial research and development', *Brookings Papers on Economic Activity*, (3), 783–831.

Malecki, E.J. (1985), 'Industrial location and corporate organization in high technology industries', *Economic Geography*, **61**(4), 345–69.

Malmberg, A. and P. Maskell (1997), 'Towards an explanation of regional specialization and industry agglomeration', *European Planning Studies*, 5(1).

Marshall, A. (1947), *Principles of Economics*, 8th edn, London: Macmillan and Co.

Martin, R. and P. Sunley (1996), 'Paul Krugman's geographical economics and its implications for regional development theory: a critical assessment', *Economic Geography*, **72**(3), 259–92.

Martin, R. and P. Sunley (2001), 'Deconstructing clusters: chaotic concept or policy panacea?', conference paper in Lund 2002 submitted to *Journal of Economic Geography*.

McKelvey, M. (1996), *Evolutionary Innovations: The Business of Biotechnology*, Oxford: University Press.

McKelvey, M. (2003a), 'Evolutionary economics perspectives on the regional-international dimensions of biotechnology innovations', paper submitted to an international journal.

McKelvey, M. (2003), 'How and why dynamic selection regimes affect the firm's innovative search activities', paper presented at Knowledge and Economic and Social Change: New Challenges to Innovation Studies, at Conference organized by ASEAT (Advances in the Economic and Social Analysis of Technology) and by the Institute of Innovation Research (I of IR) Manchester School of Management Building, Manchester, UK.

McKelvey, M., H. Alm and M. Riccaboni (2003), 'Does co-location matter for formal knowledge collaboration in the Swedish biotechnology-pharmaceutical sector?' *Research Policy*, 32, 483–501.

McKelvey, M. (2004), 'How and why dynamic selection regimes affect the firm's innovative search activities', *Innovation: Management, Policy and Practice*, **6**(1).

Mogee, M.E. (1991), 'Using patent data for technology analysis', *Research Technology Management*, **34**(1), 46–51.

Narin, F. and E. Noma (1985), 'Is technology becoming science?', *Scientometrics*, 7, 369–81.

Nelson, R.R. and S.G. Winter (1982), *An Evolutionary Theory of Economic Change*, Cambridge: Harvard University Press.

Noyons, E.C.M., M. Luwel and H.F. Moed (1998), 'Assessment of Flemish R&D in the field of information technology: a bibliometric evaluation based on publication and patent data, combined with OECD research input statistics', *Research Policy*, 27, 285–300.

OECD (2001a), 'Biotechnology statistics in OECD member countries: compendium of existing national statistics', Paris: Organisation of Economic Co-operation and Development.

OECD (2001b), 'Scoreboard 2001: The knowledge base of OECD economics', Paris: OECD.

Pavitt, K. (1985), 'Patent statistics as indicators of innovative activities: possibilities and problems', *Scientometrics*, **7**, 77–99.

Porter, M.E. (1990), *The Competitive Advantage of Nations*, New York: The Free Press.

Porter, M.E. (1998), 'Clusters and the new economics of competition', *Harvard Business Review* (November–December), 77–90.

Saxenian, A. (1990), 'Regional networks and the resurgence of Silicon Valley', *California Management Review*, (Fall), 89–112.

Saxenian, A. (1994), *Regional advantage: Culture and Competition in Silicon Valley*, Cambridge, MA: Harvard University Press.

Schmookler, J. (1971), 'Economic sources of inventive activity', in N. Rosenberg (ed.), *The Economics of Technological Change*, London: Penguin Books.

Senker, J. (1998), *Biotechnology and Competitive Advantage: Europe's Firms and the US Challenge*, Cheltenham, UK and Lyme, USA: Edward Elgar.

Soete, L.G. and S.M.E. Wyatt (1983), 'The use of foreign patenting as an internationally comparable science and technology output indicator', *Scientometrics*, **5**(1), 31–54.

Storper, M. and A.J. Scott (1995), 'The wealth of regions: Market forces and policy imperatives in local and global context', *Futures*, **27**(5), 505–26.

Suarez, F.F. and J.M. Utterback (1995), 'Dominant designs and the survival of firms', *Strategic Management Journal*, **16**, 415–30.

Trajtenberg, M. (1999), 'Innovation in Israel 1968–97: a comparative analysis using patent data', NBER working paper series, working paper, no. 7022.

Zucker, L.G., M.R. Darby and M.B. Brewer (1998), 'Intellectual human capital and the birth of U.S. biotechnology enterprises', *The American Economic Review*, **88**, 290–306.

11.   On the spatial dimension of firm formation

Annika Rickne

1. INTRODUCING THE ISSUES

The last two decades have seen an expanding life science sector in both Europe and the USA, with the growth of incumbent corporations, and a vast number of new firms being established. Interestingly, on a global arena of life sciences, US actors are remarkably influential and dominating as regards development of scientific and technological knowledge, identification and exploitation of economic opportunities as well as the ability to create economic growth. In terms of firm formation, the European life science sector showed some success during the 1990s, resulting in almost 1400 biotech start-ups between 1994 and 2000 (Ernst & Young 2001). However, there are still major differences between Europe and the USA in terms of firm growth, revenues, number of public companies as well as economic outcome created. Certainly, the European sector faces somewhat different – and some would argue much larger – challenges as compared to its US counterpart. Thus, even if the gap between the USA and Europe may be decreasing in terms of number of firms, the mechanisms and entrepreneurial innovation activities needed to transform investments in science and technology (S&T) into new firms and economic growth (Schumpeter 1934) may not be fully put into practice in the European context.

The European situation is well mirrored also in the Swedish life science sector. The Swedish challenge has been described as a 'paradox', with a competitive research sector but with difficulties in converting the scientific and technological potential into commercial success (Sörlin and Törnqvist 2000; Henrekson and Rosenberg 2000; Andersson et al. 2002; Edquist 2002; Goldfarb and Henrekson 2002). Even though entrepreneurship and other innovative activities may not be less in Sweden than in other European countries (Jacobsson 2002),[1] firm formation is undoubtedly a process in need of further understanding and improvement.

Thus, as it is clear that the European – and Swedish – challenges call for a deeper knowledge of the mechanism of firm formation, this is the focus of this chapter. On this issue, there are studies analysing the nature of S&T and the character of the innovation process, the type of technological opportunity, market characteristics and the entrepreneur, as well as the role of institutions. While these factors are all of crucial importance, this chapter focuses specifically on, and adds to, the analysis and discourse of whether or not and why the activities preceding and leading up to firm formation are spatially confined.

In particular, resting the analysis on the empirical example of a set of firms in Sweden, Ohio and Massachusetts with a core competence in biocompatible materials, the point of departure is that the particular technological regime assumes spatial importance. More specifically, in the field of biomaterials we may expect spatial concentration of innovative activity and a strong basis for the formation of groups of co-located new firms. The chapter scrutinizes *if* this is in fact true, *why* it is so and *when* co-location of young science-based firms can turn into virtuous circles and clustering effects. Three issues are discussed. The first issue is if the establishment of new ventures in biomaterials depends on access to previous scientific development and if such pre-firm activities are spatially confined. The second question is whether or not the profile of scientific work within a specific location influences the technological profile of new science-based firms. Finally, the third issue is if there are mechanisms enhancing a spatially confined firm formation process, possibly leading to clustering.

After a theoretical introduction in Section 2, a description of the research design follows in Section 3. The remainder of the chapter is structured around the three research issues, where the first two above are analysed in Section 4, and Section 5 focuses on the third research issue. Thus, Section 4 discusses if firm formation rests on access to prior scientific advance, and in order to answer to what extent the firm's technological profile hinges on the science profile in the specific location, it penetrates the potential influence of the regional science base. Mechanisms that may enhance a spatially confined firm formation processes are discussed in Section 5, with particular focus on actors engaged in developing the science and technology platform for new firms, or in providing initial capital. Finally, based on the analysis, Section 6 summarizes the results and discussed collocation as related to clustering, where the latter not only denotes spatial proximity between actors but also interdependence, synergy effects, and virtuous circles.

2. THEORETICAL UNDERPINNING

In an evolutionary view on economic development the processes of innovation, firm establishment and development, are not seen as atomistic or performed by solitary actors. Instead, these processes can be described as highly multi-technological, complex and uncertain, and therefore performed at the interface between actors, rather than being confined to the borders of a specific organization (see e.g. Freeman 1987; Håkansson 1987; Carlsson and Stankiewicz 1991; Lundvall 1992; Nelson 1993; Saxenian 1994). The continuous search and selection processes necessary for keeping up with innovation involve a diverse set of actors, all engaged in creating and diffusing knowledge (research organizations, firms that may be customers, suppliers, competitors, end-users, financiers, policy actors and so on) and many processes are under way simultaneously (Carlsson 1997; Edquist 1997). The earlier prevailing linear models of either technology push or market pull, or both, have been replaced by this view of parallel and iterative development processes, which also introduces the capability to change the process in an iterative manner (Rothwell 1994). An underlying assumption is that this is a process of coevolution. Indeed, firms and other actors, technological fields, industries, innovation systems and institutions coevolve, contrasting to a linear view of innovation (Nelson 1987). What is of particular importance in this modern view of innovation is how knowledge is formed, diffused in several directions between actors and used among actors to create economic growth (Carlsson 1997).

Thus taking a point of departure in evolutionary economics, firm formation is seen as an unfolding, cumulative process and depends on a set of interrelated factors. First, it may depend on the very nature of science and technology (S&T), and the character of the innovation process transforming scientific knowledge into commercial products and services. In general, scientific progress is seen as crucial to technological change and economic development (Krugman 1991). Nelson and Winter (1982) argue for the influence of the technological regime, and this line of thought is pursued in a set of related studies focusing on, for example, appropriability – the 'possibilities of protecting innovations from imitation and of reaping profits from innovative activities' (Breschi and Malerba 1995, p. 6) – and sectoral patterns (Winter 1984; Breschi and Malerba 1995; Malerba and Orsenigo 1997; Shane 2001, 2001a). Crucial characteristics of the innovation process include how opportunities are developed and diffused, as well as the organization of activities leading up to firm establishment (e.g. what actors are engaged, how knowledge is diffused, time frames and its spatial features). Second, features of the technological opportunity matter, including type of discontinuity and phase of the technology life cycle (Abernathy and

Utterback 1975; Tushman and Anderson 1986; Henderson 1993; Christensen 1997). Third, there are a number of influential factors relating to market structure and segmentation (Utterback 1994; Audretsch 1995; Christensen 1997). Fourth, characteristics of the entrepreneur/intrapreneur may be of value to understand firm creation. For example, Schumpeter (1934) points to the creative aspect of entrepreneurs, Roberts (1991) discusses what is driving an entrepreneur to undertake such a high-risk endeavour, and there are studies focusing on the impact of skills and experience (Zucker et al. 1998; Shane and Khurana 2000). Finally, institutions such as labour market regulations, tax rules or attitude towards entrepreneurship have a bearing on firm establishment, as may university policies and praxis on handling intellectual property.

This chapter concentrates explicitly on the first set of factors related to the nature of S&T and the character of the innovation process. As one of the main drivers of economic change processes, scientific and technological innovation and change are at the heart of an evolutionary approach to industrial dynamics. Importantly, innovation processes have varying features across technological fields and industries where the nature of the technology itself is important and the underlying technological regime influences the character and dynamics of innovation processes (Carlsson and Stankiewicz 1991; Breschi and Malerba 1995; Feldman 1999). In relation to firm formation it has been argued that this is not likely to be geographically evenly distributed, but the specific technological regime (TR) under which the young firms operate holds explanatory power both for the likelihood of firm establishment and for the spatial concentration of firms (Acs and Audretsch 1990; Breschi and Malerba 1995; Malerba and Orsenigo 1997). Certainly, the characteristics of some technological regimes render innovative activity a tendency to concentrate at certain locations in order for actors to reap benefits from spillover.[2] This chapter has as its point of departure that the particular TR of biomaterials favours new entrants, and that there is a basis for the formation of groups of co-located new firms. Based on this assumption the chapter analyses to what extent and why the activities proceeding firm formation are spatially confined.

The grounds for this assumption are as follows.[3] A technological regime can be characterized in four dimensions (Breschi and Malerba 1995): (1) the extent to which knowledge is cumulative (defined as the degree of serial correlation among innovative activities); (2) the extent and source of the technological opportunities; (3) the possibilities for appropriation by the innovator; and (4) the knowledge base, which involves both its nature and how it is created and transmitted between actors. In understanding the emergence and behaviour of young firms, the character of any specific TR has two fundamental implications for the innovation process of new companies. First, different combinations of the three first dimensions determine the scope of new firms in relation to large established firms. In some areas, the technological opportunities are small, while there is high cumulativeness, and the possibilities for appropriation are great. These features would tend to favour incumbent firms. At the other end of the spectrum – with regimes characterized by large technological opportunities, low appropriability and low cumulativeness (at the level of the firm) – a pattern of innovative activities best labelled Schumpeter Mark I may be expected, that is, with a relatively large number of innovators and entrants. Also, the way tasks are divided among firms influence the scope for new entrants. One manner for a company to cope with the often increasing multi-technological need is to source knowledge from external organizations. This induces a shift from vertical integration to external sourcing of technology, often in networking forms, opening up opportunities for new or small firms to meet this demand.

The second implication is that differences in regimes influence the importance of a spatial dimension in the innovation process. Depending on tacitness, complexity and independence of the knowledge, the spatial dimension will be more or less pronounced. When the knowledge base is tacit, complex and has strong systemic features – and/or the relevant source of knowledge is highly concentrated (e.g. in a university) – strong spatial concentration of innovative activities may occur. In such a case, and where other features of the TR are favourable to new entrants (see above), geographical agglomerations of new firms are likely to come about. The nature of the knowledge is also expected to influence the choice of the means of knowledge transmission, as well as the ease and cost of knowledge transfer. We would anticipate that the same conditions (i.e. tacit, complex and systemic features of knowledge) would tend to lead firms to source a relatively large share of their technology from a geographically limited area. Thus, in some industries, the character of the TR is such that the innovation process may have important spatial features.

Accordingly, the nature of the TR would be expected to influence both the scope of new firms in relation to older and larger firms and the extent to which the geographical dimension affects the innovation process. In turn, these two features influence the 'natality' of new firms as well as the ways in which the young companies can draw on the support of other actors in a spatially limited area (see also Audretsch and Stephan 1996).

Turning now to the specific technological regime of 'bio-compatible materials and related products' (hereafter referred to as 'biomaterials'), which is the empirical case in this study, biomaterials are defined in this study as synthetic or natural materials to be used to treat, enhance or replace human functions, and are materials that are either compatible with human tissue or mimic biological phenomena (COSEPUP 1998).[4] In this study the following sub-technologies are included into the definition: (a) the technology of osseointegration using the material titanium; (b) silicon and other related materials to be used, for example, for orthopaedic, dental, hearing and cardiovascular implants; (c) biopolymers, both biodegradable and bioresorbable, to be used in implants and wound care; (d) materials to be used to coat surfaces and in tissue regeneration; and (e) the technologies of tissue engineering employing biologic materials to be used in tissue healing and regeneration.

Analysing this field in detail has revealed that the density of technological opportunities is high, and these opportunities are generated by a variety of sources: university researchers, customers (such as medical doctors), incumbent companies, suppliers that are integrating forward and new firms. Many biomaterials technologies are pervasive in character, meaning that they may fit in a range of applications. However, most are in early phases of development and there is still competition between technological solutions. Although a prevailing design has emerged for dental and orthopaedic implants, outcome is still uncertain for most bioartificial organs, for wound care and surgical sealants and for drug delivery. Patenting is common, and appropriability seems therefore to be relatively high. However, as this is a science-based field, scientific results are also communicated and diffused through such channels as journal publications and scientific conferences. Inventions may therefore not always be easy to protect. In fact, some companies have had problems with patentability due to early publications submitted by researchers. As a result, continuous innovation and strong networks to inventors are all the more important to firms. Even if firms are highly innovative, there is not a strong dependency on specific companies to keep on innovating. Instead, there are many firms that together maintain the continuous activity of innovation. Thus, cumulativeness is low at the level of the firm but high on the system level. The features of the TR described here – the high density and many sources of technological opportunities, their pervasiveness, moderate appropriability and low cumulativeness at the firm level – make for a system that favours extensive innovative activity and high entry of firms.

As we saw above, characteristics of the TR also influence the degree of geographical concentration of innovative activity. First, as innovative activity is rather dependent on scientific knowledge, and the producers of such knowledge (universities, incumbent companies as well as new entrants) are not evenly distributed geographically, we may suppose that innovative activity is not evenly dispersed over space, but rather that an accumulation takes place on the spatial level. With university and firm location comes a labour market, enhancing spatial knowledge accumulation. In addition, network externalities are likely to enforce geographical agglomeration of innovative activities. Second, the characteristics of the knowledge base may also influence the extent of spatial concentration, and we find that the knowledge base of biomaterials has codifiable and patentable features that are transferable in the form of licence agreements, as well as tacit features that are not easily transferred between actors. This may imply that the knowledge base has a spatial dimension. Furthermore, the degree of complexity is high, as the technologies are multi-disciplinary and several competences need to be integrated (e.g. bioengineering, biology and materials science), and as a variety of different types of competences are influential (e.g. R&D, process development, production). Moreover, the knowledge needed to innovate in biomaterials has systemic features.

In summary, the characteristics described above suggest that innovative activity within the field of biomaterials has a tendency to concentrate in certain locations in order for actors to be able to reap benefits from spillover. As the TR favours new entrants, there is a strong basis for the formation of groups of co-located new firms. (Note that to be co-located here may refer to the local, regional or even national level in the case of small countries.) This also rhymes well with the wider field of life sciences, where we can also expect spatial concentration (Steinle and Schiele 2002). Certainly, in the USA innovative activity has often centred in locations with strong research, especially where universities are situated, and new firms frequently locate close to a 'parent organization' (Enright 1994; Saxenian 1994; Zucker et al. 1995, 1998).

With the point of departure in the proposition that the specific technological regime assumes spatial importance in the process of firm formation, this chapter aims to analyse if and why the activities preceding firm formation are spatially confined. Three particular sub-questions are in focus:

1. Does the establishment of new ventures in biomaterials depend on access to previous technological development and are such pre-firm activities spatially confined or globally distributed?
2. Does the profile of scientific work within a specific location influence the technological profile of new science-based firms?
3. Are there mechanisms that make for a spatially confined firm formation process?

3. THE RESEARCH DESIGN

This section discusses the choice of methodology, indicators and empirical field. The object of study in an analysis of firm formation is the young firm,

and in this study the focus is on particular science-based firms, that is firms
having a strong scientific and technological focus. In this definition, all new
entities are included independent of origin (spin-off, spin-out, independent
start-up). With the aim of comparing the firm creation process between
firms, a case study approach was chosen. While a substantial number of
cases was deemed necessary for the comparison, at the same time the data
had to be descriptive, explanatory and longitudinal in character. Given
these objectives with the study, the aim was to collect extensive data on the
firm formation process of approximately 50 firms.

Another important feature of the research design in this chapter is that
not only are comparisons made between firms, but are also made between
different locations. It is important to note that terms such as 'spatial' or
'localized' are used rather than local/regional/ national to denote that we
are not interested in the specific local/regional level but in the phenomenon
of spatial limits of a process. Such boundaries may coincide with local,
regional or national ones depending on size of region/nation, but this
varies.[5] Based on the discussion of the European – and Swedish – challenges of firm formation, the aim is to compare the Swedish situation to
that of the USA. In doing so, two regions of the USA were selected: one
which in general is perceived as high performing in terms of firm formation
and growth, namely Massachusetts; and one which is similar to Sweden in
terms of size, industry structure and renewal problems, namely Ohio.[6] It is
important to note that this study is not only concerned with the 'rare and
exclusive' type of clusters such as those in Silicon Valley. Instead it deals
with the more 'ordinary and common' phenomenon of spatial importance
in innovation processes – here particularly in firm formation processes –
that may or may not lead to clustering effects (Rosenfeld 1997).[7] As such,
it links to the current discourse on when and why spatial proximity is of
importance, and how clusters emerge and function (Porter 1990; Jaffe et al.
1993; Saxenian 1994; Audretsch and Feldman 1996; Swan et al. 1998;
Zucker et al. 1998a; Feldman 1999; Harhoff 1999; Niosi and Bas (2001);
Orsenigo 2001; Zeller 2001).

All three locations have strong scientific activities and output in the field
of biocompatible materials, thereby having one crucial 'prerequisite' for the
possibility of firm formation in the specific, highly science-based, field (see
Rickne 2000 for further discussion).[8] However, as is evident from Table
11.1, the scientific strength is considerably higher in Massachusetts than in
the other two locations, irrespective of whether we measure in terms of
number of universities awarding science and engineering doctorates,
number of doctorates in science and engineering or number of publications
in biomaterials. Sweden and Ohio display a similar position, even though
Sweden is stronger in terms of publications per capita. In general terms,

Table 11.1 Comparison of number of universities, doctorates and
publications in the three regions

	Number			Number/million capita		
	Sweden	Ohio	Mass.	Sweden	Ohio	Mass.
Number of universities awarding science and engineering doctorates	11	16	19	1.3	1.4	3.2
Total number of doctorates awarded in science and engineering, 1996	743	1094	1575	85	98	262
Total number of publications in biomaterials issued to research organizations, companies or individuals	12182	8595	22328	1400	796	3721

this comparison between Sweden and the two rather different regions within the high-performing USA renders the possibility of understanding the varieties of the firm formation process.

With the aim of capturing a set of firms working under the same technological regime, a specific field within life science was chosen. All firms included have a common knowledge base within biocompatible materials as related to medical applications (as defined above).[9] The area of biocompatible materials relate to three sets of general competence areas: materials science, biomedical engineering and biotechnology, and this means that the actors involved in technology and product development – firms as well as research groups – need to master a multitude of competence sets, with innovative activities often being performed at the crossroads between these fields. It is a science-based field still in an emerging phase, involving extensive experimenting, implying scientific, technological, regulatory and market-related uncertainties, and often leading to long-term and costly product development and market introduction. Despite having been active for several years, not all firms analysed yet have products on the market (compare Cortright and Mayer 2002).

Biomaterials-related products can be categorized in any of three sets of industries: biologics, pharmaceuticals versus medical devices and implants.[10] Biologics mainly involves tissue engineering products (design, growth and manipulation of tissue) and bioartificial organs (e.g. pancreas, liver, kidney, artificial skin and synthetic blood). Tissue engineering is emerging as an interdisciplinary technology that requires competences in materials technology, molecular biology, chemistry (growth factors, perfusion) and medicine.

Within this field the biomaterial itself may be produced and sold to other companies for use in their products, or to research organization for development work or clinical trials. In addition, several companies work on complex product development, addressing end-user needs and some firms attend to the cell culturing process itself. A new approach to materials science is providing customers with biomaterials through the development of what is called 'biomimetic' materials, which imitate the microstructure and mechanical features of living cells and tissues. Applying such structures to biopolymers, bone, collagen, synthetic polymers, ceramics, metals and so on, biocompatible and bioresorbable scaffolds for orthopaedic applications, cartilage repair or synthetic ligaments can be developed. While there is a range of implants, artificial 'human spare parts' and tissue-based products in development laboratories, there is still not much on the market. With rather few industrial players so far, tissue engineering firms suffer from an extremely uncertain innovation process, with long time periods in the laboratory, a regulatory vacuum, a hesitant market and the need for durable financing.

Biomaterials are also important in the pharmaceutical industry, for products such as drug delivery or surgical procedures for wound care, adhesion and surgical sealants. Biopolymers, for instance, can be used in surgical sealing to prevent adhesion, the scar-formation process that takes place after open surgery, where a barrier of degradable gel can be placed between the tissue ends to give the tissue time to heal. Another example are proteins, which can be used for human tissue regeneration and repair. For example, the particular protein OP-1 is produced in the kidney and in the brain, and functions as a trigger to differentiate which organ the body forms (bone, cartilage etc.). It can be used in a matrix of collagen for orthopaedic applications, to stabilize the kidney function, and for neurological disorders (such as Parkinson's disease, stroke), since it helps to grow dendrites that handle the connections in the brain.

The third group of industries where biomaterials play a pertinent role is medical implants, which have a vast range of uses, for example orthopaedic (knee repair, artificial joints, hip replacements etc.), dental, or cardiovascular (vascular grafts, artificial hearts etc.). In these cases, biomaterials come into fairly mature industries, possibly renewing product development and adding to product performance. For example, the industry for dental implants curing dental decay has gone through a number of product failures, but has been renewed in the last 20 years by new technologies such as osseointegration and biodegradable dental barriers. Currently, while most biomaterials are well established within medical implants, new development is also under way.

Given a case study approach, the comparison between three different locations and the choice to focus on the field of biocompatible materials, the entire population of biomaterials-related young firms, established between 1975 and 1998, in the specific locations were traced. Certainly, to capture issues of firm establishment, it was important to cover a fairly long time span. Therefore, in addition to a core competence base within biocompatible materials, an additional criterion for inclusion in the population was establishment within the period 1975–98, amounting to a total of 73 firms.[11] Full data were obtained for 51 of these firms, and the analysis below builds mainly on these 51 cases. The firm cases include data on pre-firm activities, firm establishment, product development, market activities, critical incidents, partnerships,[12,13] resource flows as well as innovative and economic performance.[14] Several different sources for information were used including extensive written company information, external company and industry analysis, a directory of formal alliances, and, most important, interviews with the new firms, interviews with most inventors of the firms' initial technology and interviews with some of the firms' partners. In total 129 actors have been interviewed, and the data was carefully codified to be able to perform both quantitative and qualitative analyses. While the limitations of the research design include its retrospective character, extensive data collection and analysis underlies the results.

4. THE INFLUENCE OF THE REGIONAL SCIENCE BASE

In general, science-based firms are, particularly at the onset, highly connected to a scientific community be it within universities, research institutes, clinical settings or firms (Acs and Preston 1997). Our first research issue asked if the establishment of new ventures in biomaterials depends on access to previous technological development, and if such pre-firm activities are regionally confined or globally distributed. The empirical analysis shows that the biomaterials firms' initial science and technology platform – that is the set of competences and artefacts the firm possesses at firm formation that are necessary for developing products to a market[15] – had most often been developed externally prior to firm establishment. The external S&T provided an opportunity for commercialisation, and was brought into the company through the founder, a licence agreement, recruiting or acquisition. Thus, most companies did not start with a blank slate with regard to technology, but were significantly influenced by S&T developed externally.

Public as well as private S&T activities can shape virtuous circles, spurring knowledge formation and creating spillover (Winter 1984; Jaffe et al. 1993). More specifically, several studies point to the relation between regional investments in S&T and innovative activities (see e.g. Feldman

1994). Given the scientific connectivity of young firms and the external pre-firm development of S&T competences, the formation of new firms may thereby be particularly dependent on regional S&T activities (Libeskind et al. 1996), and our first research question asked if this is the case in the field of biomaterials. Indeed, scrutinizing the data, the origin of the external pre-firm S&T development in fact proves to be fairly spatially confined. The vast majority (88 per cent) of the initial S&T opportunities for this population of new firms came from a source within the same geographical area.

Thus, in light of the importance of regional inventions and pre-firm development, we can anticipate the degree of experimentation and invention in a specific location, as well as the profile of scientific work, to influence the formation of science-based firms. More specifically, the remainder of this section aims to discuss whether or not the profile of scientific work within a specific location (in this case within Sweden, Ohio and Massachusetts, respectively) influences the technological profile of new science-based firms.

To consider this issue, we first calculate the scientific specialization of the regions, that is the revealed scientific advantage.[16] This is done by comparing all publications in biomaterials-related knowledge areas issued to research organizations, companies[17] and individuals located in each of the three locations. It is clear from the left-hand column in Table 11.2 that Sweden has a comparatively high level of publications in scientific fields related to bone-anchored implants (osseointegration), which is where Ohio and Massachusetts have their greatest weakness. Sweden is also strong with

Table 11.2   Revealed scientific advantage and revealed comparative advantage of biomaterials firms in the three regions

	Revealed scientific advantage in biomaterials			Revealed comparative advantage of firms*		
	Sweden	Ohio	Mass.	Sweden	Ohio	Mass.
Tissue engineering	0.960	0.888	1.065	0.487	0.507	1.724
Drug delivery	0.843	0.858	1.140	0.834	1.159	1.042
Bioartificial organs	0.218	2.250	0.947	0.000	1.738	1.390
Biopolymers	0.983	1.179	0.941	1.168	0.649	1.071
Cardiac assist devices	1.194	1.000	0.894	1.095	1.014	0.912
Bone-anchored implants	2.671	0.439	0.306	1.460	1.622	0.243
Biocompatible surfaces	1.408	1.245	0.684	1.947	1.352	0.000

Notes:
$N_{Total} = 73$, $N_{Sweden} = 25$, $N_{Ohio} = 18$, $N_{Massachusetts} = 30$
* Each firm can be active in more than one field.

regard to biocompatible surfaces and cardiac assist devices. The scientific field where Sweden lags is bioartificial organs, the very area where Ohio has its strength. Ohio is also fairly strong in biocompatible surfaces, and in bio-polymers. Massachusetts displays a somewhat different profile, with a higher revealed scientific advantage in tissue engineering and drug delivery, and a low profile for biocompatible surfaces.

Second, to display the specialization of new firms, the revealed comparative advantage (RCA) of biomaterials companies in the three regions is calculated.[18] The right-hand column of Table 11.2 clearly exhibits that the RCA differs between the groups of young firms. The Swedish profile of bio-materials companies is concentrated in four areas of application: biocompatible surfaces, implants based on either osseointegration (mainly dental and orthopaedic) or biopolymers, and cardiac assist devices. The Ohio-based companies have a profile skewed towards bioartificial organs, bone-anchored implants, biocompatible surfaces, drug delivery and cardiac assist devices. Thus, Sweden and Ohio have their vigour in two of the same areas. Both are comparatively weak in the area of tissue engineering, but differ with regard to biopolymers (a strength for Sweden) as well as drug delivery and artificial organs (strengths for Ohio). Massachusetts-based firms have strengths in tissue engineering and bioartificial organs (such as skin, pancreas, liver), as well as in biopolymers and drug delivery, whereas the weaknesses of Massachusetts are within bone-anchored implants and biocompatible surfaces.

The task was to scrutinize if there is a relation between the profile of the science base in each of the regions, and the profile of new firms formed. Ocular inspection of the table suggests that there may be a relation between the two. Definitely, it is within Sweden's fields of scientific strength – bone-anchored implants, biocompatible surfaces and cardiac assist devices – that most Swedish companies are active. Likewise, the young firms in Massachusetts focus on application areas where that region has its strength: tissue engineering, drug delivery, bioartificial organs and biopolymers. However, in Ohio, the pattern is ambiguous. While several new firms are active within the fields of scientific strength (bioartificial organs and biocompatible surfaces), others develop products where the region does not have a strong scientific position (bone-anchored implants and drug delivery).

A statistical analysis, however, gives a strong relation between regional publications and firm formation in the Swedish case only. The Spearman correlation between RSA and RCA for each specific location was tested, and of the three tested regions, only Sweden showed some significant correlation between the RSA and RCA. The value of the correlation was 0.89, with a p-value of 0.02, which allows us to conclude that there is some association between the RSA and RCA in that specific location. For Massachusetts the correlation was 0.75 (*p*-value = 0.07), and the association is thus weaker than in Sweden. There was no evidence of an association for Ohio, where the correlation was 0.21 (*p*-value = 0.66).

In conclusion, there is some support for the idea that the profile of scientific work within the specific location influences the technological profile of new science-based firms (see also Harhoff 1999). Even though additional analysis of this type of issue on complementary data sets is necessary, these findings lead to two propositions. First, it seems as if firms to a high degree are built on regional activities, and a fairly large percentage of new firms focus on areas where the regional science base is strong. Therefore, to ensure sufficient firm formation within a geographical location, it may be of importance to have a strong regional science base within such scientific areas that holds promise for future economic growth. Second, given that the correlation is significant in Sweden and discernible in Massachusetts, but not in Ohio, we may speculate that the Ohio firms will be less attached to regional actors for their further innovation process. Indeed, the study proposes that when the degree of experimentation and invention, and the profile of scientific work in a specific location, is allowed to influence the formation of science-based firms, this is one basis for the formation of a strong cluster.

5.   PRE-FIRM ACTIVITIES

In order to capture the very process of firm formation, this section discusses pre-firm activities – that is the activities that underlie firm creation – and focuses principally on the development of the S&T platform and the compilation of the initial capital. In particular, we are interested to analyze if these are mechanisms that enhance the *spatial* character of the firm formation process.

5.1   Actors Developing the Science and Technology Platform

The issue in focus for this section is what actors are engaged in developing the science and technology platform discussed above. The empirical data shows that both academic researchers, technology transfer units, users, firms and venture capitalists were involved in forming the initial science and technology opportunity leading to firm formation. Without doubt, a wide variety of actors engaging in innovative activities seem to be a prerequisite for firm establishment. Many of these are located within the same geographical area, even though international connections naturally are indispensable as will be noted below. Importantly, it is clear that it is not, as such, the spatial proximity to sources of S&T that matters, but the very knowledge and resources flows that are entertained through these connections (Cockburn and Henderson 1998).

There is, thus, a multitude of actors engaged in the process. Due to the science-based character of biomaterials, universities, medical schools and hospitals – regional, national as well as foreign ones – have played a crucial role in the formation of the biomaterials field. Lending the terminology of Cooke (2002), these actors are truly involved in exploration, examination as well as exploitation of knowledge. In fact, for 71 per cent of the firms the initial S&T emanates from research teams from any, or a mix, of these three types of organizations (see Table 11.3). Importantly, academic research of substance to this field is performed at universities of natural sciences or technology (e.g. Massachusetts Institute of Technology or Gothenburg University), at medical schools (e.g. Harvard Medical School) as well as in more clinically oriented organizations such as hospitals (e.g. The Cleveland Clinic or Uppsala University Hospital). There are several forms of cooperation between these types of organizations, and frequently a diverse set of actors are involved in joint projects. In addition, there is much cooperation on the individual level, and in any specific research team there are often researchers from various types of organizations engaged. As a consequence, it is not always obvious to distinguish what organization did what part of the contribution,[19] but the combination of the three types of actors seems important for innovation and firm formation.

Table 11.3   *The source of the initial technological opportunity (percentage of firms)*

	Sweden %	Ohio %	Massachusetts %	Total %
University, medical school or hospital	73	83	63	71
Company	40	8	42	33
Venture capital company/business angel		8	17	10
Founder		8	8	6
Military		8		2

Notes:
The sum of each column may exceed 100 per cent as the opportunity may have more than one origin.
$N_{TOT} = 51$, $N_S = 15$, $N_O = 12$, $N_M = 24$.
Some of the percentages are based on small numbers of observations, where for example 8 per cent in the Ohio case denotes one firm only.

Without doubt the most frequent source of initial inventions, the importance of universities, medical schools and hospitals nevertheless varies
between the locations, with Ohio especially, but also Sweden having a larger
share of their new firms based on university research (see Table 11.3).

In one-third of the firms the initial S&T opportunity originates from
another company, making the new firms in some sense corporate spin-offs.
For firms located in Sweden and Massachusetts this is an important source
of initial technology, while it is of minor relevance in Ohio (see Table 11.3).
There are two types of 'parent companies' involved. First, several new companies (14 per cent) use S&T developed in an incumbent company. Often
the pre-firm development has taken place within a relatively large life
science company, but it may also be a firm in a completely different sector.
There is a considerable competence spillover between various industry
sectors, and regional diversity of industrial specialization seems to be beneficial to create technological opportunities. This type of new firm formation may come about as a strategic managerial decision to diversify into a
new application through the formation of a new firm, or a result of former
employees basing a new firm on S&T not exploited within the incumbent.
Second, a number of young firms are spin-outs from the relatively young
and small biomaterials firms established earlier in the period 1975–98
within these geographical areas, that is these new firms could be called
'second-generation' biomaterials companies. In this respect the regions
differ. While S&T diffusion from one biomaterials firm to another is
common and important in Sweden and Massachusetts – 27 per cent of the
Swedish biomaterials firms are second-generation spin-outs, and 25 per
cent of the Massachusetts firms – this is not the case in Ohio where there
are no such second-generation firms. We will return to a possible explanation of this finding later in the chapter.

Interviews show that there are two mechanisms by which biomaterials
firms diffuse their technological knowledge to a new firm. The most
common is that the founder (often a university researcher) of the 'parent'
company sell or leave the earlier firm and start a new company, bringing
technological assets along. Thus, biomaterials-related companies have split
into two companies at an early stage focusing on different applications or
becoming competitors. Another mechanism for this technological diffusion
is that the existing company sells or by cooperative agreements conveys
technological assets to the newly established firm, for whom these assets are
seen as necessary to be able to pursue the application envisioned. For
example, a significant part of the initial technology platform of Alkermes
Inc. in Massachusetts was acquired from another young regionally located
biomaterials firm, Enzytech Inc. For any given location, this technological
diffusion – and a certain 'multiplying process' of new firms – seems to be a

feature of strength, given that one views experimentation and increased diversity as advantageous for regional and national development.

For a small part of the US ventures (10 per cent), the initial technology on which they based their business idea had been developed by the individuals in the venture capital company that financed the start-up. Interestingly, the venture capital firm actually held competence enough to be active in technological development, presumably due to long experience from similar start-ups, and the technological learning that resulted. Finally, a very small number of firms were based on the founder's inventions or on technology developed within the US military.

In conclusion, it is clear that the establishment of new ventures in biomaterials depends on access to previous technological development. Certainly, cutting-edge activities in universities and other research institutes, but also in companies, are important to firm formation. The study proposes that, for firm establishment to come about, it is crucial to have a wide variety of actors who engage in innovative activities. Moreover, as many of these actors are located in geographical proximity to the new firm, the set up of the regional/national industry structure is of significance.

5.2 Actors Providing Financial Means

A number of studies have pointed out the problems experienced by young or small firms in raising money for start-up or development, taking this fact as a partial explanation of lack of firm formation or inhibited growth (Landström and Winborg 1995; Binks and Ennew 1996). Our next question concerns what actors are engaged in providing the initial capital market.

In a definition of a capital market all the actors that may provide the young firm with financial assets for its establishment and development must be included, as well as the structure of that market and the institutions guiding it. In this broad sense, it is not only the traditional actors of capital markets, such as banks, managers of investment funds and venture capital companies, that may be of importance in providing capital, but also, research foundations, founders, other organizations (such as technology-transfer units, bridging organizations, consulting firms, military units, R&D organizations), life science companies, parent firms or new owners, the public market and customers.

Our study showed that there were several, equally important, actors providing financial means for the new firms (see Table 11.4). While venture capital companies (VCC) provided one-third of the new firms with monetary resources, the importance of research foundations (29 per cent), founders' personal sources (27 per cent), other organizations (24 per cent), life science companies (22 per cent), and parent firms or new owners (22 per

Table 11.4   Financial sources for the start of the new venture (percentage of firms receiving financing from each type of source) [abc]

	Firms in Sweden %	Firms in Ohio %	Firms in Massachusetts %	All firms[d] %
Research foundation	63	33	4	29
Founder	31	75	4	27
Other organization[e]	25	58	4	24
Life science company[f]	25	17	22	22
Parent firm or new owner	38	8[g]	17	22
Public market	0	8	4	4
Customer	6	0	4	4

Notes:
[a] $N_{Total} = 51$, $N_{Sweden} = 15$, $N_{Ohio} = 12$, $N_{Massachusetts} = 24$.
[b] Traditional financial actors as banks are not included.
[c] This table gives the frequency of different types of financiers, not their relative importance.
[d] The sum of each column exceeds 100 per cent as each firm connects to several sources.
[e] Other organizations include technology-transfer units, bridging organizations, consulting firms, military units, R&D organizations.
[f] Life science company other than parent firm or owner.
[g] Some of the percentages are based on small numbers of observations, where for example 8 per cent in the Ohio case denotes one firm only.

cent) is also considerable. In initial phases neither public offering nor development with customers are common routes of financing the analyzed start-ups.

There are, however, substantial differences regarding the importance of different actors depending on where the firms are located. In Sweden, a majority of firms have received money from research foundations (63 per cent), and financing from parent firms or new owners (38 per cent), and other life science firms (25 per cent) are also important. On the other hand, VCC financing is of less importance (13 per cent). As a contrast, over 40 per cent of the US firms report initial VCC financing. In Ohio, confirming previous studies on the importance of private investments by the entrepreneurs (see e.g. Berger and Udell 1998) the founder is a major contributor of initial financial assets (75 per cent), and the military and bridging organizations (e.g. Edison Biotechnology Center) are also relatively influential. Interestingly, initial corporate financing from parent firms or other life science companies is of less importance in Ohio, possibly reflecting the less frequent connections to S&T in other companies and the fewer second-generation firms (as noted above). In Massachusetts, the dominating financial actors in these early phases are venture capital companies. Even though the percentage of VCC financing is equally large in Ohio and Massachusetts interviews show that to the Massachusetts-based firms this VCC financing is more sufficient. This is also mirrored in the relative minor importance of other financial means in Massachusetts, where, for example, research foundations bears no relevance.

In conclusion, regarding the structure of the initial capital market for young biomaterials companies, it is clear that many different types of actors are at work, and that the type of sources differs between the regions. This may be a matter of industrial structure as well as of networks. In this study it is proposed that there is some degree of substitutability between initial financial actors, but that the firm needs to be acutely aware of potential additional resources offered by each type of financier.

5.3   Mechanisms at Work

We have seen that there is a multitude of different actors involved in the formation process – both in S&T development and diffusion, and in financing – and that these are to a high degree in a geographical proximity to the new firm. The puzzle we are faced with is why this is the case. Therefore, our final issue deals with what the underlying mechanisms are that make the firm formation process spatially confined. Three different mechanisms surface in the analysis of the biomaterials firms.

First, the activities involved in establishing new firms – the pre-firm activities – seem to be organizationally concentrated in Sweden and Massachusetts. This means that for the establishment of the entire set of biomaterials firms in these two locations, the pre-firm activities are not distributed across many actors but instead concentrated in a few. These actors from which the firms spring – these 'centres of origin'[20] – are, as was seen above (see Table 11.3), mainly corporate- and/or university-based research groups. Obviously, some of these 'centres of origin' have contributed to several firms being formed. Table 11.5 displays the organizational

Table 11.5   *The organizational concentration of pre-firm activities measured by number of 'centres of origin' for the set of firms in each location*

	Sweden	Ohio	Massachusetts	Total
Number of firms in analysis	16	14	27	57
Number of 'centres of origin'	8	13	15	36

concentration of pre-firm activities measured by the number of 'centres of origin' for the set of firms in each location.

In the Swedish case, there is in this study access to detailed empirical data on the origin of 16 firms, and Table 11.5 shows that these new firms originated from only eight 'centres of origin'. Thus, the research activities in some of these centres resulted in more than one company. For the 27 firms analysed in Massachusetts there are only 15 different 'centres of origin'. It is clear that in Sweden and Massachusetts some firm formations are highly related to one another. Figure 11.1, showing the organizational origin of a part of the biomaterials firms in Massachusetts, further illustrates this. (The different symbols in the figure denote, on the one hand, biomaterials firms, and, on the other hand, the different organizational origin of these firms (numbered). There is also a time dimension in the figure, in that the stretch of the NTBF symbol illustrates the life span of the firm.[21]) Clearly, there are several examples of university research contributing to a series of related firms. For example, scientific work performed by groups at the MIT, Harvard University and Harvard Children's Hospital including researchers such as Professors Langer and Vacanti (origin number 11 in the figure) has been influential in Massachusetts in the formation of the biomaterials firms Alkermes, Enzytech, Neomorphics, Reprogenesis, Focal, Optafood and Advanced Tissue Sciences.[22] These firms are indirectly connected to one another through their 'parent organization', and by having a highly related technology platform, and some of them also have direct connections. Even though the phenomenon of the same organizational origin is most common for firms with a university background, there are examples of corporate 'centres of origin' spinning out a number of related firms, where, for example, university research in Massachusetts coupled with work at the company W.R. Grace contributed to the formation of Biohybrid Technologies and Circe Biomedical. These examples were from the Massachusetts case, but the situation is similar in Sweden.

Interestingly, such agglomeration of research activities is not prominent in Ohio, but most new firms have their own 'centre of origin'. This means that the new firms in Ohio have little relation to each other at the time of their foundation, and, in addition, interviews show that links between these companies are seldom formed in later phases. The conclusion to be drawn is that concentrated organizational origin is one important mechanism leading to spatially confined firm formation.

Interviews suggest that one of the reasons for the establishment activity to be organizationally concentrated is the existence of serial entrepreneurs, either coming from a university, a clinical, a corporate or a financial setting. The empirical data demonstrates that repeated firm formation from an academic centre of origin seems to involve a long-term research endeavour

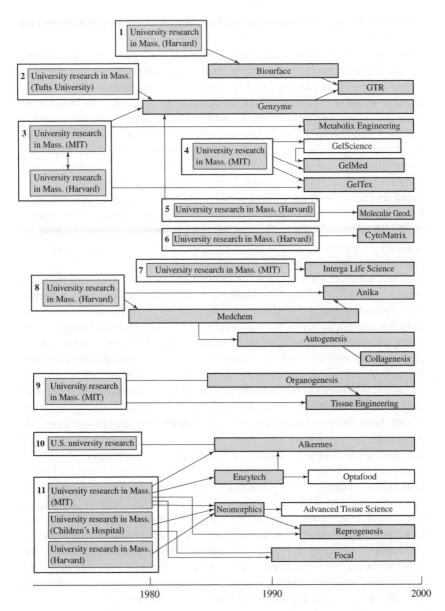

Figure 11.1   The organizational origin of a sub-set of the new biomaterials firms in Massachusetts

where the knowledge is generic in character, opening up possibilities for several applications. It is thereby important that intellectual property is wisely split into several indications to enable different firms to focus on different applications, instead of a single firm holding exclusive rights to valuable knowledge assets, blocking further firm formation. Moreover, in the case of biomaterials, serial entrepreneurship is enhanced by high labour mobility, where people move back and forth between the academic setting, industry and finance. This behaviour may lead to a diffusion of ideas and understanding of the use of technology in different industrial situations. Connections are crucial, that is strong connections between research organizations and industry as well as informal linkages between potential entrepreneurs and venture capitalists. It has also been important with incentives for serial entrepreneurship (or intrapreneurship), in terms of a committed university or corporate policy, to encourage entrepreneurial efforts, for example in providing flexible use of equipment and facilities, creating alternative career paths or making role models visible.

A second mechanism is that the interrelated new firms tend to focus on specific technological profiles or areas of application, and this enhances knowledge flows and the 'clustering effect'. This attention to certain technologies and applications has several reasons. When the educational system assures continuous development of human capital in a particular field, it may lead to a critical mass of people, and firm formation becomes technologically interrelated. For example, in Sweden, research as well as firm formation within the field of osseointegration have been built on the successful diffusion of competence through the education of PhD students, several of whom have explored different but interconnected strands of research. Moreover, mobility back and forth between the academic and industrial sector leads to a crucial knowledge spillover. Also, early division of the technological platform into several indications – each one pursued by one firm only – makes for technological interrelatedness between these firms. In this, financial providers, with know-how in the specific field, have been useful to decide on the appropriate scope of the commercial opportunity to be pursued by each firm. Finally, the focus on specific technologies or applications is partly due to the existence of tight regional networks, for example between researchers in universities and companies that allow for competence spillover. Such informal networks and information pathways are influential in creating clusters of inventions and of firms.

6. SUMMARY AND IMPLICATIONS

In order to both firmly establish and augment a firm creation processes, an understanding of the mechanisms behind such activities is necessary. This chapter has examined the spatial dimension of such processes. The chapter took as its basic assumption that the technological regime of biomaterials would influence the propensity to form new organizational entities and spur co-location of innovative activities and new firms. It is clear that if the TR is one determinant of how commercialization of S&T is exploited – in new firms or within incumbents – this might have implications for policy. In agreement with Shane (2001a), to encourage new firm formation may require a close look at the specificity of the underlying set of technologies to be able to design accurate, technology- and industry-specific, policy measures. In particular, the spatial importance may be differently pronounced in different technological regimes.

The study confirms that the early phase of firm formation truly is a geographically delimited activity. Even though international connections naturally may be – and are – important, connections and knowledge flows in geographical proximity dominate for the entire sample investigated. The analysis of biomaterials companies gave initial support for the hypothesis that the profile of scientific work within a specific location may influence the technological profile of new science-based firms. This issue needs further scrutiny, especially concerning the situations in which we may expect this type of relationship, given that an association was indicated for two locations but not for the third. Even so, from this preliminary finding it follows that the direction and volume of the science base may have a large impact, and this stresses the need for regional/national actors to focus on scientific areas that holds promise for future economic growth. Further, when the degree of experimentation and invention, and the profile of scientific work in a specific location, is allowed to influence the formation of science-based firms, this is one basis for the formation of a strong cluster of new firms. Indeed, this has been the case in Massachusetts and in Sweden. In Ohio, however, the firms' profiles do not reflect the regional science base, and the new firms are not as connected to the regional network of scientific and commercial actors – neither initially nor over time.

The study also revealed that many of the important actors – involved in S&T creation and financing – are located in geographical proximity of the new firm. One of the mechanisms leading to such spatially confined pre-firm activities is organizational concentration; the activities are not distributed across many actors but instead are concentrated in a few, where serial entrepreneurship is important. This, in turn, enhances connectedness between the new firms as well as between application areas, and paves the way for long-lasting relations between the inventor group and the new venture.

The findings lead to implications for structure as well as mechanisms on the regional or national level. As new units (or individuals within them) seem – more often than might be expected in a global technological and industrial field – to search in a spatially confined area for initial technological opportunities and financing, and as the profile of the regional science base may influence the profile of new firms, structural characteristics in the geographical setting are important.

First, a wide diversity of regional scientific activity in firms and research organizations, with much experimentation and learning between fields and actors works positively on creating many opportunities for start-up. Indeed, diversity and experimentation is key.[23] Second, a regional (or national in case of small countries) industry structure with research organizations (universities, medical schools, hospitals) is important for firm formation in biomaterials – and similar fields – to emerge. Such organizations are increasingly active, not only in creating scientific results but also in diffusing knowledge throughout society, in the early stages by linking scientific development to user needs and in commercialization processes. In this, technology transfer units are of importance, to build competences on legal arrangements for intellectual property and to create incentives for researchers to engage in industrial activities (see e.g. Shane and Cable 1998). Also, as the choice of ways to commercialize new technological opportunities is crucial, often setting the path for which applications will be focused, a competent technology licensing office can identify opportunities, split technological opportunities into several indications, make due diligence on partners/licencers, write agreements, renegotiate unfulfilled deals and so on. Third, regional companies, and their magnitude and profile of development activities, are influential for firm formation. Even though all industrial experimenting cannot lead to positive returns for the specific incumbent company, spillover effects often render regional benefits in terms of new firms. Fourth, while lack of VC access does not necessarily restrain start-up activity, as is shown in the Swedish case, VCC can, and often do, play an important role in opportunity recognition and for initial financing.

It is clear that 'co-location' of innovative activities – that is, the simultaneous location of firms and other actors – does not automatically lead to 'clustering' – that is, co-located actors that do produce synergy because of their geographical proximity and interdependence, give spillover, mutual benefits and spur each other's innovation and growth processes.[24] This is obvious by the difference between the three regions in this respect. In Sweden and Massachusetts, the regional science base influences firm formation and leads to related firms being established. There is much experimentation and strong centres of origin that leads to serial entrepreneurship. Spillover effects are reaped. Also, the spatially confined networks formed early retain their importance over time, partly due to local search processes.

As a contrast, in Ohio the new firms are not as rooted in the regional science base, and related firms are not formed as often within the field of biomaterials. The 'centres of origin' can be described as rather weak in the sense that serial entrepreneurship is uncommon, and spillover could be more pronounced. There is less scientific relation between regional actors and new firms in Ohio as compared to the other locations. The firms form no strong regional networks in pre-firm or early firm phases, but rather the young firms behave more like 'isolated islands' in relation to one another. Partly as a result the regional actors do not become crucial for further firm development, and there are no second-generation firms in Ohio. Thus, in summary, there is little clustering of innovative activities or new firms in Ohio, the new ventures are not well connected to other companies, initial corporate financing from parent firms or other biomedical companies is of less importance in the region and no second generation spin-offs are created. Even though a number of biomaterials firms are co-located in Ohio, the mechanisms important to create a 'cluster' are not present.

Thus, clusters need to be firmly rooted in the regional structure with regard to actors, profile and networks. To policy actors as well as to other organizations (firms, universities etc.) this means that, while the biomaterials activities and actors need to be highly connected to the global life science sector (Cooke 2002), regional/national experimentation should be supported and encouraged both with regard to long-term, undirected research, commercialization efforts by incumbents, entrepreneurship activities and also experimenting by financial actors.[25] To encourage serial entrepreneurship and organizational concentration of pre-firm activities within universities, where several commercial opportunities are born and spun out, some specific aspects should be considered: a continuous development of human capital in focused fields; division of technological opportunities by application; generous university policy when it comes to career paths; incentives for entrepreneurship; strong connections to industry and financiers; diffusion of best practice through role models and by mobility of people. Moreover, it is imperative that researchers/ inventors have the possibility to keep long-term relations with the firm commercializing the technological opportunity, in order to keep the vision of development and customer needs alive. This is especially true within life science, where researchers often have clinical experience and thereby understand market needs. Serial entrepreneurship and organizational concentration of pre-firm activities are also highly dependent on the behaviour of firms, where it is important with tight regional networks that allow for competence spillover, incentives for experimentation and intrapreneurship and mobility of people.

In conclusion, an important contribution of this study was to shed light on whether or not pre-firm activities are spatially confined, and which mechanisms are involved in such firm formation processes. The findings were that even though international components are important the process is indeed spatially skewed. Our discussion has centred on how to spur 'clustering' of firm formation as opposed to mere 'co-location' of new units, where spurring mechanisms include organizational concentration, application interrelatedness and a spatially confined search process.

ACKNOWLEDGEMENTS

The author would like to thank professors Staffan Jacobsson, James M. Utterback, Richard Nelson and Bo Carlsson for useful comments on earlier presentations of this study. Appreciation also goes to professors Maureen McKelvey and Jens Laage-Hellman, as well as to three anonymous reviewers. Financial support from NUTEK and Kristina Stenborg's Research Foundation is gratefully acknowledged.

NOTES

1. The indicators and measures underlying this apparent paradox has been questioned by Jacobsson (2002), and the issue is further analysed by Jacobsson and Rickne (2003).
2. In addition to knowledge spillovers there exist several other explanations as to sources for agglomeration economics; a pooled labour market, proximity to customers and suppliers, economics of scale and specialization, economies of scope and diversity (Marshall 1949; Glaeser et al. 1992).
3. This section draws on Jacobsson and Rickne (1996) and Rickne (2000).
4. Biocompatible materials are composed of several sub-technologies, forming what Lundgren (1994) calls a 'technical system', that is, technologies that are linked by being built on closely related competences and being interconnected in products.
5. Indeed, in this study, the spatial limits of the firm formation process coincide with the regional borders in Ohio and Massachusetts, but with national borders in Sweden (as this is a smaller country than the USA, and is comparable in size to regions/states in the USA).
6. Firms in the entire geographical area of Sweden, Ohio and Massachusetts, respectively, are included. In practice, the firms are not evenly distributed within the locations. Instead, in Sweden firms are mainly located by Stockholm/Uppsala, Gothenburg and Lund; in Ohio around Cleveland and Cincinnati; in Massachusetts in the Boston area.
7. The term 'cluster' here refers to concentrations of firms and other actors that are able to produce synergy because of their geographical proximity and interdependence (elaboration on Rosenfeld 1997).
8. While both Sweden and the Boston region have been identified as biotech 'hotspots', Ohio has not (see e.g. Cortright and Mayer 2002).
9. In choosing to focus on a technological field (instead of an industry), we can capture processes where technological development and strategic actions transform various industries.

10. The relation between the technological competence areas and the industries is not one to one; a mix of competences is often needed for each specific product, and there are no clear borders between the sets of industries but they overlap, and classifying a specific product is not straightforward.

11. To identify the entire population of biomaterials-related young firms in the specific regions, three complementary methods were used: a) identification of existing or potential products and of firms developing such products through corporate and product directories, b) verification of the S & T content of the firms by analyzing the firms' patent profiles, and c) verification and expansion of the population by interviews with researchers, firms and experts.

12. Our focus here is on innovation processes in the life science sector and the actors range from policy units, universities, individual researchers (in universities, institutes or companies), hospitals and clinicians, users, to life science firms of various kinds (large pharmaceuticals companies, firms focusing on devices, specialized biotechnology firms etc.) All these may be partnering and collaborating with the young firm.

13. Note that no geographical restrictions are put on the analysis, but, for example partnerships are analysed independent of location of partner.

14. This particular study is part of a larger study involving issues of firm development and performance, and the data collection is therefore substantial.

15. The S&T platform is indicated qualitatively by scrutinizing firm and technology descriptions, and by interviewing the firm about their S&T competences and artefacts and how and where these were sourced.

16. This indicator is calculated by $(PxR/\Sigma x)/(PTotR/\Sigma Tot)$, where PxR is the number of publications (P) in field x in a specific location (R), Σx is the total number of publications in field x in all three regions, $PTotR$ is the number of publications (P) in biomaterials in a specific location (R) and ΣTot is the total number of publications in biomaterials in all three regions

17. Naturally, all companies are included, not only young science-based firms.

18. This indicator is obtained by the equation $(NxR/Sx)/(NTotR/\Sigma Tot)$, where NxR is the number of firms (N) in field x in a specific location (R), Sx is the total number of firms in field x in all three regions, $NtotR$ is the number of firms (N) in biomaterials in a specific location (R) and $STot$ is the total number of firms in biomaterials in all three regions.

19. Due to frequent cooperation over organizational borders, and thereby difficulties of distinguishing various contributions, this study groups these organizations under the heading 'universities'.

20. A 'centre of origin' is defined as the group of actors (research organizations, firms, venture capital companies or a combination hereof) that develop the firms' initial S&T platform.

21. Note, however, that the size of the symbol for origin (university, company, user) does not denote volume or time span of the pre-firm efforts. The firms in uncolored boxes are startups not included in our sample, but created by the same 'center of origin'.

22. Located in California.

23. Admittedly, there needs to be a balance between specialization and diversification, where specialization is needed to ensure a sufficient volume of actors, but may hamper opportunity creation.

24. Naturally, the firms are not created in a vacuum. Location within an already well-functioning location in terms of favourable structural characteristics, high connectedness between actors and diverse and frequent knowledge flows (Porter 1990; Rosenfeld 1997), may enhance knowledge spillover by early, fast and timely knowledge transfer and may perhaps lead to higher innovativeness and competitiveness (Audretsch and Feldman 1996).

25. There may be a risk of an exaggerated policy focus on the regional/national level, and, therefore, the international character of the field must simultaneously be stressed (van Geenhuizen 2003).

REFERENCES

Abernathy, W.J. and J.M. Utterback (1975), 'A dynamic model of product and process innovation', *Omega*, 3, 639–56.

Acs, Z. and D. Audretsch (1990), *Innovation and Small Firms*, Cambridge, MA: The MIT Press.

Acs, Z. and L. Preston (1997), 'Small and medium-sized enterprises, technology and globalization', *Small Business Economics*, **9**, 1–6.

Andersson, T., O. Asplund and M. Henrekson (2002), *Betydelsen av innovationssystem: Utmaningar för samhället och politiken, En fristående studie utarbetad på uppdrag av Näringsdepartementet och Utbildningsdepartementet* (in Swedish), VFI 2002:1, Stockholm: Vinnova.

Audretsch, D. (1995), *Innovation and Industry Evolution*, Cambridge, MA: The MIT Press.

Audretsch D. and M. Feldman (1996), 'Knowledge spill-over and the geography of innovation and production', *American Economic Review*, **86**(3), 630–40.

Audretsch, D.B. and P.E. Stephan (1996), 'Company-scientist locational links: the case of biotechnology', *American Economic Review*, **86**(3), 641–52.

Berger, A.N. and G.F. Udell (1998), 'The economics of small business finance: the roles of private equity and debt markets in the financial growth cycle', *Journal of Banking and Finance*, 22, 613–73.

Binks, M.R. and C.T. Ennew (1996), 'Growing firms and the credit constraint', *Small Business Economics*, **8**, 17–25.

Bower, D.J. and A. Young (1995), 'Influences on technology strategy in the UK oil and gas-related industry network', *Technology Analysis and Strategic Management*, **1**(4).

Breschi, S. and F. Malerba (1995), 'Sectoral innovation systems technological regimes, Schumpeterian dynamics and spatial boundaries', paper presented at the Conference of the System of Innovation Research Network, Soderkoping, Sweden.

Brooks, H. (1993), 'Research universities and the social contract for science', in L. Bramscomb (ed.), *Empowering Technology*, Cambridge, MA: The MIT Press.

Carlsson, B. (ed.) (1997), *Technological Systems and Industrial Dynamics*, Boston and Dordrecht: Kluwer Academic Publishers.

Carlsson, B. and R. Stankiewicz (1991), 'On the nature, function and composition of technological systems', *Journal of Evolutionary Economics*, **1**(2), 93–118.

Christensen, C. (1997), *The Innovators Dilemma*, Boston, MA: Harvard Business School Press.

Cockburn, I. and R. Henderson (1998), 'Absorptive capacity, coauthoring behavior and the organization of research in drug discovery', *Journal of Industrial Economics*, **2**, 157–82.

Cooke, P. (2002), 'Regional drug design, the knowledge value chain and bioscience megacenters', paper presented at the International Workshop on Clusters in High-technology: Aerospace, Biotechnology and Software Compared, University du Québec á Montréal, November.

Cortright, J. and H. Mayer (2002), *Signs of Life: the Growth of Biotechnology Centers in the US*, The Brookings Institution Center on Urban and Metropolitan Policy, Washington, DC: The Brookings Institution Press.

COSEPUP (1998), 'Materials science and engineering benchmarking report', Committee on Science, Engineering and Public Policy, website www2.nas.edu/cosepup.

Edquist, C. (1997), *Systems of Innovation: Technologies, Institutions and Organizations*, London: Pinter Publishers.

Edquist, C. (2002), *Innovation Policy for Sweden: Objectives, Rationales, Problems and Measures* (in Swedish), VFI 2002:2, Stockholm: VINNOVA.

Enright, M. (1994), 'Regional clusters and firm strategy', The Prince Bertil Symposium, Stockholm, Sweden.

Ernst & Young (2001), *European Life Sciences 2001*, London: Ernst & Young International.

Feldman, M. (1994), *The Geography of Innovation*, Boston: Kluwer Academic Publishers.

Feldman, M. (1999), 'The new economics of innovation, spillovers and agglomeration: a review of empirical studies', *Economics of Innovation & New Technology*, **8**(1/2), 5–26.

Freeman, C. (1987), *Technology Policy and Economics Performance: Lessons from Japan*, London: Frances Pinter.

Glaeser, E., H. Kallal, J. Scheinkman and A. Schleifer (1992), 'Growth in cities', *Journal of Political Economy*, **100**, 1126–52.

Goldfarb, B. and M. Henrekson (2002), 'Bottom-up vs. top-down policies towards the commercialization of university intellectual property', mimeo, Department of Economics, Stanford University and Department of Economics, Stockholm School of Economics.

Håkansson, H. (1987), *Industrial Technological Development: A Network Approach*, London: Croom Helm.

Harhoff, D. (1999), 'Firm formation and regional spillovers: evidence from Germany', *Economics of Innovation & New Technology*, **8**(1/2), 27–56.

Henderson, R. (1993), 'Underinvestment and incompetence as responses to radical innovation', *RAND*, **24**(2), 248–70.

Henrekson, M. and N. Rosenberg (2000), *Akademiskt entreprenörskap: Universitet och näringsliv i samverkan* (in Swedish), Stockholm: SNS Förlag.

Jacobsson, S. (2002), 'Universities and industrial transformation: an interpretative and selective literature study with special emphasis on Sweden', *Science and Public Policy*.

Jacobsson, S. and A. Rickne (1996), 'New technology-based firms and industrial renewal: an analytical framework', Department of Industrial Management and Economics, Chalmers University of Technology, Gothenburg, Sweden.

Jacobsson, S. and A. Rickne (2003), 'The Swedish "academic paradox" – myth or reality? How strong is really the Swedish "academic sector?"', presented at the Conference in Honour of Keith Pavitt 'What Do We Know About Innovation', The Freeman Centre, University of Sussex, England, November 13–15.

Jaffe, A., M. Trajtenberg and R. Henderson (1993), 'Geographic localization of knowledge spill-over as evidenced by patent citations', *Quarterly Journal of Economics*, **108**(3), 577–98.

Krugman, P. (1991), 'Increasing returns and economic geography', *Journal of Political Economy*, 99, 483–99.

Landström, H. and J. Winborg (1995), 'Small business' managers attitude towards and use of financial sources', conference proceedings of the 15th Annual Babson Kaufman Entrepreneurship Research Conference, Babson College, Massachusetts.

Libeskind, J., A. Oliver, L. Zucker and M. Brewer (1996), 'Social networks, learning and flexibility: sourcing scientific knowledge in new biotechnology firms', *Organizational Science*, **7**(4), 428–42.

Lundgren, A. (1994), *Technological Innovation and Network Evolution*, London: Routledge.
Lundvall, B.-Å. (1992), *National Systems of Innovation: Toward a Theory of Innovation and Interactive Learning*, London: Pinter Publishers.
Malerba, F. and L. Orsenigo (1997), 'Technological regimes and sectoral patterns of innovative activities', *Industrial and Corporate Change*, 6, 83–117.
Marshall, A. (1949), *Principles of Economics*, London: Macmillan.
Nelson, R.R. (1987), *Understanding Technical Change as an Evolutionary Process*, Amsterdam: Elsevier Science Publichers B.V.
Nelson, R.R. (1993), *National Systems of Innovation: A Comparative Study*, New York: Oxford University Press.
Nelson, R.R. and S.G. Winter (1982), *An Evolutionary Theory of Economic Change*, Cambridge, MA: Harvard University Press.
Niosi, J. and T.G. Bas (2001), 'The competencies of regions – Canada's clusters in biotechnology', *Small Business Economics*, 17(1–2), 31–42.
Orsenigo, L. (2001), 'The (failed) development of a biotechnology cluster: the case of Lombardy', *Small Business Economics*, 17, 77–92.
Porter, M. (1990), *The Competitive Advantage of Nations*, New York: The Free Press.
Rickne, A. (2000), *New Technology-Based Firms and Industrial Dynamics – Evidence from the Technological System of Biomaterials in Sweden, Ohio and Massachusetts*, Doctoral thesis, Department of Industrial Dynamics, Chalmers University of Technology, Gothenburg, Sweden.
Roberts, E. (1991), *Entrepreneurs in High Technology: Lessons from MIT and Beyond*, New York: Oxford University Press.
Rosenfeld, S. (1997), 'Bringing business clusters into the mainstream of economic development', *European Planning Studies*, 5(1), 3–24.
Rothwell, R. (1994), 'The changing nature of the innovation process: implications for SMEs', in R. Oakey (ed.), *New Technoloy-based Firms in the 1990s*, London: Paul Chapman Publishing Ltd.
Saxenian, A. (1994), *Regional Advantage: Culture and Competition in Silicon Valley and Route 128*, Cambridge, USA and London, UK: Harvard University Press.
Schumpeter, J.A. (1934), *The Theory of Economic Development*, Cambridge, MA: Harvard University Press.
Shane, S. (2001), 'Technological opportunities and new firm creation', *Management Science*, 47(2), 205–20.
Shane, S. (2001a), 'Technology regimes and new firm formation', *Management Science*, 47(9), 1173–90.
Shane, S. and D. Cable (1998), 'Social relationships in the financing of new ventures', Massachusetts Institute of Technology working paper, Cambridge, MA.
Shane, S. and R. Khurana (2000), 'Bringing individuals back in: the effects of career experience on new firm founding', paper presented at the Academy of Management Conference, Washington, DC, August.
Sörlin, S. and G. Törnqvist (2000), *Kunskap för välstånd: Universiteten och omvandlingen av Sverige* (in Swedish), Stockholm: SNS Forlag.
Steinle, C. and H. Schiele (2002), 'When do industries cluster?', *Research Policy*, 31(6), 849–58.
Swan, P., M. Prevezer and D. Stout (1998), *The Dynamics of Industrial Clustering*, New York and Oxford: Oxford University Press.
Tushman, M. and P. Anderson (1986), 'Technological discontinuities and organizational environments', *Administrative Science Quarterly*, 31, 439–65.

Utterback, J. (1994), *Mastering the Dynamics of Innovation: How Companies Can Seize Opportunities in the Face of Technological Change*, Boston, MA: Harvard Business School Press.

van Geenhuizen, M. (2003), 'How can we reap the fruits of academic research in biotechnology? In search of critical success factors in policies for new-firm formation', *Environment & Planning C: Government & Policy*, **21**(1), 139–56.

Winter, S. (1984), 'Schumpeterian competition on alternative technological regimes', *Journal of Economic Behaviour and Organization*, 4, 287–320.

Zeller, C. (2001), 'Clustering biotech: a recipe for success? Spatial patterns of growth of biotechnology in Munich, Rhineland and Hamburg', *Small Business Economics*, 17, 123–41.

Zucker, L., M. Darby, M. Brewer and Y. Peng (1995), *Collaboration Structure and Information Dilemmas in Biotechnology*, Cambridge: NBER.

Zucker, L., M. Darby and M. Brewer (1998), 'Intellectual human capital and the birth of US biotechnology enterprises', *American Economic Review*, **88**(1), 290–305.

Zucker, L., M. Darby and J. Armstrong (1998a), 'Geographically Localized Knowledge: Spillovers or Markets?, *Economic Inquiry*, **XXXVI**, January, 65–86, reprinted in P.E. Stephan and D.B. Audretsch (2000), *The Economics of Science and Innovation*, Volume II, The International Library of Critical Writings in Economics 117, An Elgar Reference Collection, Cheltenham, UK and Northampton, MA, USA: Edward Elgar, pp. 400–21.

12. Examining the marketplace for ideas: how local are Europe's biotechnology clusters?*

Steven Casper and Fiona Murray

1. INTRODUCTION

Competitive advantage in the biotech industry depends on several key factors: the existence of vibrant academic institutions (Kenney 1986), venture capital and other forms of risk financing (Lerner 1995; Powell et al. 1996; Sorensen and Stuart 2001), managerial talent, (Gulati and Higgins 2002) and the embeddedness of firms in collaborative inter-firm networks (Powell et al. 1996; Carlsson 2002). Such factors certainly provide considerable explanation for the explosive development of the biotech industry in regions of the USA such as San Francisco and Boston. However, the slower growth of regions such as New York, Chicago and several European biotech clusters requires a more detailed and nuanced understanding of the emergence of local biotech clusters.

In this chapter we propose the concept of a marketplace for ideas: the social context in which ideas are brought together with human and financial capital to build a vibrant cluster. The notion of the marketplace highlights both the key resources that provide the foundation for the formation and growth of biotech firms and the more social network-like setting for these interactions in contrast to a pure neoclassical market. This chapter uses the marketplace conception of biotech cluster development to examine the following question: Why do some clusters develop superior capabilities to commercialize science than others? In particular we ask whether or not this marketplace is geographically bounded and operates within a local context or alternatively if firms are clustered locally but draw on the resources of a global marketplace. This perspective follows the traditional 'varieties of capitalism' approach to explore cross-national differences through an understanding of different institutions and the systems of incentives and actions that these institutions establish (Hall and Soskice 2001). But at the same time, we draw on the regional development literature to appreciate how local institutional distinctions lead to micro-mechanisms that can lead to dramatic variation within a given national context (Saxenian 1994; Almeida and Kogut 1999).

With this lens on the development of biotechnology we propose that differential cluster development can be thought of as arising from variation along two key dimensions: the degree to which firms can access key local marketplace resources and the extent to which firms move beyond their local context and access resources and build network connectivity outside the local cluster. We describe the marketplace for ideas of biotechnology across two important European settings – the UK around the academic institutions of Cambridge and Germany around the academic institutions of Munich – and we then compare these European clusters with the biotech marketplace that has been established around the Boston-area institutions of MIT and Harvard.

The comparison of the European biotech experience with that of Boston is of particular interest because while Europe, like the USA, has made extensive investments in basic biomedical research and has a strong biotech industry, there is also considerable evidence to suggest that at least in some locales, the industry in the USA has a superior performance record (Ernst & Young, 2001). Indeed careful comparison centred on the relative entre-preneurial activity generated from a comparable investment in basic life science research suggests that Boston's biotech cluster is currently larger in terms of firms, products in development and employment than the compar-able UK biotech industry (Casper and Murray 2002).[1]

Our research follows Powell and others in highlighting the heterogeneity of actors contributing to the biotech sector. However we attempt to move beyond the more traditional approaches that establish and count ties among actors. Rather, in framing our analysis as a marketplace and elaborating on the interactions among actors within the marketplace we hope to provide a more nuanced understanding of commercialization dynamics. Building on these basic comparisons, we examine firms in different biotech clusters along two key dimensions: The source of their core technology together with the nature and extent of their ongoing scientific linkages, and their access to labour markets – dimensions that are closely tied to our conception of the marketplace as a setting in which firms must bring together key resource inputs – ideas, management and capital – for future development and success.[2] In particular we focus on salient differences in the extent, variety and geographic scope of linkages that firms have been able to establish with university science. These include using universities as a source of ideas for start-ups, scientific collaboration between firms and laboratories, the role of scientists on the scientific advisory boards of firms, and the role of univer-sities in supplying firms with a labour market for talented scientists.

Our empirical evidence suggests that the marketplace institutions in the UK are relatively strong, creating incentives for people, ideas and finance to come together to commercialize science in the Cambridge cluster. Moreover, firms in the Cambridge cluster tap into national resources in addition to the local resources of the Cambridge region. Marketplace development in the UK may therefore be viewed as having established an 'under-expressed' version of the US marketplace with the architecture of the surrounding institutions having basically emerged to support cluster development, although with some limitations in the current sector-specific policies that relate to structuring technology licensing from universities.

In contrast Germany is more orchestrated and as one might expect from comparative institutional analysis, the government has attempted to shape the institutions surrounding the marketplace for science (Fligstein 2001). The outcome is more local in terms of the resources and the connections that are made and while a biotech cluster has been created, marketplace architectures are underdeveloped, particularly the national labour markets for scientists. Furthermore, while science-linkages across Munich firms are strong – stronger in fact than their Cambridge counterparts and dominated by Munich's core academic institutions to a degree beyond the scope of the University of Cambridge's role in the Cambridge cluster – their ability to develop capabilities needed to commercialize their ideas is weak.

In conclusion, we argue that the marketplace metaphor may lead to a different policy agenda – though in many respects a complementary one – than much existing research on clusters. In particular we have two main conclusions: the first is that the incentives for interaction and collaboration have been underemphasized in policy formulations that attempt to bring actors together. The second is that while the emphasis in biotech policy has often been on local collaboration and networking as key drivers of the success of clusters we argue that in the case of Cambridge, the cluster is shaped around a national dynamic with the majority of scientific collaborations being with laboratories other than at the University of Cambridge; moreover, Cambridge scientists do not dominate the scientific advisory boards of firms. We conjecture that clusters of the scale of Cambridge or Munich are too small to support purely local cluster development and thus the transplantation of US networking metaphors to the small European clusters may lead to hasty policy conclusions – seen in particular in recent years in Germany's biotech clusters. Instead, we show how a local marketplace metaphor with firms operating in a national and international context may lead to a useful policy agenda that could lead to the improved performance of Europe's biotech clusters.

The structure of the remaining chapter is as follows: first we substantiate the notion that there is in fact an empirical puzzle to be solved by examining the regional differences in European biotechnology; second we examine the scientific engines of the marketplace for ideas in Cambridge, Munich and Boston. We then provide a comparative overview of scientific linkages that characterize these clusters including the origins, collaborations and advisory networks of the firms. Our last section of empirical evidence is a comparative analysis of the labour market dynamics across the clusters. We end with our policy conclusions and the implications of our findings for European policymakers whose attention is focused on biotechnology.

2. THE EMPIRICAL PUZZLE

Whereas there was virtually no biotech industry in Germany pre-1995, several hundred biotech companies were formed in the latter half of the 1990s (for more details see Ernst & Young 2001). While many of these firms are tiny companies or firms specializing in very low-risk activities such as the supply of specialist supplies or contract research activities, several dozen companies including Apovia, Cellzome and others have gained critical mass in technologically intensive areas focused on therapeutic research. This flurry of activity can be connected to German federal and state government spending of several hundred million euros promoting the development of biotech clusters centred around German universities that underscores a stated governmental goal to establish Germany as Europe's leading biotech marketplace by the turn of the new millennium. This aim was particularly striking as prior to the mid-1990s the dearth of biotech firms in the country arose in a large part due to strong social and regulatory restrictions on recombinant DNA research but also due to a dearth of economic institutions supporting entrepreneurial start-ups (see Casper 2000).

In comparison, the biotech industry in the UK has its origins in the early development of monoclonal antibodies and the formation with government support of Celltech in 1980 through the National Enterprise Board, the establishment of British Biotechnology in 1986 and the formation of Cambridge Antibody Technology (Senker 1996). However, with the by now well-known refusal of the Medical Research Council to patent Kohler and Milstein's monoclonal antibody research, the foundation of the industry faltered and it was only in the early to mid-1990s that the industry restarted (see Murray and Kaplan 2003 for a more detailed analysis of the waves of firm formation in the biotech industry). By the late 1990s, however, the success of firms such as Oxford Glycosciences, PowderJect and Shire Pharmaceuticals established the UK as Europe's largest biotech sector, with over 40 public companies dedicated to human therapeutic biotechnology and dozens of privately held start-ups located in university-centred clusters in Cambridge, Oxford, the Dundee area in Scotland and London (see Ernst & Young 2001 for more details). Unlike the Munich case, government policy towards the UK industry has been relatively passive, with no large-scale subsidy programmes aimed at the industry. Nevertheless, the UK government has focused explicit policies on promoting the commercialization of university research with considerable investment in the University Challenge funds, seed funds and inducements to early-stage funding of novel biomedical ideas.

The differences in the German and UK biotech clusters do not stop with a simple numerical comparison. While there are notable exceptions, in particular the small cluster of firms surrounding the Max Planck Institute (MPI) for Biochemistry in Munich and a small number of firms spun-out of the European Molecular Biology Laboratory (EMBL) in Heidelberg, the vast majority of German biotech companies currently face severe difficulties, if not bankruptcy, as initial funding begins to expire. Very few German firms have been able to develop promising therapeutic products (Ernst & Young 2001). Instead, most have moved into lower-risk platform technologies (see Casper and Kettler 2001); segments which generally depend more on downstream relationships with industrial collaborators rather than upstream relationships with basic research laboratories. In comparison the UK biotech firms have been more successful in transforming basic scientific discoveries into therapeutics for human health care – ranging from monoclonal antibodies to advances in gene therapy and population genetics – probably the most potentially profitable sector of the biotech industry. The UK biotech industry has thus significantly overtaken its German counterpart at least in terms of therapeutic drug discovery productivity. Table 12.1 dramatically demonstrates this claim through comparing therapeutic pipelines of publicly listed companies in the two countries. As of 2001, the ratio of UK to German compounds in established pipelines was about 20:1.

There are many important differences between the broader German and

Table 12.1   Therapeutic pipelines of German and UK public biotech companies, mid-2001

	Preclinical	Phase I	Phase II	Phase III	Total
UK	32	37	46	13	128
Germany	2	2	1	1	6

Source: Ernst & Young (2001).

UK industrial landscape (see Hall and Soskice 2001) which should make one wary about quick comparisons across the two countries even within an individual sector. However, following in the tradition of comparative industry-based research, we believe that deep empirical analysis, grounded in theory, will reveal some of the policy choices and dynamics that have led to the relatively poor performance of the German biotech sector in comparison to its British counterpart. Furthermore, we believe that an analysis of the UK–German empirical puzzle is not only of relevance to biotech policymakers but also relevant for broader studies of cluster policy.

What follows is a close analysis of biotech firms in two comparable regions of the UK and Germany, Cambridge in the UK centred on the University and the Sanger Institute for Genomics, and Munich in Germany the focal point for a series of critical scientific institutions and much of the recent entrepreneurial activity. Our comparison has important implications for economic policy particularly at the intersection of science and entrepreneurship and those policies that are intended to shape the marketplace for ideas, now considered such an important engine of growth in Europe.

3. SCIENTIFIC ENGINE DRIVING THE MARKETPLACE FOR IDEAS

Our choice of European clusters is predicated first on the Germany–UK puzzle explored above. While France is also an interesting case, our chosen comparison is particularly dramatic and illustrates quite distinctive policy interventions and outcomes in terms of resulting cluster dynamics. Nevertheless both nations and the specific clusters we have chosen to study are centred on powerful scientific institutions that are the foundation for the public scientific knowledge base in each nation (along with other institutions in both countries that together make the UK and Germany among the leading European nations for scientific excellence and productivity). The comparison of these two science-oriented clusters is part of our research design because following the notion of a marketplace for the commercialization of scientific ideas, we argue that we are controlling for the supply of high-quality ideas into the marketplace and for the *potential* for ongoing university–firm networks.

3.1 The Scientific Institutions have Broad Similarity

Munich and its environs has emerged as Germany's largest life science research complex. Research in the area is dominated by several large life science departments and teaching hospitals belonging to the University of

Munich. This includes the Munich Genezentrum, an autonomous department of the University launched in the 1980s to conduct interdisciplinary genetics research. The city is also home to the Max Planck Institute (MPI) for Biochemistry, which has a large campus in the Martinsried area employing over 800 scientists and technicians. This institute has become a leading centre for research in the area of cell signalling, an area of biochemistry with a particularly strong commercialization potential. The region also houses the GSF Institute for Environment and Health Research. The GSF has been a core coordinator of German contributions to the International Human Genome Project, and it houses the Munich Information Centre for Protein Sequences, a large bioinformatics centre.

The Cambridge scientific institutions are centred on Cambridge University. The University has a number of five-star-rated departments within the broadly defined life sciences arena and receives substantial public funding for science.[3] It has also played a strong historical role in the development of the intellectual foundations of the biotech industry as the locale for Watson and Cricks' ground breaking work on the structure of DNA some 50 years ago (Watson and Crick 1953). The University of Cambridge is also surrounded by several critical affiliated life sciences research institutes. These include two Medical Research Council institutes, the MRC Laboratory for Molecular Biology – another important contributor to the early foundations of molecular biology, research on protein crystallography and DNA sequencing techniques – and the MRC Centre for Protein Engineering. In addition, the Cambridge region has a significant presence in genomics-related research: The Sanger Institute and the European Bioinformatics Institute (EBI) are large, specialized and world-class scientific research organizations. Sanger is one of the world's largest gene sequencing centres responsible for decoding over one-third of the genome as part of the public Human Genome Project. Its director during the genome project was Dr John Sulston, recently awarded a Nobel Prize for his contributions to gene sequencing. Similarly, the EBI is located in close proximity to Sanger and is a leader in developing the software used to manage a huge volume of genetic code created by new genomics technologies.

4. AN OVERVIEW OF SCIENTIFIC LINKAGES

4.1 Origins of Firms: Laboratories as a Source of Start-ups

One of the most basic, although relatively recent, roles of universities within biotech clusters is as a source of ideas for spin-offs. To gauge the

Table 12.2   Origin of Cambridge and Munich biotech firms

	Cambridge	Munich
Local university spin-off	15 (52%)	19 (68%)
Non-local university spin-off	7 (24%)	4 (14%)
Industry spin-off	5 (17%)	5 (18%)
Unknown	2 (7%)	0 (0%)
Total	29	28

Note: Includes all known foundings, 1990–2002.

Source: Authors' research.

importance of local universities as originators for spin-offs, we collected data on biotech firms established in the Cambridge and Munich region since 1990. Though we cannot claim that this list is exhaustive, it is based on an in-depth search of a number of Internet-based biotech industry guides, surveys of local science and technology parks, and information available from technology licensing offices and regional development agencies in both areas. Table 12.2 provides an overview of the origin of 57 biotech start-ups found in either Cambridge or Munich between 1990 and 2002. Figure 12.1 displays data on the frequency of start-ups on a yearly basis. Table 12.2 shows that universities were the primary source of biotech start-ups in both regions.[4] And the strength of local academic science is strongly correlated with the founding of biotech firms in both areas. However, there are also some differences across the two areas.

Cambridge has a more diversified portfolio of companies than Munich, at least as defined by origin. Half of the Cambridge firms are spin-offs from various departments and institutes within the University of Cambridge. A wide diversity of laboratories are involved in the formation of these firms, ranging from pure science departments such as chemistry to more applied research institutes that have clear connections to biomedical research, such as the Institute for Cancer Research or the MRC Centre for Protein Engineering. A few departments have been involved in serial spin-offs. These include the Departments of Chemistry and Biochemistry and the MRC Centre for Molecular Biology, which has spun-out two of Cambridge's more high-profile biotech firms, Cambridge Antibody Technology (CAT) and

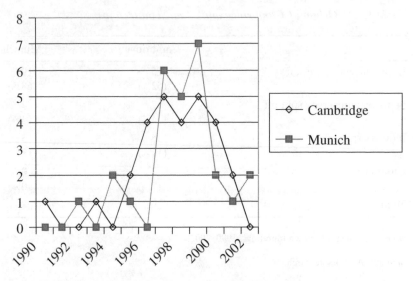

Figure 12.1 Frequency of biotech foundings, 1990–2002 – biotechnology start-ups by year: Cambridge and Munich

more recently Astex. However, relatively few Cambridge academic scientists have become serial founders (i.e. launched more than one biotech company). Our research has located only two: Tom Blundell and Craig Winter, both from the MRC Centre for Molecular Biology. Furthermore, some very significant research laboratories, such as the European Bioinformatics Institute and the Sanger Institute – both involved in high-profile genomics-related research – are absent from our biotech start-up list of originating institutions. We discuss both these issues in more detail below.

While the University of Cambridge is the dominant source of ideas for new biotech firms in Cambridge, half of the region's firms did not derive from the University. Roughly a quarter of the firms were spin-offs from universities in other parts of the UK but located in the Cambridge region. There are also a significant number of industrial spin-offs. These derive from large UK-based research and development (R&D) laboratories of Glaxo and SmithKline Beecham (before their merger), from established biotech firms, and in one case from a venture capital firm. In sum, the Cambridge region is clearly a magnet for biotech start-ups. While we discuss several problematic issues surrounding the performance of Cambridge as a biotech cluster, the willingness of a significant pool of founder teams to relocate to Cambridge is strong evidence that Cambridge, at least relative to the rest of the UK, is performing well as a biotech cluster.

During the same 12-year time frame, the Munich area fostered a similar number of firms (28 in all), but the predominance of firms originating in local university founder laboratories is stronger than in Cambridge. As in Cambridge, firms are formed based on the research of a diversity of laboratories. Most important has been the MPI for Biochemistry. One of its directors, Professor Axel Ullrich, entered the biotech industry in its very early days joining Genentech in the USA in 1979, and seems to have imported more US-style norms of academic–industry relations to Munich – he has been involved in founding four of the Munich firms. However he is the only serial founder we located in Munich. Several scientists working within the Munich Gene Centre have become involved with firms, as have professors within numerous departments of the University of Munich and its teaching hospital. While there are fewer firms in Munich originating in non-local universities, one of the areas' most prominent firms, GPC Biotech, migrated to Munich from Berlin, where it was spun-out of research associated with the German Human Genome Project carried out at the MPI for Molecular Biology. Company spin-outs include firms developed by employees of SmithKline Beecham (local Germany R&D laboratories), local management consultants and in two cases local biotech companies.

The rate of founding in Munich shows less of a smoothly increasing upward trend than in the Cambridge area – with a large number of companies developed in the 1997–1999 period. This is a reflection of strong technology policies aimed towards the formation of biotech companies in Germany during the late 1990s. Such 'orchestrated' development of biotechnology also helps account for the greater predominance of local area spin-offs in Munich than in Cambridge. As we discuss in more detail below, local actors tend to be privileged by regional cluster policies.

4.2   Collaborative Activities of Firms

Previous studies of the biotech industry have concluded that universities often spin-out companies rather than license technologies to established companies due to the inability of laboratories to codify technologies in a manner that can be easily transferred to companies (see Zucker et al. 1998; Kenney 1986). Many nascent biotechnologies include tacit elements, such as laboratory protocols, that are difficult to completely codify. Furthermore, embryonic technologies often need significant development before it becomes clear whether there is a significant value in their full commercialization. In addition to facilitating tacit knowledge transfer, by locating in close proximity to a university fledging biotech firms also provide strong incentives for the continued collaboration with inventors who participate as equity-holding founders. Firms collaborate to secure access to tacit knowledge, such as innovative laboratory methods that may exist within laboratory personnel that have not moved to the firm. Firms also routinely collaborate to validate methodologies, or more broadly to secure a reputation for scientific research prowess. Having access to world-class academic researchers becomes a key component of such strategies. Biotech clusters develop in proximity to universities in part due to this need for close collaboration.

All firms within our sample that were spun-off from university laboratories in Cambridge and Munich have also located in the region, which could facilitate close collaboration. However, the existence of relatively large numbers of firms with no founding connection to the university suggests that other drivers of clustering must exist. To add depth to this analysis we have collected material on patterns of research collaboration for 17 Cambridge biotech firms and 10 Munich companies – all firms we were able to locate that were established in roughly the last five years. Using bibliometric search engines on the Internet we examined all publications for each firm in order to determine the composition of collaborators for each publication.[5] Table 12.3 summarizes our results.

Table 12.3   Composition of research collaborations for Munich and Cambridge biotech firms

	Cambridge		Munich	
Authors include:	Number	%	Number	%
Academic founder lab	44	51	39	28
Other academic lab	32	37	112	82
Other firm	15	17	26	19
Company sole author	31	36	17	12
Total	86		137	

Source: ISI Web of Science, www.isinet.com. Individual publications can be included in multiple categories.

There are a number of similarities in publishing patterns across firms in both regions. Collaborations with academic laboratories strongly outweigh those with other companies, signifying that ongoing research links with academia are important for most companies. We also see significant ongoing relationships between firms and their founder laboratories. This relationship is particularly strong for Cambridge firms, where half of all publications have a scientist from the academic founder laboratory listed as a co-author. Relationships with other academic laboratories are also important. Over a third of Cambridge publications involve a third-party academic laboratory, while 82 per cent of all Munich firms involve this type of collaboration. Firms in both regions also routinely co-publish with scientists working in other biotech firms. Cambridge firms also have a large number of publications where scientists within the firm are the sole author; a smaller number of Munich publications are in this category.

In general, the relatively high extent of firm publishing suggests that access to basic research is important to their success. The high degree of involvement of founders suggests that publishing surrounds the ongoing transfer and exploitation of a firm's founding ideas from the academic laboratory. This evidence strongly supports the claim that access to university research is crucial for most biotech firms. However, while the founder laboratory connection appears to be a highly valued resource for most firms, geographical proximity is not a strong driver of collaborations. Table 12.4 shows the geographical location of academic research laboratories

Table 12.4   Geographic location of academic collaborators

	Cambridge		Munich	
	Number	%	Number	%
Local	73	31	90	25
National	76	33	148	42
Foreign	84	36	119	33
Total	233		357	

Source: Author calculations based on ISI Web of Science Data, www.isinet.com. Particular collaborators can appear multiple times, as their location will be counted in each publication in which they appear as a co-author.

involved in 233 research collaborators listed as co-authors in 114 publications involving Cambridge biotech firms and 357 collaborators involving 159 in Munich. In both regions local proximity of academic research laboratories does not seem to be a strong driver of the propensity to collaborate. In only 31 per cent of collaborations with Cambridge firms are co-authors local. An equal share of co-authors are located within the broader UK academic science system, while over a third of all collaborators are foreign. The Munich data shows a similar trend. Only a quarter of all collaborations involve a local author, while a third are with foreign laboratories. If we take out the founder laboratory collaborations from the data (on the basis that by construction for many of these firms these collaborations are local), the trend towards more geographically extensive collaboration reaching well beyond the boundaries of the local cluster are more pronounced.

While access to academic science appears to be important for most

Cambridge and Munich area biotech companies, this evidence does not present a clear picture on whether local relationships, presumably driven by tacit knowledge or local proximity, are a particularly important driver of collaborations between biotech firms and laboratories. Companies may seek out the best available scientific collaborators regardless of location. Given the very frequent occurrence of successful publication of research, it appears that distance-based collaboration is viable for most biotech firms. Local science networks are obviously an important focal point for firms in each region, but may not drive their pattern of research collaborations or their ultimate success. This suggests that firms do and possibly should broaden their scope of collaboration in order to successfully access key scientific knowledge that may not be locally available, particularly within smaller clusters whose breadth of local expertise is limited.

4.3 Scientific Advisory Boards

In addition to collaboration, most biotech companies invite a group of scientific experts to attend regular meetings to advise the company as to its scientific trajectory. Scientific advisory boards (SABs) usually include a firm's scientific founders and several additional experts within the firm's scientific niche. SABs, when comprising famous scientists, can also enhance the prestige of a firm and are also a good source of networking contacts for the firm (see Audretsch and Stephan 1999 for a brief discussion of SABs).

The SAB composition again demonstrates the importance of scientific connections to most biotech firms in each region. Drawing on the dataset of post-1996 start-ups, Table 12.5 denotes the affiliation of SAB members. For Cambridge firms, 73 per cent of SAB members are university faculty members, while only 12 per cent come from industry, with the remaining 15

Table 12.5   Scientific advisory board membership by affiliation

	Cambridge		Munich	
	Number	%	Number	%
University	35	73	51	81
Hospital	7	15	0	0
Biotech	3	6	6	10
Pharma	3	6	1	2
Other (VC/Consulting)	0	0	5	8
Total	49		63	

Source: Company websites.

per cent from hospitals. Munich SABs have a similar composition – over 80 per cent from universities, with the rest from industry. This again suggests that Cambridge and Munich biotech companies are strongly science-based. In other words, the advice of basic research scientists appears to be far more valuable to these firms than that of experts within the pharmaceutical industry, or from other biotech firms (though industry experts commonly sit on company boards). The low occurrence of affiliations with hospital faculty (even with those who have a joint affiliation of some sort), particularly in Munich, suggests that few of these firms have developed pronounced therapeutic research area specializations or have compounds in clinical trials; alternatively it could be interpreted that they are failing to establish critical connections to the medical community who may facilitate such commercialization.

We have also compiled a geographic breakdown of SAB members, as shown in Table 12.6. This figure complements our research on scientific col-

Table 12.6   Geographic location of SAB members

	Cambridge		Munich	
	Number	%	Number	%
Local	16	30	12	19
National	25	47	14	22
Foreign	12	23	37	59
Total	53		63	

Source: Company websites.

laborations, demonstrating that proximity probably is not the driving factor in SAB selection. In general, these results are not surprising, given the relatively arm's-length relationship between SAB members – who usually attend meetings at most on a quarterly basis – and the firm. The fact that most firms can recruit SAB members from a wide variety of geographic locations indicates the wide, international scope of their scientific networks.

In the Cambridge cluster, almost 80 per cent of SAB members reside in the UK, while only 20 per cent reside abroad (primarily in the USA). However, only 30 per cent are located in Cambridge. The fact that 70 per cent of SAB members do not reside within the local area suggests that firms attempt to secure the best scientific advice, regardless of its location. The Munich data shows an even stronger trend away from local SAB relationships. There, less than 20 per cent of SAB members are from local universities – in most cases these are the founder scientists. A similar proportion of SAB members are from the broader German science system, while close to 60 per cent of scientific advisors are from foreign institutions. Inviting prominent international scientists to sit on company advisory boards probably reflects a strategy of using the SAB to build up credibility for the firm. The ability of tiny start-ups in the Munich cluster to attract international scientific advisors may also reflect the differences in the German science system compared to those in the USA or the UK. The German science system tends to be more hierarchical, in which senior scientists enjoy long-term research funding that is commonly used to develop large laboratories. Germany's senior academic scientists may be leveraging international contacts developed within their scientific community to develop capabilities and credibility for firms they are sponsoring. We will see a similar trend across Munich firms in recruiting scientific staff.

5.   LABOUR MARKET DYNAMICS

One of the most important roles of any technology cluster is grounded in its capacity to generate a vibrant labour market for scientists and engineers. Recent research on technology clusters has linked the organization of local labour markets to the innovative capacity of firms: Saxenian's book *Regional Advantage* (1994) established the claim that the success of Silicon Valley in the semiconductor and computer industries rested in part in its establishment of extremely deep and flexible labour markets for engineers. Saxenian's ethnographic research, more recently verified with a statistical analysis by Almeida and Kogut (1999), documents frequent job-hopping among Silicon Valley's engineers. These authors have found that other regions in the USA where job-hopping is less pronounced have a lower innovative output (as defined by the level of patenting and forward citation activity).

Extensive job mobility aids in the diffusion of tacit knowledge surrounding new science and engineering techniques across a region's firms. Moreover – and particularly important for volatile industry segments such as biotechnology – the existence of a deep, flexible labour market substantially reduces the individual career risk, from the point of view of a scientist or engineer, of moving to a high-risk technology start-up. If the firm fails, new jobs can be found quickly. Firms located in deep, flexible labour markets should have an advantage in securing the employment of high-quality scientific personnel for high-risk projects. Examining the organization of scientific labour markets surrounding Cambridge and Munich biotech firms is thus an important aspect of the marketplace for ideas such as labour markets significantly enhance and complement the access to scientific ideas and ongoing scientific linkages documented above. In addition to obtaining an estimate of the extent of labour market mobility in the region, we are also interested in examining the degree to which labour markets form bridges which establish additional connections between basic academic research laboratories and commercial start-ups.

Building on the data on firm publications described earlier, we have collected data on all publishing scientists within our sample of Cambridge and Munich area biotech firms. Using publication records allows us to use bibliometric methods to construct information about the prior employment and scientific activities of scientists. Scientists fall within our dataset if they have published with a biotech firm included in our sample and have published in the course of at least one prior job. This method allows the systematic collection of career histories for publishing scientists, but excludes a relatively large number of scientists who have not published with the firm. This could create biases, particularly if strong publication records are related to job mobility – scientists who publish more may have more jobs open to them, in either academic laboratories or industry. This might also lead to inflated pictures of the amount of job mobility in a particular region. None the less, taking this possible source of bias into account, we used the 'Web of Science' to compile lists of all publications for each scientist identified as an author on each publication by all firms in our post-1996 group of Cambridge companies. We were able to trace the career histories of 83 scientists who have worked within 15 Cambridge biotech companies and 52 scientists with Munich companies.

Table 12.7 shows the type of most recent prior employment for each scientist prior to joining the current firm. It illustrates the diverse nature of the Cambridge labour market pool: 46 per cent, close to half, of the scientists

Table 12.7   Employment origins of biotech firm scientists

	Cambridge		Munich	
	Number	%	Number	%
Founder lab	15	18	26	50
Local lab	8	10	4	18
National lab	12	14	11	21
International lab	10	12	7	13
Pharma	23	28	1	2
Biotech	15	18	3	6
Total	83		52	

Source: Author calculations based on data from the ISI Web of Science.

came to their current job from another job in the life sciences industry, either a pharmaceuticals company or another biotech company. The other half came directly to the firm from an academic science laboratory. While a significant number, 15 per cent or 18 per cent, came to the firm directly from its founder laboratory, a large number of scientists came to the firm from another laboratory (36 per cent). Only 10 per cent of these people came to Cambridge biotech firms from an academic laboratory located in Cambridge itself. The rest came from abroad or elsewhere in the UK.

One interesting result from the data for Cambridge is that close to half of all scientists came to their current job from another firm rather than from academia. This shows an important degree of labour market flexibility within the region. Moreover, several scientists came to their current jobs from firms that have failed, such as Axis Genetics, or companies undergoing major reorganizations and shifts of scientific direction, such as Celltech or Cantab. The fact that scientists appear to have survived failure and gone on to find jobs in subsequent start-ups suggests that flexible labour market dynamics resembling those in Silicon Valley may be beginning to appear in Cambridge. We believe that this is a critical element in the marketplace for ideas, with firms needing to draw not only on a deep supply of novel science but flexible scientific linkages and, importantly, flexible labour markets bringing scientists with a depth and breadth of experience into the firm. This view augments prevailing views of the importance of management teams (Gulati and Higgins 2002) and extends it to the scientific team.

The data for Munich scientists indicates the existence of markedly different labour market dynamics. For over 90 per cent of scientists, the job with the current biotech firm is their first job in industry. More specifically, the primary labour market bridge connecting universities with firms appears to be the founding laboratory. Half of the Munich scientists came to their firm through previous employment in the company's academic founder laboratory. The remaining 40 per cent of these scientists were recruited from a roughly even mix of non-founding Munich area laboratories, laboratories from elsewhere in Germany, and in 13 per cent of cases international laboratories. The relatively small contingent of scientists recruited from abroad contrasts with the high percentages of SAB members and collaborators who are foreign. This implies that founding scientists have not been able to translate their foreign contacts to recruit scientists for their firms. Instead, they have primarily used their own laboratories as a core labour market pool.

Also of interest is the very low number of German scientists that were previously employed in industry – only 4 people out of 52, or 8 per cent. It is not surprising that so few Munich scientists came from biotech companies, as so few companies existed previously in the country. More surprising is the inability of Munich biotech companies to attract scientists from the country's large pharmaceuticals industry – as the Cambridge companies have done. As emphasized by varieties of capitalism research (see below), this finding is likely linked to long-term employment patterns within large German companies, limiting the market for experienced mid-career scientists. The strong founder laboratory link helps explain why the Munich biotech firms have matched the Cambridge companies in developing links with basic research scientists. However, they may also help account for the relatively poor performance of German firms in commercializing their science – few scientists with direct commercialization experience appear to be working within the Munich companies.

6. THE GEOGRAPHIC DISPERSION OF THE MARKETPLACE IN BIOTECH CLUSTERS: A PROBLEM FOR TRADITIONAL CLUSTER POLICY?

Through our conceptualization of biotechnology clusters as a marketplace which brings together ideas, scientific and managerial talent and capital, we would like to steer attention away from two prevailing theories of cluster formation, both with strong policy implications. The first relates to the importance of local, tacit knowledge in driving clusters, while the second relates to the role of local networks as the basis for cluster development.

A common formulation for the foundation of clusters formation is that they are driven primarily by proximity, or tacit knowledge (see Cooke 1999 and generally Winter 1987). According to this line of thought, an inability to codify knowledge brings people together, forcing research organizations, whether public or private, to collaborate in close proximity. A policy implication of this argument is that governments should take the initiative in facilitating the establishment of relationships between laboratories within regional universities and technology spin-offs. They typically do so through sponsoring the development of incubator laboratories and technology parks in close proximity to university laboratories.

In line with this argument we found some evidence supporting the claim that tacit knowledge drives science linkages: local universities are a major source of new firms, academic collaborations, SAB members, and, particularly in the Munich case, employees. While tacit knowledge may drive some relationships, particularly knowledge transfer between founding laboratories and firms, the convenience of proximity may also play a role in many of these other activities. However, along each of these four dimensions, national and at times international ties were also abundant, and in some cases outnumbered local ties. In these cases it seems less obvious that tacit knowledge is driving the structure of scientific linkages. Or, even if tacit issues exist, they may be overcome through modern patterns of communications or, at least in the case of SAB membership, travel. If tacit knowledge plays a lesser role in driving university–industry relationships, then it may not be the core driver of regional cluster formation. Policies aimed at facilitating local linkages through breaking down barriers to tacit knowledge exchange may be misdirected.

The second, closely related idea is that clusters are primarily driven by local networks, and that these networks may be insufficiently formed due to the heterogeneity of actors involved and due to marketplace failure. We have certainly seen a variety of laboratories, organizations and individuals that partake within the Cambridge and Munich biotech arenas. Networking policies are predicated on the notion that due to the diversity of this community many useful relationships may not be realized. Such policies can help by bringing people together and giving them resources to explore their complementary interests. This has been the driving idea behind much German cluster policy towards biotechnology, and has also begun to take hold in the UK. A recent example of such policy in the UK is the Genetic Knowledge Park Initiative; a £15 million scheme to foster interdisciplinary networks of participants in genetics based research in five UK clusters.

Again our research lends partial support to the local networking argument. Our evidence shows a great variety of organizations participating in both biotech clusters (supporting Powell et al. 1996). Moreover, networks or prior ties, not distant or unsocialized 'market' relationships underpin most activities in both clusters. Recent German cluster policies have undeniably induced scientists from most of the key biomedical research laboratories into commercialization processes. We have also seen that in Cambridge, an area in which active recruitment-oriented policies have traditionally been weaker, important institutions such as the Sanger Institute and the European Bioinformatics Institute (EBI) have not actively participated in the commercial cluster. Policies such as the Genetic Knowledge Park Initiative could usefully develop networks of relationships that could strengthen their role in commercializing Cambridge area science. However, at least among firms and their scientific collaborators and employees there seems to be little evidence for problems in local network formation.

Furthermore, the geographic dispersion of the participants in the Munich and particularly the Cambridge cluster again presents strong evidence against purely local network-driven cluster policies. Our evidence emphatically demonstrates that a core strength of Cambridge as a cluster is its ability to bring in a broad range of firms, organizations and individu-

Table 12.8   Geographic location of last academic job

	Cambridge		Munich	
	Number	%	Number	%
Local	31	38	48	46
National	30	38	27	26
Foreign	19	24	29	28

als from outside Cambridge (see Table 12.8). Such activity is less pronounced in Munich possibly leading to the cluster's poorer performance.

Cambridge is a magnet for commercial activities because of its strength as a general marketplace for biomedical research. Networks, or ties, must be important as sources of contact for individuals, collaborators or firms entering into this marketplace. But these networks most plausibly have their origin in other activities that fall outside the scope of traditional cluster policies – for example long-standing scientific research communities, or contacts forged through previous affiliations in firms or laboratories. It is this ability to act as a magnet, to draw outside actors in, that drives the agglomerations and related network effects that are widely seen as driving cluster growth (Porter 1998; Fujita et al. 1999). Cluster policies aimed at facilitating networks of actors that can exploit local tacit knowledge links may be misdirected. Non-local actors will not benefit directly from local cluster policies aimed at strengthening existing local ties.

A related problem with network-oriented cluster policy is that it assumes rather than demonstrates, that information barriers or other problems prevent fruitful relationships from emerging. Our evidence on collaborations and SAB memberships for both clusters, however, shows that dozens of laboratories and individual scientists have found commercial partners. Moreover, when there is evidence that important institutions are not participating in commercialization activities, it would be wrong to automatically assume that networking failures are the cause. For example, while it is clearly the case that the Sanger Institute and the EBI in Cambridge are not engaged in the Cambridge biotech cluster, this lack of participation can be traced to policies explicitly adopted against actively collaborating with industry. Professor John Sulston, the recently retired long-term director of the Sanger Institute, has been an active proponent of a pure diffusion model of scientific research (see Sulston and Ferry 2002). Substantial Wellcome Trust funding for the Institute's Human Genome Project research was justified in part to compete with Celera's private sequencing initiative (see Sulston and Ferry 2002). The EBI has been an active participant in the development of 'open source' bioinformatics software, again precluding strong commercial ties of the more traditional formal type, in favour of more informal networks of distant collaboration. Thus direct policy, not a lack of network opportunity, has driven this outcome.

One aspect of the marketplace we have not considered is the connection between scientific ideas and appropriate forms of capital to support their early commercialization. Of the two approaches to regional policy, the argument for tacit knowledge is not relevant to financial resources. On the other hand, it may be the case that in this dimension, network failure is a critical explanation for the underperformance of biotech regions; anecdotally it is seen as the source of some market failure in the UK while in Germany and Munich in particular, the venture capital subsidies described earlier have been an explicit policy attempt to overcome the lack of financial resources in the marketplace. Nevertheless, in line with our argument about the non-local quality of local clusters, we believe that the incorporation of financing from geographically dispersed sources is critical to cluster dynamics and is therefore often outside the scope of local networking policies.

7.   CREATING MORE EFFECTIVE POLICIES

A cluster analysis oriented around a marketplace metaphor emphasizes the development of rules, norms or frameworks governing the market. We also stress the importance of ensuring that non-market actors, particularly universities in the biotech case, have sufficient resources to play a strong role within the marketplace. Within this perspective, policies are useful to the extent that they present stable incentives to promote and sustain interaction within the marketplace and maximize the different resources on which the marketplace operates. Incentives structuring marketplace architectures can be shaped by public policy. We adopt an institutional approach to examine such policy instruments, first by examining the importance of national institutional frameworks emphasized by 'varieties of capitalism' research, then examining more sector-specific policies.

7.1   National Policies – Varieties of Capitalism

Marketplaces as a locus of economic activity are developed at a fundamental level within the context of a country's core economic institutions (Hall and Soskice 2001). Recent comparative research has emphasized that different models of capitalism exist, favouring different patterns of economic coordination. National economic institutions help frame the incentive structures facing marketplace participants and, through doing so, present an overarching framework through which sector-specific support policies can be both analysed and designed.

National institutional factors can help explain the success of several UK regions in developing biotech clusters. Comparative research on varieties of capitalism has emphasized that the USA and the UK are 'most similar' cases in developing 'liberal market economies' that privilege marketplace solutions in fostering economic activity (Hall and Soskice 2001). This research has linked the existence of deregulated labour markets and financial systems oriented around capital markets to the rapid and flexible steering of resources towards high-tech industries. The ability of many UK biotech companies to develop in the early 1990s with virtually no sector-specific government support validates the claim that national institutional variables influence patterns of industry specialization.

Varieties of capitalism research is also a good starting-point for an analysis of the German experience. A core assertion of this line of research is that 'coordinated market economies' such as Germany while, on the one hand, conducive to long-term financial and human resource commitments that may be ideal for securing investment in engineering, machine tools, and other industries with cumulative technological tracks (see Casper and Soskice forthcoming), on the other are seen as inappropriate for harnessing risky investments in less stable emerging technologies. Problems include the long-term employment practices at most firms that hinder the development of flexible labour markets that nascent firms could draw on to quickly develop, and at times shed, capabilities and the lack of capital markets capable of supporting venture capital-backed investment strategies (see Casper 2000).

To be effective, cluster policies should work within institutional constraints and opportunities generated by national institutional frameworks. Policies must be 'incentive compatible' with a country's core economic institutions. Our analysis of marketplace development in Cambridge is predicated on 'fine-tuning' marketplace rules within an appropriate national institutional architecture. We discuss such sector-specific policies below. In Germany, on the other hand, the durability, and in some cases even existence, of these core marketplace institutions is questionable. Much more extensive, heavy-handed policies may be warranted in Germany to construct fundamental marketplace institutions. However, it is not clear the extent to which sector-based public policy can provide an effective substitute for inappropriate national economic institutions.

Policy in Germany can be analysed as attempting to overcome marketplace challenges. Specifically, government-sponsored technology transfer and venture capital programmes are mechanisms to compensate for the relative inability of actors within the country to incorporate these resources into the marketplace. Creating a technology transfer infrastructure and jump-starting venture capital became the core goals of German policy towards biotechnology. As part of a 'BioRegio' programme aimed at university-based cluster development, German government officials developed schemes to recruit scientists into industry through subsidized consulting, business plan development, patenting and angel financing for nascent biotech firms. Once firms were established, they could find homes, often rent-free, in government-sponsored incubator laboratories and technology parks near universities. Financing was made available through extensive 'public venture capital', which provided government matching funds for any private investments funding a firm raised (often from both the federal and state governments) (see Adelberger 2000; Casper and Kettler 2001 for overviews). Industry surveys claim that over 400 firms were started through these schemes (Ernst & Young 2001), though far fewer (probably less than a hundred) obtained critical mass as ongoing organizations.

Have these policies created a marketplace for biotechnology in university clusters such as Munich? In one sense they have. Most German clusters have been tightly orchestrated to establish a critical mass of companies. University actors appear to be well integrated into the process, seen in our data by the relatively high number of university spin-offs in Munich and the frequent involvement of university laboratories and personnel in their activities. However, other aspects of the policies are failing.

While policies have increased the resources available in the marketplace, German venture capital subsidies have not sparked a fundamental reorientation of German financial institutions towards a capital market-based system. Venture capital subsidies are still widely available and, though not discussed here, a venture capital industry does now exist, headquartered in the Munich area (see Adelberger 2000). However, the Neuer Markt, a technology-oriented stock market created in 1997 to accommodate initial public offerings by German biotech and Internet companies, crashed as part of the decline in Internet stocks, and was formally closed in January 2003. With the initial public offering (IPO) market in Germany inactive over the last two years (not unlike other markets around the world), venture capital is increasingly inaccessible for established companies in need of additional funding but without prospects for a short-term exit, with the absence of government subsidies for most follow-on investments creating an additional disincentive.

A second aspect of marketplace failure lies in Germany's labour markets. Specifically, one clear conclusion from the data on Munich is that very few firms have been able to recruit scientists with industry experience. Rather, the engine driving most Munich-based firms is large groups of scientists that worked previously in academic laboratories. This has given these firms the ability to forge links with basic research institutes that are at a par with those established in Cambridge. However, this marketplace failure may have long-term consequences in the degree to which these firms can translate research prowess into commercial success. Such a lack of policy instruments capable of generating a flexible labour market for scientists, and particularly scientists with previous experience in industry, suggests another overall policy failure. As discussed earlier, the success of most new technology clusters is driven by the establishment of a plethora of 'weak ties' (Grannovetter 1973) across a dense network of individuals and organizations. This appears to be particularly relevant in developing dense, flexible labour markets capable of sustaining radically innovative, but risky, companies. We were surprised to discover the importance of strong ties between founder laboratories and biotech firms within the Munich cluster. Our ongoing research on the German industry has revealed a similar pattern in the other three large clusters based in Heidelberg, Cologne, and Berlin. However, founder laboratory ties alone cannot supplant the development of a labour market for experienced biotech scientists.

Over time, the German biotech industry could mature and develop more sustainable marketplaces for people, money and ideas. Furthermore, international surveys consistently point out that a small number of German firms, predominately located in Munich, appear to be world class (see e.g. Ernst & Young 2001). But simultaneously, there exists a risk that the collapse of several hundred small firms could wipe out the entire industry. One present danger in Germany is that, despite access to strong scientific resources, a multitude of failed firms could sour the investment climate for years to come, while affirming to promise scientists and managers that 'safe' careers in large companies may be preferable to a life in start-ups. It does not appear that German policies have created viable substitutes for key national institutional frameworks.

7.2 Sector-specific Policies: University–industry Relationships

Our research suggests that the establishment of effective marketplaces for ideas relies on more than supportive national frameworks. Though the Cambridge cluster is clearly performing well, there is little doubt that its performance could be improved. For example, the Boston area cluster located in the Kendall Square area surrounding MIT is clearly outperforming not just Cambridge, but the entire UK biotech industry. Recent comparisons suggest that the Boston area and the UK as a whole receive roughly similar public support for biomedical research. However, the Boston cluster is home to more public companies, has higher rates of successful commercialization, gauged through comparing the number of therapeutic compounds in various stages of development, and employs a third more people (Casper and Murray 2003). While the variation in national institutional frameworks emphasized by varieties of capitalism research clearly helps account for the performance differences across the Cambridge and Munich clusters, more sector-specific policies require a deeper understanding of how these factors play out in the particular sectoral marketplace and an understanding of the micro-dynamics of the marketplace. Sector-specific regulation and policy may strongly affect the architecture of markets through creating what Mowery and Nelson (1999) call a 'sector support system'.

In particular we note that within biotechnology the university–firm boundary is of particular significance in understanding market construction in biotechnology. Incentives structuring marketplace interaction are particularly dependent on the orientation of rules structuring university–industry relationships, as well as the resources available to university technology licensing offices. Moreover, experience in the UK and Germany stands in contrast to US settings such as Boston and the Bay Area. The discrepancy in performance between Cambridge and the Boston area may be partly explained through differences in these policies.

To demonstrate the importance of policy in this area to the effectiveness of biotech clusters, we briefly contrast the organization of Technology Licensing Office (TLO) policies surrounding the University of Cambridge and MIT in the Boston area. The efficiency of the Cambridge biotech marketplaces has been impeded by the high levels of ambiguity surrounding micro-level rules and regulations structuring the transfer of ideas from universities to laboratories. We were surprised to discover how few prominent biomedical scientists working within Cambridge University have become serial founders. During interviews with several founders, a common issue raised was the complexity and relative lack of transparency of rules governing technology licensing from Cambridge University. Clear 'standard operating procedures' governing the process are only slowly emerging in Cambridge.

Rules governing the funding relationships between laboratories and firms are one area where frameworks differ across the USA and the UK. A key difference between the University of Cambridge and MIT is that within MIT university-based scientific founders of companies must make a clear choice between taking an equity position and receiving ongoing research funding from the company. This tends to demarcate boundaries between laboratories and firms in a clearer way than seen in Cambridge, in which founders may accept both funding and equity. Inappropriate rules governing funding relationships between laboratories and firms have also reduced the incentives to start many firms. Rather, most founder scientists have established relatively dense, long-term relationships with a single firm spun-off from their laboratory. In contrast MIT is characterized by serial founders such as Professor Robert Langer who has founded more than six firms, and similarly Harvard's faculty include individuals like George Whitesides and Stuart Schreiber who have both founded firms and serve as advisory board members.

A second important difference in frameworks governing relationships between laboratories and firms is that in most US universities academic researchers are prevented from taking line positions in companies so long as they maintain their university jobs. One frequent practice in the UK, seen for example in prominent firms such as Kudos in Cambridge or Cyclacel in Dundee, is that scientific founders have become Chief Scientific Officers of a firm while retaining professorships within university departments. Firms in some respects have become extensions of basic research laboratories. While this may lend prestige to particular firms and over time channel tacit intellectual property to the firm from the laboratory, a lack of clear frameworks structuring relationships between laboratories and firms can also impede the success of firms. It is not immediately clear why a brilliant academic research scientist should necessarily be a successful line manager, particularly if dividing time between two jobs and early US experience underscores the difficulty in this approach. Furthermore, as a firm begins to mature, commercial priorities may diverge from a founder's basic research stream, leading to conflicts between the two agendas. On the other hand, the success of the US biotech industry indicates that academic scientists may serve well as scientific advisors on SABs.

Micro-level policies in these areas could significantly improve the efficiency of biomedical marketplaces in the UK. From the point of view of the broader cluster (and, universities that have equity stakes in firms), the lack of serial spin-offs from senior scientists reduces the number of firms in the area. In an industry in which only a small percentage of early-stage firms succeed, this diminishes the ability of both venture capitalists and universities to profit from portfolio strategies. Having fewer firms in an area also diminishes labour market externalities that we have linked to the innovative capacity of firms. Cluster policy in this area is clearly complex and technical. However, the difference in performance across Boston and Cambridge might be significantly explained by more effective regulatory structures governing the market for university spin-offs in Boston-area universities.

8.   CONCLUSION

Biotech clusters are initiated by scientific relationships and fostered by marketplace relationships that combine scientific, human and financial resources. In general, such relationships cannot be easily massaged through well-meaning governmental policy. Perhaps the most difficult to establish are the initial relationships between the scientific community and the marketplace itself. Our research suggests that there are three common features of these scientific linkages underpinning biotechnology clusters: the diversity of actors involved, the breadth of collaborative activities and the wide geographic scope of their location. The critical breadth and lack of the previously assumed 'localness' serves to hinder the effectiveness of local policies designed to orchestrate the construction of biotech regions through a focus on tacit knowledge and local networks. Instead, we have argued that biotech clusters should be understood as marketplaces for ideas.

The marketplace metaphor as one that is less geographically bounded may lead to a different policy agenda in contrast to much existing research on clusters. Clusters of the scale of Cambridge or Munich are too small to support purely local cluster development. The success of biotech clusters in Cambridge, and to a lesser degree Munich, may be measured by their ability to attract large numbers of firms, academic institutions and individuals from outside the area to participate in the local cluster, thus expanding its reach. The transplantation of US networking metaphors to the small European clusters may lead to hasty policy conclusions.

The marketplace metaphor, with local firms operating in a national and international context, may lead to a useful policy agenda that could lead to the improved performance of Europe's biotech clusters. Our aim is not to suggest that governments should take a 'hands-off' approach to marketplace formation. Rather, we have argued that the role of policy should be to sponsor the development of norms and frameworks governing and sustaining the market. Within the UK, rather narrow policies, focused in particular on the rules and resources surrounding technology licensing offices, might dramatically improve the efficiency of university-based clusters. When, on the other hand, national economic institutions are misaligned with the type of marketplace activities needed within a particular sector, as is the case in Germany, there may be a role for the introduction of stronger policies. German policy has created a substantial technology transfer infrastructure virtually from scratch. However, failures in the areas of entrepreneurial finance and the lack of effective policy instruments in the area of labour markets may demarcate limits in the ability of governments to construct new biotechnology marketplaces.

NOTES

* The authors would like to acknowledge the generous support for this research from the Cambridge MIT Institute and the National Competitiveness Network.
1. Matching for the level of basic research funding the appropriate scale of comparison is the entire UK biotech cluster since the UK's funding of basic research and that of the Boston area are comparable at around US$1.5 billion in 2001.
2. Note that in this chapter we pay only limited attention to issues relating to financial capital as this is a large topic and one that is extensively covered by others.
3. This rating system is applied to all UK universities that receive public funding for research and education. It is a department by department ranking of the quality of research and teaching which is used to determine subsequent allocations of funding. The highest rating is a five star. In the life sciences, relevant departments include chemistry, biochemistry, biology and so on.
4. We define university spin-offs as those firms that have emerged from the university based largely on ideas developed within the context of the university. While in many cases such firms are the recipients of licences to university intellectual property, others are founded with the extensive involvement of university academics.
5. This methodology follows the work of Murray (2003), Zucker et al. (1998), Henderson and Cockburn (1994) and others in using publications to examine the networks of collaboration that are established by firms in the biotech and pharmaceutical sectors. While such measures may undercount the full scope of collaboration, when compared across firms, it becomes a valid means of examining variation in collaboration patterns.

REFERENCES

Adelberger, K.E. (2000), 'Semi-sovereign leadership? The state's role in German biotechnology and venture capital growth', *German Politics*, **9**, 103–122.

Almeida, P. and B. Kogut (1999), 'Localization of knowledge and the mobility of engineers in regional networks', *Management Science*, **45**, 905–17.

Arthur Andersen (1998) *Technology Transfer in the UK Life Sciences*, UK: Arthur Andersen/Garrets/Dundas & Wilson.

Audretsch, D.B and Stephan P.E. (1999), 'Knowledge spillovers in biotechnology: sources and incentives', *Journal of Evolutionary Economics*, **9**(1), 97–107.

Carlsson, B. (ed.) (2002), *Technological systems in bio industries: an international study*, Boston, MA: Kluwer Academic Publishers.

Casper, S. and F. Murray (2002), 'Marketplace for ideas: biotech', paper presented at the Cambridge–MIT Initiative Competitiveness Summit, London.

Casper, S. and F. Murray (2003), 'Scientific labor markets and the building of biotechnology firms: a comparative analysis', paper presented at the 2003 Academy of Management Conference, Seattle, USA.

Casper, S. (2000), 'Institutional adaptiveness, technology policy, and the diffusion of new business models: the case of German biotechnology', *Organization Studies*, **21**, 887–914.

Casper, S. and H. Kettler (2001), 'National institutional frameworks and the hybridization of entrepreneurial business models: the German and UK biotechnology sectors, *Industry and Innovation*, **8**.

Casper, S. and D. Soskice (forthcoming), 'Sectoral systems of innovation and varieties of capitalism: explaining the development of high technology entrepreneurship in Europe', in F. Malerbo (ed.), *Sectoral Systems of Innovation: Concepts,*

Issues and Analyses of Six Major Sectors in Europe, Cambridge University Press, Cambridge.

Cooke P. (1999), 'Biotechnology clusters in the UK: lessons from localisation in the commercialisation of science', Cardiff: Centre for Advanced Studies, Cardiff University.

Ernst & Young (2001), *European Life Sciences 2001*, London: Ernst & Young International.

Fligstein, N. (2001), *The Architecture of Markets: An Economic Sociology of Capitalist Societies*. Princeton, NJ: Princeton University Press.

Fujita, M., P. Krugman and A. Venables (1999), The Spatial Economy.

Granovetter, M. (1973), 'The strength of weak ties', *American Journal of Sociology*, **78**, 1360–80.

Gulati R. and M. Higgins (2002), 'When do ties matter? A contingent model of the implications of interorganizational partnerships for IPO and post-IPO success', *Strategic Managaement Journal*, **24**(2), 127–44.

Hall, P. and D. Soskice (2001), 'Introduction', in P. Hall and D. Soskice (eds) *Varieties of Capitalism*, Oxford: Oxford University Press.

Henderson, R. and I. Cockburn (1994), *Strategic Management Journal*, **15**, 63–84.

Kenney, M. (1986), *Biotechnology: the University–industrial Complex*, New Haven: Yale University Press.

Lerner, J. (1995), 'Venture capitalists and the oversight of private firms', *Journal of Finance*, **50**, 301–18.

Mowery, D. and R. Nelson (eds) (1999), *The Sources of Industrial Leadership*, Cambridge: Cambridge University Press.

Murray, F. (2002), 'Innovation as overlapping scientific and technological trajectories: exploring tissue engineering', *Research Policy*, **31**.

Murray, F. and S. Kaplan (2003), 'Evolution and evidence: the social construction of the discontinuity in biotechnology', Sloan School of Management working paper.

Porter, M. (1998), 'Clusters and the new economics of competition', *Harvard Business Review* (November–December), 77.

Powell. W., K. Koput and L. Smith-Doerr (1996), 'Inter-organizational collaboration and the locus of innovation: networks of learning in biotechnology', *Administrative Science Quarterly*, 41, 116–145.

Saxenian, A. (1994), *Regional Advantage*, Cambridge, MA: Harvard University Press.

Schitag Ernst & Young (1998a), *Germany's Biotechnology Takes Off in 1998*, Stuttgart: Schitag Ernst & Young.

Schitag Ernst & Young (2000), *German Biotechnology Report* (German language version), Stuttgart: Schitag Ernst & Young.

Senker, J. (1996), 'National systems of innovation, organisational learning and industrial biotechnology', *Technovation*, **16**(5), 219–29.

Sorensen, O. and T. Stuart (2001), 'Syndication networks and the spatial distribution of venture capital investments', *American Journal of Sociology*.

Sulston, J. and G. Ferry (2002), *The Common Thread: A Story of Science, Politics, Ethics and the Human Genome*, London: Bantam Press.

Watson, J. and F. Crick (1953), 'A Structure for Deoxyribose Nucleic Acid', *Nature*, 25 April, 171, 737–8.

Winter, S. (1987), 'Knowledge and competence as strategic assets', in D.J. Teece (ed), *The Competitive Challenge: Strategies for Industrial Innovation and Renewal*, Cambridge, MA: Ballinger.

Zucker, L., M. Darby and M. Brewer (1998), 'Intellectual human capital and the birth of U.S. biotechnology enterprises', *American Economic Review*, **88**(1) 290–306.

13. Creation and growth of high-tech SMEs: the role of the local environment

Corinne Autant-Bernard, Vincent Mangematin and Nadine Massard

1. INTRODUCTION

The aim of this chapter is to understand the determinants of the creation and growth of high-tech small to medium-sized enterprises (SMEs) at a regional level as industry matures. The biotech sector is examined to understand the role of local environment in the economic valorization[1] of scientific results. We specifically investigate the influence of two determinants of the creation and growth of biotech SMEs: a scientific and technical profile of a region and the size of the local market for biotech products and services.

In the USA, the biotech sector has developed around poles of scientific excellence (Audretsch and Stephan 1996; Zucker et al. 1995; Zucker et al. 1997). The movement of researchers from academic laboratories towards the private sector has provided a vehicle for the diffusion of knowledge and has been a powerful force behind the creation of start-ups (Almeida and Kogut 1999; Catherine et al. 2004). Economic and tax policies to support entrepreneurs in the creation of high-tech firms have had impressive results in Quebec, now North America's third region with regard to the creation of biotech firms (Niosi 2000). In Europe, different public policies have been implemented and the results in terms of creation and growth of biotech SMEs show a contrasting picture. The UK, Germany and Sweden are performing better than France and southern Europe (Senker et al. 2001).

In France, national and regional policies are combined to support the economic development of high-tech sectors. National policy-makers define economic and tax policy for all entrepreneurs and not specifically for high-tech firms, while the regions have various incentives (incubators, science parks, regional funding and so on) to promote local economic development and the emergence of new technologies, especially biotech and information technologies. Two effects are underlined: economic wealth and creation of employment through the development of new economic activities and cluster effects which increase the attractiveness of the region for firm locations because of expected spillovers and positive externalities (Cooke 2001). Regions are thus competing to attract start-ups, with a view to generating increasing returns to location (cluster effects) in their territory (Krugman 1991). Until the 1990s the authorities facilitated technology transfers between academic laboratories and industry via large firms. However, the development of 'biotechnopoles' around large firms was only moderately successful (Genet 1997). Since the early 1990s the number of biotech SMEs has grown rapidly and firms' location seems to depend on multiple factors: scientific excellence, regional specialization in scientific sub-fields, regional industrial specialization, public incentives but also the size of the local market as the biotech sector matures.

During the last 10 years, public policy-makers have encouraged economic development through firm start-ups. The determinants of creation and growth of biotech SMEs since the 1990s are thus examined to understand the role of local environment in the economic value attached to scientific results. We explore relations between the scientific and technical profiles of a region, the size of the local market for biotech products and services, and economic development viewed as start-up creation and growth. Thus, the definition of local environment is not limited to the local technological externalities and public research–industry links, as are most analyses of biotech locations in the USA (Zucker et al. 2002). Apart from academic research, the size of the local market is also taken into account as a potential factor in the development of high-tech industries. Set between the national and local (*departement*) levels, the regional level is considered as relevant in identifying the effects of location. Moreover, statistical series on research and development (R&D) firms and public sector research organizations are available at the level of aggregation.

As all public and private actors recognize the industrial and commercial potential of the life sciences, location of firms in any region seems to be one of the most important issues in regional policy. Either the dynamics of self-reinforcement in highly concentrated scientific and technological poles dominate processes of location, in which case the role of public intervention is severely limited; or the weight of determinants such as demand links with related industries or the challenges of commercialization in the recent development of these activities paves the way towards a new dynamics of location, sources of original regional development trajectories.

In order to understand the determinants of the local dynamics of firm creation and growth, Section 2 considers the main elements in the theoretical literature on which we base our analysis of geographic determinants in the creation and growth of biotech SMEs. It concludes with a number of testable hypotheses. Section 3 presents the data and methods used in our econometric estimations. It also provides a statistical image of scientific, technological and industrial profiles of French regions in the biotech domain. The results of estimations are interpreted in Section 4, and conclusions on regional public policy concerning innovation are presented in Section 5.

2.   DETERMINANTS OF THE CREATION AND GROWTH OF BIOTECH START-UPS

In France, most biotech firms are young and have an average of about 50 to 60 employees (Lemarie et al. 2001). The creation of SMEs and their growth is thus one of the best indicators for measuring the development potential of a geographic area in biotechnology. Age and size make it possible to identify a region's capacities both to attract investments in biotechnology and to offer a stimulating environment for firms established in the area to grow. After an exploration phase in which scientific progresses have been screened to identify potential innovations, the biotech sector enters into a maturing phase in which innovations are developed and marketed (Afuah and Utterback 1997; Feldman and Francis 2001; Mangematin and Nesta 2002). In order to understand spatial trends in highly innovative science-based sector, three groups of studies can be mobilized. The first includes work on the 'geography of innovation' which seeks to explain the concentration of innovative activities by measuring the spatial dimension of knowledge externalities. Here the focus is on the advantages of being located near sources of knowledge. The second group illuminates the role of local environment in firms' capacities to absorb knowledge. Studies in this group aim to identify local organizational profiles best suited to benefit from the externalities produced. The third group focused less on innovation and technological externalities and identifies broader determinants of the location of industrial activities borrowed from economic geography: increasing returns, pecuniary externalities and market structures.

2.1   Proximity to Sources of Knowledge

Even if spatial economics focused on the role of technological externalities on spatial concentration of activities for a long time, empirical tests appeared only recently. Empirical analyses of the mechanisms of spatial concentration of innovative activities often proves to be difficult at an econometric level (Autant-Bernard and Massard 1999), and suggests it is risky to infer questions of localization from indicators such as patents, the number of innovations or even relations between geographic areas and R&D expenditures (Anselin et al. 1997).[2] We can nevertheless identify a number of conclusions of such research:

- Innovation in a given region is closely related to public and private sector research expenditure in that region (Feldman 1994), including in sectors that are not research-intensive (Mangematin and Mandran, 2001).
- Innovation in a given region is not only related to public and private R&D expenditure but is also related to the region's entire technology transfer infrastructure (presence of technological centres, technology transfer agencies etc.) (Feldman 1994). Thus, the presence of complementary activities generates more spillovers and reduces costs and risks related to firms' innovation.
- There are no eviction effects between public and private R&D expenditure, that is public and private R&D expenditure seems to be more complementary than substitutable. They enhance each other to create areas of expertise (Jaffe et al. 1993). Although academic research is often a major source of local knowledge externalities (especially in the USA), universities are not the only generators of externalities, and externalities very often spread within industries.

The last three points emphasize the role of research infrastructure on the firm creation. The regional research infrastructure includes public sector research (universities) as well as private firm research laboratories.

H1: Biotech SMEs are set up in centres of excellence. They are localized nearby public academic research.

H2: Biotech SMEs are set up in the regional neighbourhood of major firms with substantial research potential, which are expected to produce knowledge externalities.

2.2 Local Environment and Absorptive Capacity

Traditional explanations in terms of spillovers remain unsatisfactory. Breschi and Lissoni (2001) provide a critical reassessment of the theoretical conception of localized knowledge spillovers and econometric literature based on the knowledge production function. They consider that the insistence on knowledge as a pure local public good and the concept of

knowledge spillovers as the major analytical category to explain the localized nature of innovative activities has diverted research away from examining other mechanisms governing knowledge flows. In fact, economic
theory remains indecisive as to the ability of SMEs to capture externalities.
Whereas conceptual studies emphasize the concept of absorptive capacity
based on local interactions, empirical work describes a correlation between
the intensity of presence of university research in a given geographic area
and the propensity to innovate, irrespective of the sector concerned. It
sheds little light on the verified presence of local externality effects.
Audretsch and Feldman (1996a, 1996b) and Audretsch and Stephan (1996)
made a specific analysis of high-tech sectors and showed that, in sectors
where innovation is based on science, geographic links are weaker. Seventy
per cent of formal relations between biotech firms and universities are not
based on geographic proximity. Mangematin and Nesta (1999) showed that
the higher the absorptive capacity of the firm, the fewer the university–industry relationships based on local interactions. Studies of relations
between biotech firms and universities are not, however, transposable to
choice of location when firms set up. In this phase, relations between start-
ups and firms in their 'natural' network are fundamental and location is
often in the entrepreneur's 'natural' environment (Liebeskind et al. 1996;
Steier and Greenwood 2000). On an infra-regional scale, Acs and Preston
(1997) highlight the effects of local spillovers owing to the presence of a
large university and reputable research departments.

It seems, therefore, that the results of the analysis are very different at the
time of start up, when the survival and development of the firm depends
on the founder's close network of relations (Baum et al. 2000), and later,
when the firm is established and builds sound relations in the same scientific, productive and commercial network (Pisano 1991). One of the main
explanations for these differences relates to firms' absorptive capacities.
According to Cohen and Levinthal (1990), in order to capture technological externalities, firms need to have their own adequate knowledge and
competences (e.g. a high level of research, a variety of available competences). Firms therefore try to develop their absorptive capacities, in other
words, to master specific knowledge in order to be able to identify and
exploit new knowledge available in their environment. But, as Autant-
Bernard (2001) notes, it is likely that the level of research and its degree of
diversity affect not only the level of externalities captured in a local environment but also their geographic origin. The fact of having a high level of
varied in-house competences seems decisive in the ability to take advantage
of distant sources of knowledge. Conversely, highly specialized emergent
firms with weak internal resources need a healthy local environment to help
them to take advantage of neighbouring sources of externalities. Although

the question of absorptive capacities defined at the level of a geographic zone is still seldom addressed in the literature, there are some bases on which to found hypotheses, including not only traditional determinants such as level of research done locally and level of knowledge produced (publications, patents etc.), but also conditions of transmission of knowledge. Proximity to sources of knowledge is not a favourable factor as such; it has to be associated with the establishment of real links between the institutions through which the knowledge can flow. A high level of collaboration between researchers enhances firms' absorptive capacities. The results of Zucker et al. (1998) and Cockburn and Henderson (1998) suggest that firms benefit from public research only if they collaborate with a university researcher. The simple fact of being located near a university is not enough.

Such relations nevertheless often imply a common language and codes. The existence of a degree of specialization is thus a condition for the diffusion of knowledge. Yet, if use of a common internal language is all that is emphasized, the firm could fail to capture external knowledge because it is too specialized. We have here the well-known debate on the comparative advantages of diversity or specialization, found in studies of the geography of innovation (Feldman and Audretsch 1999). Specialization is necessary to master a set of common knowledge, and is particularly advantageous in the start-up phase. By contrast, diversity enables a firm to be more sensitive to a wider range of information. Based on a study of large pharmaceuticals firms during the 1990s, Arora and Gambardella (1994a) distinguish between specialized competence that makes it possible to evaluate and run a specific research programme in a narrow domain, and the architectural competences used to combine different domains of specialized competences. They show that in the short term, specialized competences are correlated with a better innovative performance in the firm, whereas in the medium term, the ratio is inverted and architectural competences predominate in firms with the best performance. Moreover, the importance of innovation through a recombination of knowledge in high-tech sectors, and the difficulties of coordination and transfer between individuals with different specialities, explain the advantage of proximity. Firms therefore try to set up in areas with varied competences. Feldman (1999) confirms this point by showing how the presence of diverse but complementary knowledge is capable of creating economically beneficial local externalities. In this sense, the diversity of knowledge should facilitate innovation and the growth of existing SMEs by allowing individuals to establish new connections between knowledge.

This work enables us to formulate the following hypotheses:

H3: A region's capacity to promote scientific collaboration helps to develop local SMEs' absorptive capacities. Local interactions are particularly enriching in the start-up phase.

H4: A region's specialization in a specific area of knowledge (scientific specialization) can enhance the absorptive capacities of start-ups and thus facilitate their establishment.

2.3   Local Industrial Environment and Agglomeration Dynamics

It is generally considered that biotechnology is an activity with a 'high potential for concentration' (Steinle and Schiele 2002). Yet, even if the link with basic research and the role of 'star scientists' is unquestionable in the early development phases of biotech firms (especially in the USA), the characteristics of their development in France and, above all, the recent shift to a new phase of maturation, profoundly modify this initial model. Links with basic research and perhaps even with the research capacities of large groups (Cooke 2002) are losing weight as relations with customers and users become more important. Certain factors explaining the location of industrial activities, inspired by economic geography, can prove decisive. Based on the existence of increasing returns, the size of the local market is a key element of attraction and of dynamic self-reinforcement. In the biotech business this market is measured not so much in terms of size of regional population as in terms of presence of user industries, since biotech firms are primarily suppliers of other firms in the life sciences. The main user industry of biotechnology are firms, hospital and university laboratories involved in life sciences, that is agriculture, agro-food, pharmacy, medicine, veterinary and biotechnology (human, vegetal and animal health and the production of research tools and devices). Feldman and Ronzio (2001) suggest that service companies in biotechnology set up close to major production centres rather than to universities, even if in some areas they are close to both.

   The size of the market is the first dimension. However, the market structure is not neutral. In their explanation of agglomeration processes, economic geography models most often emphasize the role of diversity, whether it concerns consumer preferences (Krugman 1991) or available intermediate goods (Krugman and Venables 1995). The combination of economic geography and endogenous growth (Martin and Ottaviano 1999) presents a far more complex picture of the role of diversity, in which the impact of the effects of diversity depends on the region's specialization in the production of technological goods (Massard and Riou 2001). Spillovers from related industries increase the attractiveness of a region for firm creation. Contributions focused on biotechnologies highlight the advantages of specialization during the set-up phase of SMEs, based on the advantages of specialization in terms of identification of a region in relation to its competences, available skilled manpower or outlets for specialized products/services.

Finally, the work of Krugman (1991) also highlights the dispersing effects of competition. The presence of too many firms in an area strengthens competitive pressure, both in the search for consumers and in the job market. This pressure is a dispersing force because in certain conditions it can discourage firms from setting up in the area. Even if the market of biotech is hard to describe in a statistical way, we consider that biotechnology firms are providing the whole life science industry (pharmacy, agriculture, agro-food, environment) with new products or services as well as research organizations involved in life sciences and biotechnology. Thus, markets of biotechnologies include products and services, which focus industry as well as academic institutions.

Based on these theories concerning determinants of industrial agglomeration, the following hypotheses can be put forward:

H5: The larger the market, the greater the number of start-ups.

H6: Specialization of a region in a sector related to the life sciences (biomedical, agro-food etc.) stimulates the creation of SMEs.

H7: The intensity of competition in related industries reduces firms' propensity to create.

The creation of wealth in a region can be attributed primarily to firms. Hence, the location of firms and especially high-tech firms likely to create jobs and wealth in the future is a major challenge for the regions. Yet start-up is just a stage in the life of a firm, which has to grow. Do the regions play a key part in the start-up phase? Do they impact on the subsequent growth of firms?

In the explanation of business creation we find a mix of attraction effects (choice of location) and positive effects of local environment in the emergent sector. In the explanation of growth, attraction effects disappear (due to the irreversibility of location created when the firm is set up), and other local development factors appear which can determine start-ups' abilities to survive and grow. The difficulties encountered in reaching a clear conclusion in the debate on specialization and diversity can perhaps be explained by the confusion of its mechanisms in most empirical studies carried out to date.

To analyse the role of local environment in the economic valorization of scientific results, and to define public policy to support business, it is necessary to analyse not only the creation phase but also determinants of growth. Thus, hypothesis H7 is worth testing.

H8: The characteristics of the regional environment influence SMEs' growth potential.

3. DATA AND METHOD

To analyse the determinants of creation and growth of biotech SMEs, several databases have to be connected to perform econometric analysis.

3.1 Econometric Models

The empirical tests that we propose are based on two types of model: a regional SME creation model and a biotech SME growth model. The former explains a region's capacity to stimulate the creation of biotech SMES in relation to four types of regional characteristic:

1. characteristics of public sector research;
2. characteristics of private sector research which give an overall picture of research in biotech firms;
3. characteristics of the market, describing the region's profile in related industries;
4. organizational characteristics, in the scientific area as well as market area.

The creation model is as follows:

$$NBCREA_{it} = \alpha + \beta_1\,RPUB_{it} + \beta_2\,RPRIV_{it} + \beta_3\,MARCH_{it} + \beta_4\,ORG_{it} + \varepsilon_{it}$$
(13.1)

with:

$NBCREA_{it}$ = the number of new SMEs set up in region i during year t,
$RPUB_{it}$ = the matrix of variables characterizing the public sector research potential in region i in year t,
$RPRIV_{it}$ = the matrix of variables characterizing the private sector research potential in region i in year t,
$MARCH_{it}$ = the matrix of variables characterizing the market in related industries in region i in year t,

ORG_{it} = the matrix of variables characterizing the organization of research (openness, collaboration, specialization etc.) in region i in year t,
E_{it} = a random term.

The second model considers that the level of growth of a biotech SME is explained both by a set of variables characterizing the firm itself and by the characteristics of its regional environment. It thus takes the following form:

$$MAGR_{ki} = \alpha + \beta_1 \, RPUB_i + \beta_2 \, RPRIV_i + \beta_3 \, MARCH_i + \beta_4 \, ORG_i + \beta_5 \, ENTRE_k + u_{it} \tag{13.2}$$

with:

$MAGR_{ki}$ = mean annual growth rate of firm k situated in region i,
$RPUB_i$, $RPRIV_i$, $MARCH_i$, ORG_i = the variables of the characteristics of the region to which k belongs, calculated either for the initial year or as a mean of the period under consideration,
$ENTRE_k$ = all the variables characterizing firm k,
u_{it} = a random term.

3.2 Data and Variables

Five databases were mobilized for this study. Table 13.1 presents the different variables used as well as the main sources of data.

- Data on SMEs were drawn from the database created by UMR GAEL[3] on the basis of a survey carried out in 1999 by the French research and technology ministry (website www.biotech.education.fr). The data were updated and enriched by the UMR GAEL team in Grenoble (http://www.grenoble.inra.fr). Two hundred and fifty independent SMEs created between 1960 and 2002 are recorded in the database. We extracted from this base the number of start-ups created every year in each French region (except Corsica) between 1993 and 1999 (NBCREA variable) that is to say 165 creations. We also calculated the mean annual growth rate in terms of employee numbers (MAGR) of SMEs existing during the period 1996–99. One hundred and twenty-two firms are present throughout the period and are correctly reported. The variables characterizing firms are described in Table 13.1.
- Patent data concerning patents filed by SMEs were supplied by the INPI[4] (French national patent office), while those concerning

Table 13.1 List of variables used

Type	Name	Definition
Biotech SMEs	MAGR	Mean annual growth rate of employee numbers of SMEs (between 1996 and 1999)
	SME	Number of SMEs in the region (in 1996)
	TO	Annual turnover of SMEs (1996–99)
	EFF	Annual employee numbers of SMEs (1996–99)
	PATENT	Annual number of patents registered by SME (1996–99)
	NBCREA	Annual number of SMEs created (1993–99)
Public research in life sciences	PUB	Annual number of publications
	PhDS	Annual number of life science PhDs. Proxy of public research inputs in biotechnology per region
	SPE_PUB	Index of specialization of publications (variance of the revealed technological advantage – RTA, calculated on the basis of the different scientific disciplines)
Private sector biotech research	RE_BIO	Number of researchers in firms doing biotech research
	PART_DERD_BIO	Private R&D expenditure by firms doing biotech research, in relation to all R&D in the region
	CONCENT_BIO	Index of geographic concentration of R&D (Herfindhal index on departmental R&D expenditure)
	SPE_BIO	Index of sectoral specialization of R&D (RTA variance calculated on the basis of sectoral DERD)
	PATENT_BIO	Number of biotech patents of which at least one inventor lives in the region

Category	Variable	Description
Biotech market: related industries	INV_RI	Investments of related industries
	NB_ETS_RI	Number of firms in industries related to the biotech sector
	CONCURR_RI	Average size of establishments in related industries
	CONCENT_RI	Index of geographic concentration of the biotech market (Herfindhal index on employee numbers per department in related industries)
	SPE_RI	Specialization index of the biotech market (RTA variance calculated on the basis of employee numbers per sector in related industries)
	PART_EFF_RI	Share of employee numbers in related industries in total number of employees in industry
Organization of biotech activities	TXCOPUB	Share of co-publications in the total number of life science publications
	TXCOPEXT	Share of publications co-authored with researchers from outside the region in the total number of publications
	TXCOPINT	Share of publications co-authored with foreign researchers in the total number of publications
	COPUBPP	Number of publications co-authored by public- and private-sector researchers
Overall regional characteristics	PUBTOT	Number of publications in all disciplines put together
	DERDTOT	Domestic R&D expenditure in all sectors put together
	INVTOT	Industrial investments in all sectors put together
	PATENTREG	Patents filled by actors within the region

numbers of patents filed in each region were drawn from the OST/OEB database of patents aggregated by geographic area. The inventor's address was used for location (PATENTREG variable).

- Public research is evaluated in two ways, the first of which is inputs. The only available reliable variable broken down into both region and year is the number of biotech PhDs, provided by a survey by the Ministry of Research (DGRT). This variable gives a good approximation of existing research capacities (PhD variable). The second indication is output, estimated in terms of publications signed by authors belonging to institutions in the region (PUB variable). This measurement represents the role of public research in the production and diffusion of new knowledge. Publication data are drawn from the Science Citation Index (SCI) and the Biotech Citation Index (BCI) and are analysed in terms of institutions' addresses, since the final figures relate to participation in articles by institutions located in a particular region.

- Data drawn from the research ministry's annual survey are used in the construction of variables concerning the analysis of the regions' private sector research potential (national expenditure on R&D and employee numbers). The selection of biotechnologies was based on the biotech code that appeared in the 1999 and 2000 surveys.

- Finally, data concerning the customer industries of biotech firms, primarily in sectors related to the life sciences (human, animal and plant health, environment, agriculture and agro-food) that we term related industries (RI) are drawn from the SESSI[5] CDRom, 'SESSI-region', on which data are regionalized on the basis of the EAE[6], the annual survey on enterprises.

Table 13.1 also presents a set of organization variables calculated for each domain: scientific (publications), technological (patents and private R&D expenditure) and industrial (employee numbers in related industries). Internal concentration within a region is measured in terms of Herfindhal indexes on the basis of distribution between departments and regions[7]. Sectoral specialization is measured by the classic 'revealed technological advantage' (RTA) indicator (Frost 2001; Nesta 2001). If P_{if} is the activity of region f in domain i, for a given year, the region's technological advantage in relation to other regions in domain i is defined by:

$$RTA_{if} = \frac{P_{if} \big/ \sum_i P_{if}}{\sum_f P_{if} \big/ \sum_{if} P_{if}}$$

The measurement of specialization in a domain is defined as the ratio of two proportions. The first is internal, in the region (the numerator) while the second is relative to all actors active in the biotech domain at national level (the denominator). For domain *i*, if the proportion of the region's activity in this domain is superior (inferior) to that of all actors in biotechnology, the indicator is greater (less) than the unit. The variance of this indicator measured for each domain within a region gives an overall degree of specialization for the region. Specialization is calculated on the basis of a nomenclature of publications in terms of theme, R&D expenditure in sectors of research, patents in technological domains and employee numbers in related industries in main sectors of users. The level of competition in related industries is measured by the average size of firms present in the region. A large size thus represents a low level of competition.

Last, the degree of interactivity of regional institutions is measured in terms of the data on co-publications (COPUB). They reveal the existence of effective scientific collaborations (McKelvey et al. 2002). Contrary to citations, which generally do not imply effective collaborations, co-publications signal collaborations among co-authors. It seems to be a good indicator of the regional interactions among laboratories and researchers. Different variables have been defined to describe collaborations, that is public sector or public/private sector collaboration (COPUBPP), and whether they are purely internal to the region or involve institutions outside the region. All these COPUB variables are described in Table 13.1.

3.3   Empirical Results

Table 13.2 presents the main characteristics of the regions as regards firm creation and firm growth. In France the biotech sector was still expanding fast in the late 1990s. The number of institutions' participations in scientific publications in this domain rose by 64 per cent from 1993, up to 5879 in 1999. R&D expenditures by private sector firms increased by 49.6 per cent during the same period – a higher growth rate than that of all other sectors. Hence, the relative share of biotech R&D expenditures compared to all R&D expenditures rose by over 20 per cent. Employee numbers in related industries also show the strong growth of these industries since they rose by nearly 3.5 per cent between 1993 and 1999 and increased their relative share in the total number of employees in industry. However, biotech SMEs remain small and recently set-up.

Table 13.3 provides a synthetic view of the regional distribution of biotech activities in France. Four facts emerge strongly:

Table 13.2   Regional characteristics of SMEs

REGIONS	# of firm creation NBCREA93-99	Average number of employees EFF96	Mean Annual Growth Rate MAGR96-99
Ile de France	53	110.0	5.5
Champagne–Ardenne	1	119.0	−3.9
Picardie	0	13.0	17.4
Haute-Normandie	2	53.5	13.5
Centre	1	34.0	1.9
Basse-Normandie	0	28.0	−9.1
Bourgogne	2	33.6	8.9
Nord-Pas-de-Calais	3	108.0	−0.2
Lorraine	4	6.4	0.6
Alsace	11	76.0	3.6
Franche-Comté	0	22.0	2.2
Pays de la Loire	10	6.7	6.7
Bretagne	6	40.8	7.7
Poitou-Charentes	0	–	39.1
Aquitaine	13	31.9	7.6
Midi-Pyrénées	15	13.4	15.5
Limousin	1	–	–
Rhône–Alpes	10	29.6	6.4
Auvergne	14	7.7	6.9
Languedoc–Roussillon	9	9.2	15.2
PACA	10	34.7	8.2
FRANCE	165	55.2	7.1

Source: INRA database.

1. Despite supportive public policies in favour of biotechnology and innovation in general (creation of knowledge base and commercialization), the results in terms of employment creation are disappointed and public authorities seem to justify their investments by the expected indirect returns.
2. The strong concentration of activity in the Ile de France region, with the Rhône–Alpes region following far behind. Forty-two per cent of publications, 58.6 per cent of national R&D expenditure and 13.1 per cent of related industries were situated in Ile de France in 1999. Ten per cent of publications, 15.6 per cent of national R&D expenditure and 10 per cent of related industries were located in Rhône–Alpes, which consistently increased its share in national activity since 1993 while Ile de France maintained its position with regard to R&D expenditure and

Table 13.3 Regional distribution of biotech activities in France between 1993 and 1999

Region	Publications 1993	Rank	%	Publications 1999	Rank	%	Growth	1993 RI empl. nos.	Rank	%	1999 RI empl. nos.	Rank	%	Growth 93–99	Biotech DERD €93	Rank	%	Biotech DERD 99 (€)	Rank	%	Growth 93–99
Ile de France	1713.22	1	47.9%	2474.19	1	42.08%	44.41%	73.771	1	13.1%	78.516	1	13.47%	6.43%	741819	1	58.55%	1111303	1	58.63%	49.80%
Champagne–Ardenne	13.95	19	0.39%	44.82	18	0.76%	221.25%	17.678	15	3.13%	15.684	16	2.69%	−11.27%	5920	16	0.46%	9853	16	0.51%	66.44%
Picardie	27.25	15	0.76%	45.27	17	0.77%	66.14%	31.248	6	5.55%	29.437	6	5.05%	−5.79%	14806	11	1.16%	11484	15	0.6%	−22.43%
Haute-Normandie	23.45	16	0.66%	45.95	15	0.78%	95.98%	27.691	8	4.91%	25.69	10	4.40%	−7.22%	32321	7	2.55%	37089	7	1.95%	14.75%
Centre	52.11	12	1.46%	105.10	13	1.78%	101.67%	26.692	10	4.74%	26.93	9	4.62%	0.89%	40340	4	3.18%	54950	6	2.89%	36.21%
Basse-Normandie	11.83	20	0.33%	26.45	20	0.45%	123.60%	16.715	16	2.96%	16.201	15	2.78%	−3.07%	8451	15	0.66%	7873	18	0.41%	−6.83%
Bourgogne	46.43	14	1.3%	89.75	14	1.52%	93.27%	17.826	14	3.16%	18.306	12	3.14%	2.69%	4294	18	0.33%	6123	19	0.32%	42.60%
Nord-Pas-de-Calais	86.18	9	2.41%	201.25	8	3.42%	133.50%	40.09	4	7.12%	40.262	5	6.91%	0.42%	21204	8	1.67%	18891	10	0.99%	−10.90%
Lorraine	95.96	8	2.69%	118.59	11	2.01%	23.58%	18.45	12	3.27%	17.067	14	2.92%	−7.49%	1085	20	0.08%	3671	20	0.19%	238.41%
Alsace	253.56	3	7.10%	358.21	5	6.09%	41.27%	24.444	11	3.34%	24.356	11	4.18%	−0.36%	35498	6	2.8%	57492	5	3.03%	61.95%
Franche-Comté	15.86	18	0.44%	30.15	19	0.51%	90.04%	8.342	20	1.48%	8.213	20	1.4%	−1.54%	5306	17	0.41%	13438	14	0.7%	153.24%

Table 13.3 (continued)

Region	Publications 1993	Rank %	Publications 1999	Rank %	Growth	1993 RI empl. nos.	Rank %	1999 RI empl. nos.	Rank %	Growth 93–99	Biotech DERD €93	Rank %	Biotech DERD 99 (€)	Rank %	Growth 93–99
Pays de la Loire	69.32	11 1.94%	138.98	10 2.36%	100.48%	36.075	5 6.4%	43.896	4 7.53%	21.67%	18798	9 1.48%	20333	9 1.07%	8.16%
Bretagne	79.61	10 2.23%	208.77	7 3.55%	162.24%	49.386	3 8.77%	59	3 10.12%	19.46%	10944	13 0.86%	22826	8 1.2%	108.58%
Poitou-Charentes	22.45	17 0.63%	45.71	16 0.77	103.64%	15.375	17 2.73%	14.766	17 2.53%	-3.96%	2315	19 0.18%	8221	17 0.43%	255.09%
Aquitaine	113.01	7 3.16%	158.01	9 2.68%	39.81%	26.91	9 4.77%	27.834	8 4.77%	3.43%	15707	10 1.23%	17723	11 0.93%	12.83%
Midi-Pyrénées	184.14	6 5.15%	294.09	6 5%	59.70%	17.869	13 3.17%	17.964	13 3.08%	0.53%	69857	3 5.51%	101119	3 5.33%	44.75%
Limousin	4.41	21 0.12%	22.58	21 0.38%	411.34%	4.565	21 0.81%	5.013	21 0.86%	9.81%	2450	21 0.01%	3058	21 0.16%	1148.47%
Rhône-Alpes	288.39	2 8.07%	607.99	2 10.34%	110.82%	57.383	2 10.19%	59.998	2 10.29%	4.55%	175462	2 13.85%	297178	2 15.68%	69.36%
Auvergne	50.17	13 1.4%	105.66	12 1.79%	110.58%	11.982	18 2.12%	12.986	18 2.22%	8.37%	9899	14 0.78%	17633	12 0.93%	78.13%
Languedoc-Roussillon	217.80	4 6.1%	363.32	4 6.17%	66.80%	11.369	19 2.01%	11.228	19 1.92%	-1.24%	12625	12 0.99%	16938	13 0.89%	34.16%
PACA	203.43	5 5.69%	394.36	3 6.7%	93.85%	29.155	7 5.17%	29.225	7 5.01%	0.24%	39878	5 3.14%	57966	4 3.05%	45.35%
Total	3572.61	100%	5879.30	100%		563016	100%	582572	100%		1266774	100%	1895163	100%	

related industries but regressed with regard to scientific publications (48 per cent in 1993).

3. Alongside these two large leading regions, another group of regions has a strong potential in biotechnology but with specific trajectories. From the point of view of scientific potential, the PACA (Provence–Alpes–Côte d'Azur) region is currently very well placed, followed by Alsace and Languedoc–Roussillon. Another point these regions have in common is a lack of buoyant markets for biotechnology since they have few related industries established in the region. By contrast, Brittany is a typical example of a region with a high level of industrial activity in which biotechnology is used, while its capacities for publication and research are more limited (even if these seem to have been growing rapidly since 1993). The Nord-Pas de Calais and Pays de Loire regions are in the same category. Finally, Midi-Pyrénées is in a different category since it is in third position with regard to R&D expenditure, despite a far weaker scientific and industrial potential.

4. Finally, despite a few noteworthy developments, the relative positions of the regions appear to be relatively stable during the 1993–99 period. This probably reveals the weight of structural determinants.

No obvious relation between the scientific presence of a specific region and firm creation appears. An understanding of the mechanisms influencing the creation and growth of SMEs is indispensable for gaining further insight into regional location and its dynamic evaluation. A total of 165 biotech firms were created in France during the period under study, 22 of which were founded in 1993 and 34 in 1999. Whereas the average number of employees of biotech SMEs in 1996 was 58, their mean annual growth rate from 1996 to 1999 was 7 per cent.

4. RESULTS OF ECONOMETRIC ESTIMATIONS

To understand the relative effects of the different regional variables on the firm creation and growth, two different models are tested: the first one concerns the ability of regions to attract firm creation (panel data on region) and the second one analyses the firm growth (panel data on firms).

4.1 Estimation Method

The data used for the dependent variable of the first model are count data. They are therefore positive wholes with a large number of zero values. We therefore prefer a Poisson model which seems well adapted for business creation data at a regional level. On this scale, the number of zero observations is very high, or else observations take low values (in our sample, the mean is 1.12 and the max is 12). Yet the dispersion level of distribution can sometimes prove to be higher than the mean (3.94 in our sample). Negative binomial model is appropriate (Greene 2000). By introducing 'alpha' parameter, the generalization of the Poisson model takes into account the heterogeneity of the dependent variable. The estimation and significance of this parameter indicates whether this model has to be preferred to the traditional specification of the Poisson model.

The backward elimination method was used to select the variables.[8] We present only models with significant variables (Table 13.4). For information, the results of the global model are also provided in Table 13.5.

Table 13.4   Results of estimations – creation model

Dependent variable: NBCREA
Number of observations: 147

	Poisson regression	Neg. binomial regression
Constant	2.98***	2.98***
	(0.88)	(0.74)
SPEPUB	−0.38***	−0.38***
	(0.08)	(0.08)
CONCENT_BIO	−1.55***	−1.55***
	(0.56)	(0.55)
CONCENT_RI	1.54**	1.54***
	(0.61)	(0.53)
TXCOPEXT	−3.81***	−3.81***
	(1.27)	(1.10)
COPUBPP	0.23E-01***	0.23E-01***
	(0.69E-02)	(0.83E-02)
DERDTOT	0.37E-07**	0.37E-07*
	(0.18E-07)	(0.21E-07)
Log-likelihood	−161.10	−160.97
χ^2	220.76***	0.26 ns
Pseudo-R^2	0.63	
Alpha		0.02 ns

Notes: The figures between brackets are standard deviations. The significance thresholds are indicated by *, ** and *** which signify 10 per cent, 5 per cent and 1 per cent, respectively.

The 'alpha' parameter estimated in the regression on the negative binomial model does not appear significant. Poisson model is preferred to the negative binomial model.

A relatively low number of variables significantly influences SME creation in each region. This is hardly surprising, given the strong correlations that generally link this type of variable. Thus, with regard to the influence of global 'weight' of the region, a single variable emerges significantly: the total amount of R&D expenditure in the region (all sectors taken into account) is more important than biotech R&D expenditures. It appears that this variable alone reflects all the quantitative effects of agglomeration that prompt firms to set up more easily in areas that already have a strong scientific, technological and industrial potential. Yet the influence here seems much too weak to effectively validate this hypothesis.

Moreover, the absence of the significance of potential variables measuring the relative weight of biotech capacities in the regions proves that even if the effect of overall size is relevant, relative specialization in the biotech sector has no particular attractive effect. Hypotheses H1 and H2 therefore receive little confirmation in this study. One of the explanations can be found in the generic nature of biotechnology that irrigates different economic sectors, just like information technologies.

All the other significant variables are, in our own terms, organizational variables. Both the industrial organization and the organization of research within a specific area influence firm creation. The linkages between public sector research and industry increase the regional propensity to attract firms. The intensity of competition within the biotech sector decreases the attractiveness of the region while the geographic concentration of user industries increases it. In this respect our results strengthen the validity of various theories that show that apart from effects of 'automatic' self-reinforcement that give a systematic advantage to large size, the organizational capacities of the regions are essential. The creation of new biotech firms therefore seems to be a particularly relevant field for illuminating this type of regional determinant.

4.2 The Positive Effect of Capacities for Private/Public sector Local Interaction

The results obtained from co-publication data measure the impact of the level of scientific interactions characterizing the region. The significant positive effect of the public–private co-publications variable (COPUBPP) validates our hypothesis H3. We show here that the presence of public research is not in itself beneficial. Based on co-publications, McKelvey et al. (2002) show the importance of geographic proximity in public/private scientific collaborations in biotechnology. However, collaborations among organizations do not depend on geographic proximity. The value of the existence of real, identifiable collaboration between local public and private sector institutions is highlighted primarily when the creation of SMEs is being promoted. On the other hand, the level of collaboration between local actors with institutions outside the region (TXCOPEXT) has a very strong negative effect on the propensity to create SMEs within the region. This difference of results can be explained by the particular characteristics of processes of new business creation which prefer attraction effects in a region that immediately offers much potential for local public/private sector interaction, whereas the effects of openness onto the outside immediately seem far more beneficial for public sector research itself than for firms. Local public authorities have to note that the level of extra-regional collaborations (TXCOPEXT) seems to have a negative effect on creation. Another explanation can be found in the characteristics of firms: only a minority of firms have a worldwide standing in terms of scientific advances and competences. Thus the effects of openness of the collaboration patterns cannot be detected through statistical analysis. On the contrary, the majority of SMEs are service-oriented and target local markets, at least in their early stages. Thus, the density of linkages among local actors plays an important role.

4.3   The Positive Effects of the Diversity of Scientific Competences

A particularly robust result of our regressions highlights the negative role of regions' scientific specialization. The beneficial effects expected from specialization, especially the facilitation of exchanges of knowledge in the emergent phases of activities Hyphothesis (H4), are not confirmed. In fact, this specialization acts negatively, thus accrediting the favourable effect on diversity in firms' start-up phase. Measured in terms of publications, specialization of public research describes the more or less limited scope of scientific fields of competence available in a region. The analysis of the role of scientific knowledge in the technological development of biotechnology can help us to interpret this result. First, the technological maturation of a sector transforms the modalities of related scientific research. The production of new knowledge involves more and more recombination of existing modules and less and less highly specialized original research (Arora and Gambardella 1994b). Second, and more specifically, the biotech sector has the characteristic of providing generic tools to a wide range of related industries. Generic orientation, linked to scientific and technical trends towards oneness of the living world, generates economic opportunities in different sectors. A diversified regional research framework would therefore increase start-up firms' opportunities for finding the varied scientific competences they need.

This interpretation in terms of diversity is particularly appropriate here since our measurement of specialization concerns basic scientific competences. Its negative effect therefore reflects no competitive effect. In many studies on the role of local sectoral structures in the development of innovative agglomerations, measurement of diversity concerns the characteristics of local employment or private sector research. The negative effect in this case cannot lead directly to the conclusion that diversity is an advantage, without directly measuring the impact of diversity. It may well result from effects of competition. Even if a specialized local structure logically allows many beneficial interactions between firms in the same sector, it is not a key resource for firms, especially in the appropriation of knowledge. This argument is developed by Feldman and Audretsch (1999) who also observe that specialization is a relatively strong curb on innovation in local areas in the USA. By contrast, in our results the variables directly intended to measure the impact of the degree of specialization in related industries and/or of the degree of competition (Hypotheses H6 and H7) reveal no significant effects.

4.4 Influence of Intra-regional Spatial Structuring of Activities

The variables of spatial structuring within regions have a strong albeit contrasting influence. The degree of intra-regional spatial concentration of biotech research expenditure (CONCENT_BIO) proves to have a negative influence on business creation, whereas the spatial concentration of demand (CONCENT_RI) has a positive impact. The former point can be interpreted in two ways. First, this variable may reflect a competition effect not revealed elsewhere. However, observation of the data suggests another interpretation. It is often the very 'small' regions in terms of research potential that provide the strongest concentration indicators. Hence, when this potential is very weak (e.g. in terms of research institution), dispersion is impossible. It is therefore essentially an effect of size that we have here, due to the reduced attractiveness of areas with very little biotech research potential.

By contrast, intra-regional spatial concentration of related industries has a strong beneficial effect on the creation of new firms. Interest in a proximity effect therefore seems to concern above all a link to the market. Thus, rather than the size of the market itself (Hypothesis H5), it is the ability to benefit from agglomerations of customer industries that is sought. Studies in geography of innovation explain the advantages of proximity by the attraction of basic public research and centres of scientific excellence. Our study underlines the role of proximity in the linkages between start-ups and their potential consumers (other companies and public laboratories). This confirms our hypothesis on a clearer orientation of SMEs towards the market, marking a new phase in the maturity of the biotech sector in the 1990s (Mangematin et al. 2003).

The main conclusions of this estimation of a regionalized model of SME creation attest to the complexity of the processes at play. The determinant expected attraction effect of public sector research is not the size *per se* but the diversity of the knowledge base. Note, however, that this absence of effect may result from the type of data used to characterize public research. Publication data describe a local productive capacity but one whose use is not limited in space, unlike the number of researchers which reflects not only the level of available resources in public research but also a capacity for local diffusion. Future research is needed to explore this aspect, especially the role of tacit knowledge in science production.

The importance of organizational variables is the other obvious result of these estimations. It is clear that apart from the resources mobilized locally, the organizational capacities likely to help local firms to take full advantage of those resources are decisive. Results clearly emphasize the role that local public authority can play to encourage firm location, using economic tools like science parks, incubators and so on.

In light of these observations, we understand the problems involved in defining regional policies in this context, even if conclusions can already be drawn. Capacity for local public/private sector interaction, diversity of available scientific competences, and concentration of market potential are all assets for new business creation. The theoretical literature nevertheless suggests the need for caution with regard to these diverse influences, by considering that anything applying to the explanation of the number of start-ups in a region does not necessarily apply to the analysis of the influence of regional environment on the growth of incumbent firms in an area.

4.5   Results of the SME Growth Model

The model presented in the Equation (13.2) presents an estimation of the growth rate of SMEs during the early years of creation. It links the mean annual growth rate of biotech SMEs between 1996 and 1999 to two sets of variables: the first describes the intrinsic characteristics of the firm itself while the second defines the scientific, technological and industrial characteristics of the region in which this firm is established. Several studies have tried to identify the main internal determinants of the growth of biotech firms (e.g. Prevezer 1997; Niosi 2000; Mangematin and Mandran 2001). Estimations of the influence of variables describing regional environment have been performed with control variables on the intrinsic characteristics of firms. The model is estimated by MCOs with robust standard deviations in order to correct the heteroscedasticity. The results are presented in Table 13.5. Only the best specification of the model is presented.

Table 13.5   Results of estimations – growth model

Dependent variable: MAGR
Estimation: MCO with robust standard deviations
Number of observations: 122

Variables	Coefficients
Constant	−1.958**
	(591)
SARL (limited responsibility firm)	−6.31
	(3.81)
CREATION	0.98***
	(0.29)
RESNET96	−0.002***
	(0.000)
EFF96	−0.02***
	(0.008)
R^2	0.109
F-test	3.80***

Notes: The figures between brackets are standard deviations. The significance thresholds are indicated by *, ** and *** which signify 10 per cent, 5 per cent and 1 per cent respectively.

Four internal variables show significant effects. The introduction of dummies representing the legal structure of the firm enables us to show the positive influence of SA structure (i.e. limited liability versus SARL structure which is preferred for family business) on a firm's growth. Year of creation (CREATION) also has a significant favourable effect, indicating that the most recent firms have the highest growth rates. SMEs therefore seem to have difficulty getting over the exhaustion of opportunities that characterizes the end of the exploitation of an initial idea. Likewise, the negative coefficient of the employee variable at the beginning of the period (EFF96) shows the absence of increasing returns and suggests the eventual exhaustion of growth potential. Finally, the firm's net income at the beginning of the period has a significant negative effect even if it is very weak (RESNET96).

None of the variables characterizing regional environment emerges significantly. While the choice of a new firm's location is influenced by regional factors, the development dynamics of established firms relates entirely to their own specific characteristics. Interpreting this type of result is complex. Further research would be needed to complete these preliminary results. However, at this stage we can venture three comments likely to guide future research:

1. The first comment concerns the geographic level of reference. The regional level was chosen because it represents a relevant level for science and technology public policy in France. This level seems to be relevant regarding firm creation because specific public policy measures have been defined at this level. However, the growth of firms may not be influenced by regional variables but rather national ones. Regional level may not be the relevant level to study firm's growth. Most theories on the geography of innovation or the analysis of proximity effects situate the relevant local framework on a smaller geographic scale (districts, clusters or even metropolitan areas). A true measure of the influence of local environment probably does need to be considered at more detailed infra-regional levels. This would tend to mean that even if a region's overall resources have no direct influence, the characteristics of the intra-regional spatial organization of those resources are likely to be important. Basic analyses of intra-regional concentration are probably not enough to make such influences fully evident.

2. The second comment is related to the evolution of firms. Mustar (1998) showed that the development of high-tech firms follows different stages. The support of the local environment is necessary at the early stages of the firm. But, its development requires finding clients and support outside the local environment: national or international venture capitalists, national or international clients or partners and so on. Thus, while the local environment (region) plays a key role at the beginning, the firm must enlarge its network when it is developing. The regional level is thus not relevant in explaining the growth.

3. The third comment provides a more fundamental explanation based on the possible interdependence of individual and regional effects. Can we account for the effects of regional structures independently of the characteristics of firms in the region? While regional characteristics can create a certain potential for externalities, for example, the exploitation of that potential will mostly result from active strategies by the firms. Externalities, whether of public or private sector origin, are perceived less and less as phenomena to which firms are subjected. Rather,

firms are seen as playing an active part by trying to appropriate the knowledge they produce, as far as possible, and, above all, by capturing knowledge produced elsewhere. As a consequence, externalities are not inherent characteristics of knowledge but refer to the way firms manage knowledge (Breschi and Lissoni 2001). The creation of an absorptive capacity and the establishment of effective relations, especially with external sources of knowledge production, cannot result exclusively from regional structural characteristics. In that sense, it is worth noting that firms do not all search for the same resources in the regions. Mangematin et al. (2003) have revealed the existence of different patterns of development of biotech firms. Some SMEs aim for a world market to industrialize their innovation while others are not designed to experience exponential growth but choose to target local markets. These categories must certainly be taken into account in measuring the influence of regional environment on SME's growth.

5.   CONCLUSION

In this chapter our aim has been to provide quantified elements for assessing the role of regional environment in the dynamics of creation and growth of biotech SMEs in France in the 1990s. Apart from the particular context under study, we provide elements of empirical refutation of theories on the location of innovative activities and economic geography. Five databases have been used and a large number of variables created to describe this regional environment. Even if methodological improvements are now envisaged (finding better proxies for public sector research, using the panel method), the first results obtained are fairly clear. The existence of a minimum level of activity within a region is necessary for new business to establish a positive trajectory. The dynamism of this positive trajectory nevertheless seems to depend strongly on organizational factors. Rather than the quantitative potential of public and private sector research in the region, it is the diversity of available scientific competences and the capacity to develop public/private interactions that favour the establishment of biotech start-ups in the region. Likewise, it is not so much the size of the regional market itself that is important, as the opportunity for the firm to exploit agglomeration effects in such markets.

Hence, in the biotech domain, regional policies cannot count only on the accumulation of regional resources. It is not simply the fact of belonging to the same region that facilitates transfers and stimulates phenomena of externalities. In this sense regional technology policy is far more than support for R&D expenditures or for the direct production of artefacts. Its role is also to set up and support the variety of mechanisms that allow firms to capture and absorb external knowledge. In biotechnologies, diffusion of technological knowledge is complex; hence, the need for a varied institutional infrastructure. Promoting diversity, facilitating the establishment of relations and cooperation between actors with varied competences, belonging to networks or institutional systems all marked by their own culture, using the agglomeration to encourage producer-user relations: all these are examples of the 'organizational capacities' that can distinguish regions.

The results obtained on our growth model prompt us to conclude with a final comment. Even if the characteristics of the regional environment can, in themselves, have attraction effects for investments in biotechnology, they are insufficient to determine the growth potential of firms in the region. These characteristics do determine a more or less favourable context, but the growth dynamics of firms result above all from their ability to take advantage of that context through the implementation of active strategies *vis-à-vis* their environment.

ACKNOWLEDGEMENTS

Financial support from Ministry of Research, Research Statistics Bureau is highly acknowledged. E. Weisenburger also helped us to understand and to use national statistics on research. We appreciate his contribution.

This contribution largely benefits from the help of Lionel Nesta, Clementine Body and Roger Coronini to collect and organise data. We greatly acknowledge their help. This study also benefits from helpful comments from the participants of the Biotech Workshop in Gothenburg 25–27 September 2002, organized by Chalmers University of Technology. Special thanks to Jens Laage-Hellman, Maureen McKelvey and Annika Rickne, for the comments on the earlier versions. Usual caveats apply.

NOTES

1. By economic valorization, we mean creating economic value from scientific results.
2. Acs et al. (2002) nevertheless show the relevance of econometric studies that use data on numbers of innovations and patents at detailed levels of location.
3. Website www.grenoble.inra.fr, available on request.
4. INPI: *Institut National de la Propriété Industrielle.*
5. SESSI: *Service d'Etudes des Stratégies et des Statistiques Industrielles.*
6. EAE: *Enquête Annuelle d'Entreprise.*
7. In France, departments are administrative units within regions.
8. Step-by-step introduction yields similar results.

REFERENCES

Acs, Z. and L. Preston (1997), 'Small and medium-sized enterprises, technology, and globalization: introduction to a special issue on small and medium-sized enterprises in the global economy', *Small Business Economics*, **9**, 1–6.

Acs, Z., L. Anselin and A. Varga (2002), 'Patents and innovation counts as measures of regional production of new knowledge', *Research Policy*, **31**(7), 1069–85.

Afuah, A. and J.M. Utterback (1997), 'Responding to structural industry changes: a technological evolution perspective', *Industrial and Corporate Change*, **6**(1), 183–202.

Almedia, P. and B. Kogut (1999), 'Localization of knowledge and the mobility of engineers in regional networks', *Management Science*, **45**(7), 905–18.

Anselin, L., A. Varga and Z. Acs (1997), 'Local geographic spillovers between university research and high technology innovations', *Journal of Urban Economics*, **42**, 422–48.

Arora, A. and A. Gambardella (1994a), 'Evaluating technological information and utilizing it: scientific knowledge, technological capability, and external linkages in biotechnology', *Journal of Economic Behavior and Organization*, **24**(1), 91–114.

Arora, A. and A. Gambardella (1994b), 'The changing technology of technological change: general and abstract knowledge and the division of innovative labour', *Research Policy*, **23**, 523–32.

Audretsch, D. and M. Feldman (1996a), 'Knowledge spillovers and the geography of innovation and production', *American Economic Review*, **86**(3), 630–40.

Audretsch, D. and M. Feldman (1996b), 'R&D spillovers and the geography of innovation and production', *American Economic Review*, **86**(3), 630–40.

Audretsch, D. and P. Stephan (1996), 'Company scientist locational links: the case of biotechnology', *American Economic Review*, **86**(3), 641–52.

Autant-Bernard, C. (2001), 'Science and knowledge flows: evidence from the French case', *Research Policy*, **20**, 1069–78.

Autant-Bernard, C. and N. Massard (1999), 'Econométrie des externalités technologiques et géographie de l'innovation: une analyse critique', *Economie Appliquée*, **4**.

Baum, J.A.C., T. Calabrese and B.S. Silverman (2000), 'Don't go it alone: alliance network composition and startups' performance in Canadian Biotechnology', *Strategic Management Journal*, **21**, 263–94.

Breschi, S. and F. Lissoni (2001), 'Knowledge spillovers and local innovation systems: a critical survey', *Industrial and Corporate Change*, **10**(4), 975–1006.

Catherine D., F. Corolleur, M. Carrere and V. Mangematin (2004), 'Turning scientific knowledge into capital: the experience of biotech start-ups in France', *Research Policy*.

Cockburn, I.M. and R.M. Henderson (1998), 'Absorptive capacity, coauthoring behavior, and the organisation of research in drug discovery', *The Journal of Industrial Economics*, **XLVI**(2), 157–82.

Cohen, W.M. and D.A. Levinthal (1990), 'Absorbtive capacity, a new perspective of learning and innovation', *Administrative Science Quarterly*, **35**, 128–52.

Cooke, P. (2001), 'Biotechnology clusters in the UK: lessons from localisation in the commercialisation of science', *Small Business Economics*, **17**(1–2), 43–59.

Cooke, P. (2002), 'Rational drug design, the knowledge value chain and bioscience megacentres', paper presented at the Clusters in High-Technology Conference, Montreal, Canada, 22–24 September.

Feldman, M. (1994), 'Regional innovative capacity', *The Geography of Innovation*, Boston, MA: Kluwer Academic Publishers, pp. 77–91.

Feldman, M. (1999), 'The new economics of innovation, spillovers and agglomeration: Review of Empirial Studies', *Economics of Innovation and New Technology*, **8**(1), 5–25.

Feldman, M.P. and D.-B. Audretsch (1999), 'Innovation in cities: science-based diversity, specialization and localized competition', *European Economic Review*, **43**, 409–29.

Feldman, M. and J. Francis (2001), 'Entrepreneurs and the Formation of Industrial Clusters', paper presented at the Complexity and Industrial Clusters, Milan.

Feldman. M. and C. Ronzio (2001), 'Closing the innovative loop: moving from the laboratory to the shop floor in biotechnology manufacturing', *Entrepreneurship and Regional Development*, **13**, 1–16.

Frost, T.S. (2001), 'The geographic sources of foriegn subsidiaries' innovations', *Strategic Management Journal*, **22**(2), 101–23.

Genet, C. (1997), 'Quelles conditions pour la formation des biotechnopoles: une analyse dynamique', *Revue d'Economie Régionale et Urbaine*, (3), 405–24.

Green, W.H. (2000), *Econometric Analysis (4th edition)*, Upper Saddle River, New Jersey: Prentice-Hall.

Jaffe, A.B., M. Trajtenberg and R. Henderson (1993), 'Geographic localization of knowledge spillovers as evidenced by patent citations', *Quarterly Journal of Economics*, **108**(3), 577–98.

Krugman, P. (1991), 'Increasing returns and economic geography', *Journal of Political Economy*, **99**(3).

Krugman, P. and A. Venables (1995), 'Globalization and the inequality of nations', *Quaterly Journal of Economics*, **110**.

Lemarie, S., V. Mangematin and A. Torre (2001), 'Is the creation and development of biotech SMEs localised? Conclusions drawn from the French case', *Small Business Economics*, **17**(1–2), 61–76.

Liebeskind, J.P., A.L. Oliver, L. Zucker and M. Brewer (1996), 'Social networks, learning, and flexibility: sourcing scientific knowledge in new biotechnology firms', *Organization Science*, **7**(4), 428–42.

Mangematin V. and N. Mandran (2001), 'Innovation without internal research: spillovers from public research or from other firms? The case of the agro-food industry', in A. Kleinecht and P. Mohnen (eds), *Innovation and Firm Performance: Econometric Explorations of Survey Data*, London and Basingstoke: Palgrave.

Mangematin, V. and L. Nesta (1999), 'What kind of knowledge can a firm absorb?', *International Journal of Technology Management*, **37**(3–4), 149–72.

Mangematin, V., S. Lemarié, J.P. Boissin, D. Catherine, F. Corolleur, R. Coronini and M. Trometter (2003), 'Sectoral systems of innovation, SMEs development and heterogeneity of trajectories: the case of biotechnology in France', *Research Policy*, **32**(14): 621–38.

Martin, P. and G. Ottaviano (1999). Growing locations: industry location in a model of endogenous growth', *European Economic Review*, **43**, 281–302.

Massard, N. and S. Riou (2001), 'Specialization and diversity: the debate on the nature of innovative agglomerations', *Third Congress on Proximity, New Growth and Territories*, Paris.

McKelvey, M. H. Alm and M. Riccaboni (2002), 'Does co-location matter for formal knowledge collaboration in the Swedish biotechnology-pharmaceutical industry?', *Research Policy*.

Ministry of Research (DGRT) (2000), 'Rapport sur les e'tudes doctorales 1999. Paris: Ministère de l'Education nationale, de l'enseignement superieur et de la recherche.

Mustar, P. (1998), 'Partnerships, configurations and dynamics in the creation and development of SMEs by researchers', *Industry and Higher Education*, 217–21.

Nesta, L. (2001), 'Cohérence des bases de connaissances et changement technique: une analyse des firmes de biotechnologie de 1981 à 1997', *department of Economics*, Grenoble: Université Pierre Mendes France.

Nesta, L., and V. Mangematin (2004), 'The dynamics of innovation networks', paper presented at the Academy of Management, New Orleans, LA, 6–11 August.

Niosi, J. (2000), 'Strategy and performance factors behind rapid growth in Canadian biotechnology firms', in J. de la Motte and J. Niosi (eds), *The Economic and Social Dynamics of Biotechnology*, Kluwer: Boston.

Pisano, G. (1991), 'The governance of innovation: vertical integration and collaborative arrangements in the biotechnology industry', *Research Policy*, **20**, 237–49.

Prevezer, M. (1997), 'The dynamics of industrial clustering in biotechnology', *Small Business Economics*, **9**(3), 255–71.

Senker, J., P. Van Zwanenberg, C. Enzing, S. Kerns, V. Mangematin, R. Martinsen, E. Monoz, V. Diaz, S. O'Hara, K. Burke, T. Reiss and S. Wörner (2001), *European Biotechnology Innovation System*, Brussels: EC.

Steier, L. and R. Greenwood (2000), 'Entrepreneuship and the evolution of angel financial networks', *Organization Studies*, **21**(1), 163–92.

Steinle, C. and H. Schiele (2002), 'When do industries cluster? A proposal of how to assess an industry's propensity to concentrate at a single region or nation', *Reseach Policy*, **31**(6), 849–58.

Zucker, L., M. Darby, M. Brewer and Y. Peng (1995), 'Collaboration structure and information dilemmas in biotechnology: organisational boundaries as trust production', Cambridge, MA: NBER.

Zucker, L.G., M.R. Darby and J. Armstrong (1998), 'Geographically localized knowledge: spillovers or markets?', *Economic Inquiry*, **36**(1), 65–86.

Zucker, L.G., M.R. Darby and J. Armstrong (2002), *Commercializing knowledge: university science, knowledge capture, and firm performance in biotechnology*, *Management Science*, **48**(1), 138–53.

Zucker, L.G., M.R. Darby and M. Torero (1997), 'Labor mobility from academe to commerce', NBER working paper no. 6050.

APPENDIX

*Table 13.A.1   Regional characteristics of public research in life sciences,
1993–1999*

REGIONS	PUB	SPEPUB	PhD
Ile de France	15 343	0.11	734.0
Champagne-Ardenne	205	4.03	3.3
Picardie	256	4.98	36.7
Haute-Normandie	246	6.82	19.9
Centre	611	2.56	32.1
Basse-Normandie	167	7.69	11.1
Bourgogne	493	3.67	34.6
Nord-Pas-de-Calais	1 007	1.56	67.7
Lorraine	678	8.08	85.6
Alsace	2 156	0.59	90.6
Franche-Comté	186	4.97	15.0
Pays de la Loire	677	2.15	35.1
Bretagne	1010	2.13	67.4
Poitou-Charentes	234	5.23	20.0
Aquitaine	1 032	1.12	80.9
Midi-Pyrénées	1 764	1.86	77.3
Limousin	109	8.48	2.1
Rhône-Alpes	3 047	0.43	199.4
Auvergne	544	3.00	37.1
Languedoc-Roussillon	2 171	1.21	133.4
PACA	2 240	0.45	131.4
FRANCE	34 178	3.39	1 914.7

Source: Biotechnology Citation Index

Table 13.A.2   *Regional characteristics of private research in biotechnology,*
1993–1999

REGIONS	CH BIO	SPE BIO	CONCENT BIO
Ile de France	2710	27.6	0.25
Champagne-Ardenne	40	10.9	0.97
Picardie	49	2.2	0.35
Haute-Normandie	85	0.2	0.71
Centre	117	2.8	0.45
Basse-Normandie	30	4.4	0.78
Bourgogne	30	1.3	0.77
Nord-Pas-de-Calais	60	8.1	0.81
Lorraine	10	2.6	0.66
Alsace	167	25.5	0.57
Franche-Comté	32	0.5	0.54
Pays de la Loire	86	5.6	0.36
Bretagne	97	9.1	0.44
Poitou-Charentes	9	24.1	0.60
Aquitaine	67	18.1	0.36
Midi-Pyrénées	347	24.4	0.41
Limousin	25	0.1	0.83
Rhône-Alpes	756	14.7	0.65
Auvergne	50	13.7	0.60
Languedoc-Roussillon	79	11.0	0.92
PACA	240	3.5	0.49
FRANCE	5089	10.0	0.60

Source: R&D Survey, Ministère de la Recherche et de la Technologie.

Table 13.A.3   Regional characteristics of markets, i.e. related industries,
1993–1999

REGIONS	EFF IR	SPE IR	CONCURR IR	CONCENT IR
Ile de France	77625	1.17	81.7	0.23
Champagne-Ardenne	16527	0.31	75.7	0.82
Picardie	30200	0.03	127.0	0.63
Haute-Normandie	26125	0.57	107.3	0.52
Centre	27407	1.36	93.9	0.35
Basse-Normandie	16167	0.38	92.9	0.42
Bourgogne	18368	0.24	88.4	0.53
Nord-Pas-de-Calais	39983	0.14	98.9	0.52
Lorraine	17611	0.30	89.5	0.56
Alsace	24453	0.09	100.9	0.57
Franche-Comté	8417	0.40	68.1	0.41
Pays de la Loire	39712	0.49	88.8	0.27
Bretagne	54020	0.57	101.2	0.30
Poitou-Charentes	14982	0.30	72.8	0.37
Aquitaine	26620	0.04	71.1	0.39
Midi-Pyrénées	17348	0.03	59.7	0.55
Limousin	4785	0.38	62.2	0.64
Rhône-Alpes	58433	0.29	85.4	0.66
Auvergne	12158	0.11	63.4	0.82
Languedoc-Roussillon	19978	0.10	60.9	0.61
PACA	28519	0.67	71.6	0.31
FRANCE	570441	0.38	83.9	0.50

Source: Enquête Annuelle d'Entreprises du Ministère de l'Industrie (SESSI).

Table 13.A.4 Regional characteristics of the organization of research in biotechnology, 1993–1999

REGIONS	TXCOPUB	TXCOPEXT	COPUBPP
Ile de France	58.2	54.1	134.5
Champagne-Ardenne	58.3	73.2	3.3
Picardie	53.3	66.7	2.9
Haute-Normandie	61.3	66.3	2.9
Centre	57.7	73.2	4.5
Basse-Normandie	49.6	61.5	2.3
Bourgogne	42.6	68.6	5.6
Nord-Pas-de-Calais	63.5	59.6	9.3
Lorraine	46.4	63.8	9.2
Alsace	47.8	76.1	18.1
Franche-Comté	60.2	57.5	3.9
Pays de la Loire	60.4	64.5	11.3
Bretagne	57.7	62.3	11.5
Poitou-Charentes	52.2	71.1	3.6
Aquitaine	52.8	58.2	9.7
Midi-Pyrénées	47.8	63.8	22.2
Limousin	69.3	69.7	0.9
Rhône-Alpes	55.8	68.5	37.9
Auvergne	51.1	60.5	5.9
Languedoc-Roussillon	54.3	67.1	27.1
PACA	58.3	66.2	21.6
FRANCE	55.2	65.4	348.3

Source: Science Citation Index.

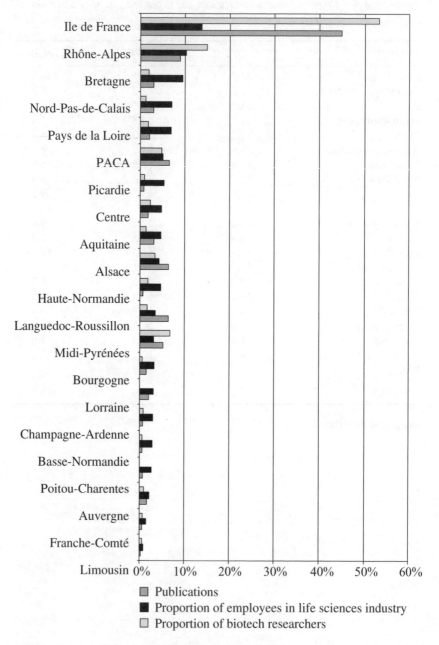

Figure 13 A.1   Regional profile of biotech activities

Table 13.A.5   Creation model, results of estimation

		Poisson regression	Neg. Binomial regression
	Constant	2.64 (1.68)	2.65 (1.71)
Public research	PUB	0.69E-02 (0.52E-02)	0.69E-02 (0.72E-02)
	PhDS	−0.40E-02 (0.50E-02)	−0.40E-02 (0.61E-02)
	SPE_PUB	−0.34 (0.11)	−0.34 (0.13)
Private research	RE_BIO	−0.13^E-04 (0.17E-02)	−0.13^E-04 (0.29E-02)
	PART_DIRD_BIO	−6.49 (5.20)	−6.49 (6.10)
	CONCENT_BIO	−0.85 (0.71)	−0.85 (0.74)
	SPE_BIO	−0.33E-03 (0.40E-02)	−0.33E-03 (0.51E-02)
	PATENT_BIO	0.14E-01 (0.19E-01)	0.14E-01 (0.26E-01)
Market for biotech i.e. interrelated industry	INV_IR	0.27E-06 (0.38E-06)	0.27E-06 (0.40E-06)
	NB_ETS_IR	0.46E-02 (0.21E-02)	0.46E-02 (0.26E-02)
	CONCUR_IR	−0.37E-02 (0.10E-01)	−0.37E-02 (0.10E-01)
	CONCENT_IR	1.73 (0.85)	1.73 (1.02)
	SPE_IR	−0.22 (0.57)	−0.22 (0.66)
	PART_EFF_IR	−0.47E-01 (0.33E-01)	−0.47E-01 (0.42E-01)
Organisation of biotech research	TXCOPUB	0.11 (1.53)	0.11 (1.90)
	TXCOPEXT	−3.44 (1.92)	−3.44 (2.04)

Table 13.A.5   (continued)

		Poisson regression	Neg. Binomial regression
	TXCOPINT	1.90	1.90
		(2.40)	(2.53)
	COPUBPP	0.27^{E}-01	0.27^{E}-01
		$(0.20^{E}$-01)	$(0.28^{E}$-01)
Regional characteristics	PUBTOT	-0.27^{E}-02	-0.27^{E}-02
		$(0.23^{E}$-02)	$(0.36^{E}$-02)
	DERDTOT	-1.08^{E}-06	-1.08^{E}-06
		$(0.82^{E}$-07)	$(0.12^{E}$-06)
	INVTOT	-0.21^{E}-06	-0.21^{E}-06
		$(0.12^{E}$-07)	$(0.14^{E}$-07)
	Log-likelihood	-161.10	-160.97
	X^2	220.76	0.26ns
	Pseudo-R^2	0.65	
	Alpha		0.02ns

PART V

Conclusions

14. Reflections and ways forward

Hannah Kettler, Maureen McKelvey and Luigi Orsenigo

1. PARADOXES AND WAYS FORWARD

Having addressed many aspects of the economic dynamics of knowledge, this chapter concludes the book by offering some reflections and ways forward. The current section returns to the four paradoxes raised in Chapter 1, in order to reflect on the contributions of this book as well as to suggest many interesting avenues for further social science research. These topics are also relevant to the policies and strategies of governments, universities and firms. Section 2 highlights one such neglected avenue, namely biotechnology for the health care needs of developing countries. This is such an important topic for further research and recommendations, that a whole section is devoted to it. Section 3 concludes by discussing challenges to conceptual understanding that have been raised in studies of modern biotechnology, including this book.

The first paradox referred to controversies over the negative versus positive societal impacts. On the one hand, modern biotechnology is claimed to be as crucially important to many industrial sectors, to large as well as small firms, and as useful to address basic human needs. On the other hand, modern biotechnology is also the centre of controversies about the possible negative impacts related to modification of nature, food safety, animal welfare, environmental protection and global exploitation of the poor.

Chapters within this book have raised issues relevant to the government's role in regulation, handling of risks, the effects of university–industry relationships and so on. On the one hand, without commercial success, society will not fully benefit from the large public and private investments being made in biotech research and development (R&D). On the other hand, it remains to be seen to what extent these results will prove profitable to investors and beneficial to society as a whole. Specific research results, services and goods products may also prove to have unintended negative consequences. Still, the evidence suggests that governments, universities and firms experiment with various incentives, solutions and organizational structures to redistribute risks and benefits across actors.

Future research could follow these avenues opened up in preceding chapters, in understanding modern biotechnology as an economic process, which involves many diverse actors around the globe. Another aspect would be a more systematic discussion of actual and potential future consequences. This would be useful, if it explicitly considers the economic trade-offs involved, in a dynamic perspective.

The second paradox was that definitions, methodology and data about modern biotechnology are currently unsatisfactory. The boundaries between firms and actors have also been shown as highly porous, with the movement of persons, ideas and resources. Still, progress is being made by national agencies as well as by social science researchers, working individually and in teams. Many avenues are possible to solve these challenges – and those results are needed.

As shown in Chapter 2, progress is being made for the public sources of quantitative data – despite the problems of definition, sampling strategies and coverage.[1] Statisticians have taken major steps forward, particularly in countries such as Canada, that are alert to the need for better innovation-related data. The situation with modern biotechnology is similar to that of other emerging technological areas – where it is common to find a plethora of approaches and methods, leading ultimately to the convergence of international standards for statistics that permit some real comparability. That convergence is clearly on its way, but future research and policy initiatives will be important. Even with additional work, it will be some time before we can really generalize with any certainty about the quantitative dimensions – and before we can link quantitative analysis to qualitative data.

Chapters 3 to 13 have also made progress in capturing certain aspects of the economic dynamics of knowledge, using both quantitative insights and qualitative data about modern biotechnology. Much of this empirical work is unique and represents significant investment by the authors into designing, obtaining and interpreting the empirical results. They move our conceptual understanding into new areas, such as scientific networks within the food industry and into risk-taking behaviour related to genetic information for human health care. Additional research of this type will help us re-conceptualize modern biotechnology, by allowing us to identify and discuss aspects related to challenging the existing and forming the new.

The third paradox portrayed modern biotechnology as being at once fundamentally global in terms of knowledge flows of skilled persons, ideas, services and products – yet also being extremely local in terms of co-located actors, sometimes within geographically situated agglomerations.

Chapters within this book have contributed by questioning the extent to which regional relationships really matter. In some cases, regional and highly personal contacts clearly matter – but in other cases and in other phases, these are of minor importance, as compared to being part of global market and knowledge processes. Chapters have examined different organizational structures, as related to issues of process of acquisition and development of new competences.[2] Large pharmaceutical companies have undergone long and often painful periods of transformation, with regard to their knowledge bases, their strategies and their organizational structures. The reactions of firms within this sector have been highly heterogeneous and largely influenced by their past histories. Similar characteristics in the adaptation and strategies of firms are observable within other sectors.

More work needs to be done to adequately address how and why companies and networks in different countries and across the globe are working together via alliances, mergers, contracts and so on. Better understanding of such issues is likely to impact our thinking on the models and institutional settings within specific countries as well as the sources of sustainable competitive advantages for countries, regions and firms. Another topic for research is the extent to which specific actors are able to join – or are excluded from – networks of collaboration as well as competitive markets.

The fourth paradox is that modern biotechnology has seemed for several decades to be primarily a US phenomenon – with the rest of the world lagging behind. Certainly, debates have raged about whether in fact – and if so, for what reasons – the USA has had a competitive advantage.

Chapters within this book have contributed by highlighting the diversity within Europe – by sector, by nation, by phenomena under consideration. There is not one Europe – but many examples of what is happening within Europe. Change over time also matters, as in the example when the development of science and technology opens up markets for ideas and for the formation of new companies. Concluding that European investment in R&D is lagging behind the USA and suffers from fragmentation, the Commission of the European Communities (2002) has launched The Sixth Framework Programme with the primary aim to restore Europe's leadership in life sciences and biotechnology research. As the European Union (EU) expands to include more and more countries, important research questions arise about the impact of the relative diversity and unevenness across these countries as well.

With regard to Europe's performance, some existing literature could serve as a starting point for future research. Allansdottir et al. (2002) have benchmarked the European situation in biotech-pharmaceuticals, by examining the relative performance of firms, their innovativeness and performance, the issue of competition in different national markets and so on. Still more could be done along these lines, because it is not clear whether the lag behind the USA is due mainly to size differences, fragmented strategies to support scientific research or to other fundamental factors. Moreover, future research could directly question the assumptions underlying international policy advice, such as the need for increasing investment in relevant science for immediate effects on firms and regional growth; for supporting regional agglomerations and local ties; and for restructuring intellectual property rights.

Hence, many relevant and fundamental problems of conceptualization and measurement remain, highlighting avenues for future research and for advice to policy and strategy. In this context, developing countries are one neglected area, a subject to which we now turn.

2. HIGHLIGHTING THE IMPORTANCE OF BIOTECHNOLOGY FOR DEVELOPING COUNTRIES

In the context of this book and in associated research communities, developing countries have been a relatively neglected topic of research. Hence, a number of justifications can be given for highlighting their importance in this section, including the global magnitude of these pressing issues as well as the overwhelming need for interactions among the developing countries and global actors to tackle these problems. Firms situated in Europe and the USA are already involved in many initiatives for developing countries. Because this book does not address developing countries in previous chapters, this section addresses some issues in some detail, to point towards an avenue for future social science research and for recommendations.

While Europe appears to be lagging behind – but catching up with – the USA, the EU is still clearly far advanced as compared to the knowledge creation and exploitation in the developing world. Currently the majority of the wealth of the global infrastructure and capabilities in biotechnology rests with 10 per cent of the global population in the developed world, leaving the developing world dependent on international product and knowledge flows. Effective and sustained transfer of these products as well as new capacity building is hindered by significant challenges to policy, practical action and debates. Critical questions are raised about affordable access to products in the developing world; about the globalization and enforcement of intellectual property rights (IPR); and about the neglect of research in 'global diseases of poverty', that is diseases such as tuberculosis and malaria that afflict predominately poor populations in the poorest regions of the world. Still, in some areas, the developing countries hold a wealth of genetic information about plants and animals, that may prove valuable one day.

One reason for flagging the importance of future social science research from this perspective for developing countries is that modern biotechnology already has – and will have additional – societal impacts that reach far beyond those clustering of firms and scientists in a few key regions of the developed countries. All countries stand to benefit from advances in medical and agricultural applications, even though modern biotechnology has been singled out as vital to meeting the medical and nutritional societal needs in developing countries. The developing countries have particular problems – but solving them will likely involve diverse actors around the world.

From an individual company's perspective, the costs and risks associated with developing and launching products that are targeted at developing country markets – be they biomedical or agricultural – far exceed the likely returns. The discussion that follows will be limited to modern biotechnology in relation to human health care. Given the essential role that biotech and pharmaceutical companies play in the commercialization and development of research, one problem is caused by the inability of patients and potential consumers to afford new products. Why spend the company's money on research and innovation for such products, if those costs are not likely to be recuperated, much less result in profits? In the absence of additional incentives, companies have little interest in investing their time and money in the development of products targeted at small patient groups (as in the case of 'rare' diseases) or targeted at millions of people without any money (as in the case of diseases of poverty).

Some policy instruments for drugs have been effective in giving companies incentives to innovate for products towards previously neglected patient groups. Governments' recognition of this lack of incentives to innovate in certain areas is reflected in the establishment of the Orphan Drug Act in the USA, Japan and the EU. Orphan Drug Act policies combine cost-reducing policies – such as grants and tax credits for clinical trials in the USA; fast track-product approval in all three policy areas – with revenue-enhancing policy, achieved by providing a company with a certain number of years of market exclusivity for any approved orphan-designated product.

Evidence from the USA suggests that the Orphan Drug Act has been a success from the standpoint of encouraging new R&D into rare diseases and of encouraging biotech companies which, in particular, have responded to these incentives. Since 1983, 227 new products have been approved for orphan diseases, as compared with 53 such products in the 20 years prior to the Act. Over 900 substances have been designated orphan status. Over 70 per cent of those designations have gone to biotech companies and 50 per cent of all biotech products approved by the Food and Drug Association (FDA) are in orphan diseases. That said, only 25 of these 9000 designations have been in diseases of poverty resulting in 12 approvals. Hence, despite qualifying as orphan diseases, the policy has not worked to draw new players into global diseases of poverty – but the policy instrument has been useful to change incentive structures for some types of company innovation.

Indeed, the pursuit of government policy instruments to direct biotech research towards global diseases is under way on a number of fronts. In terms of incentive policies, the UK introduced a 50 per cent tax credit in 2002 for R&D on malaria, TB and HIV while the EU is considering a broader package of 'push' and 'pull' incentives.

Other initiatives link government policy with a broader range of actors. A significant amount of money and resources have been allocated towards new initiatives – public–private partnerships (PPPs) – that focus on R&D in specific diseases. These include the Malaria Vaccine Initiative (MVI), Medicines for Malaria Venture (MMV), International AIDS Vaccine Initiative (IAVI), Institute for One World Health (IOWH) and the Global Alliance for TB Drug Development (GATB). Key founders and funders of these PPPs include the Gates Foundation, the Rockefeller Foundation, the World Health Organization (WHO), the British Pharmaceutical Association, Medicins sans Frontieres (MSF), as well as select governments. A number of biotech companies have contributed to these partnerships, as is illustrated in Table 14.1.

Table 14.1 shows the pharmaceutical companies in bold typescript, and the biotech firms in normal typescript, along with the respective public–private partnership for combating such diseases. By 2003, some of these initiatives had led to the development of early stage clinical testing of vaccines and drugs designed to boast immune systems, such as for malaria.

Further research is needed to build on the existing literature about how to most effectively engage and use research and biotech companies in these partnerships. Questions include the value of the biotech companies' contributions, as compared to that of large pharmaceutical companies. Moreover, one could ask what impact does working in these different networks have on the biotech companies' core business strategies, for example non-diseases of poverty. Hence, the avenues for future research introduced in Section 1 are also relevant in relation to developing countries.

In terms of the perspective of innovation systems of diverse actors and linkages among them, dedicated biotech firms are only one of a set of actors required to meet the problems of innovating and delivering new products and new tools to the developing world. Modern biotechnology in itself will not solve the fundamental problems. Even if all the development of modern biotechnology in the whole world were turned to the task of solving problems relevant to developing countries, this would not be enough. Diffusion of modern biotechnology into products and services affecting society requires changes in many other areas as well. Areas

Table 14.1   Companies contributing to select PPPs

• Chiron, GATB	• Apovia, MVI
• **GSK, MMV, MVI**	• Progen Industries, MVI
• **Roche, MMV**	• Bharat Biotech, MVI
• AlphaVax, IAVI*	• Oxxon Pharmaceuticals, MVI, IAVI
• Maxygen, IAVI	• **Schering Plough, MMV**
• Targeted Genetics, IAVI	• **Bayer, MMV**
• Therion, IAVI	• Jacobus Pharma, MMV
• Berne Biotech, IAVI	• Sequella Inc.
• **Japanese Pharma, JPMW**	• Celera, IOWH
• Ranbaxy, MMV	• Korea Shing Poong Pharma, MMV
• Immtech Int'l, MMV	• Biotech Australia, MVI
• Adprotech, MVI	• Gropep Industries, MVI
• **Novartis Pharma, MMV**	• **Bristol Myers Squibb, MMV**

Notes:
*Contract was not renewed in 2002.
Pharmaceutical companies are in bold.

Source: PPP websites, annual reports, personal communications.

include the health care system, agricultural practice, public health care measures such as vaccines, basic scientific research, company R&D efforts to continue developing vaccines and so on. Hence, a broader innovation system perspective could be useful both for explaining why a 'technological solution' is not a simple thing, but instead linked to more fundamental change, as well as for giving recommendations for policy and strategy.

From the perspective of the global society, the cumulative costs of not addressing these problems – including the costs of poverty, sickness and political destabilization – may already exceed the costs of acting now. Given the mismatch between societal need and supply provided by private actors, it suggests that assertive public action could be undertaken, be that in the form of policy incentives or direct investments. However, as is usually the case, many questions arise about who has the incentives and resources to lead such assertive public actions for the developing countries.

As donors, policy-makers, researchers and producers, Europe has the potential to lead the 'transfer' of biotech research as well as to influence the future trajectories of knowledge development to make it relevant to the developing world. Clearly, there are difficulties associated with achieving such a complex political goal, particularly one playing out on the other side of the globe. One starting strategic focus could be on the narrowing of the gap between the real costs and the uncertain benefits of innovation, as compared to the potential societal benefits. Using policy incentives to realize this goal may encourage companies to pursue global health and nutrition improvements as part of their private business strategy. Note that sustained results are more likely if individual firms purse potentially rewarding innovation opportunities in direct collaboration with informed and committed public, public–private and non-profit organizations.

Thus, one question for the future is whether the EU – with all the diversity of included institutions, industrial structures, policies, business models, resources and so on – will draw on existing practice to rise to the challenges of developing countries. Is the EU in a position to lead in the global health arena, a move that would bring not only political applause but potentially also real business benefits? What contributions can future social science research make, both in terms of identifying and explaining the crucial issues, as well as providing guidelines for policy and strategy?

3.   CONCEPTUAL UNDERSTANDING

This section concludes the book by reflecting on our conceptual understanding. It could be argued that studies of modern biotechnology have pioneered – and in any case contributed heavily to – the development of new concepts and methodologies that are now becoming state of the art in the analysis of industrial change and innovation. Within that wider debate, this book aims to contribute by offering new empirical evidence, raising more fundamental theoretical questions, and providing insights of relevance to governments, universities and firms. Each chapter contributes to the literature in some way – by arguing for new perspectives, by questioning existing assumptions, by providing unique data and so on. Many chapters address research questions related to how and why existing industries, networks, firms and organizations are challenged as well as new ones formed.

Modern biotechnology is a rather amorphous concept, including diverse products, sectors and knowledge bases. It is therefore pervasive across society, leading to heated societal debates as well as opening up new business opportunities. As such, analyses of the emergence, development and impacts of modern biotechnology pose fundamental problems. These conceptual problems may appear extreme in this emerging technological area and difficult to resolve. Even so, researchers, analysts and decision-makers must tackle them and also consider novel interpretations.

First, the case of modern biotechnology illustrates the usefulness and power of an evolutionary approach – however loosely defined – to the study of industrial change and innovation. Different varieties of evolutionary metaphors are present in the literature, even appearing at times within neo-classical economics. For this book, we mean an evolutionary approach in the sense of an emphasis on heterogeneous agents, which have limited capabilities and which change over time. Such agents affect the transformation of industrial structures and organizational forms, as driven by processes of learning and by multiple forms of selection.

Second, modern biotechnology raises a number of problems for the conventional understanding of the relationships between innovation and the development of science and technology. While there is little question that scientific knowledge increases the productivity of industrial R&D – by providing new opportunities and by helping the researchers select lines of research – science does not simply come as a free good, nor may its impact on R&D and innovation be simply described as a linear process.

The empirical evidence goes against the notion that scientific research should in principle be abstract, codified and public in such a way as to be instantaneously available to anyone in the world. The very process of creating new, science-based firms can be interpreted as a consequence of the partially tacit nature of even scientific knowledge which, especially in its early stages of development, is embodied in specific individuals and research teams. Moreover, the increasing number of ways for university and industry to interact must be seen as attempts to privately appropriate the potential economic benefits stemming from such knowledge. If anything, the stylized facts of innovation processes in modern biotechnology (see Chapter 3) and the evidence in Chapters 4 to 13 demonstrate how complex innovation processes and such interactions among diverse actors are when we analyse them cognitively, organizationally and institutionally.

Third, these complex processes of corporate and industrial evolution as linked to knowledge are tightly coupled to organizational issues. The processes of creating, transforming and using scientific and technological knowledge has implied the development of complex and sophisticated organizational structures. These range from inter-organizational collaboration and the creation of markets for technology to networks and regional agglomerations. Access to strategic relationships and access to new sources of knowledge may turn out to be two of the most important strategic assets within modern biotechnology. These organizational devices involve a variety of economic and institutional actors, such as governments, universities, firms, bridging institutions and so on. Moreover, the conventional functions of such actors tend to overlap and blur as they interact over time.

Fourth, firms are complex organizations, which make decisions, act, follow routines and operate in ways that involve complex organizational, technological and management problems. Firms do not act alone, nor do they act only based on internal factors. Firms are also partly dependent on their networks with other firms and organizations, as well as on the broader socio-political-economic context.

Fifth, within this broad context, modern biotechnology challenges the standard, simplified accounts of the patterns of industrial evolution and life cycles. The standard account is one in which the early days of an industry are characterized by high rates of entry, and this new breed of innovators gradually supplants the incumbent firms. These newer firms then grow and become the incumbents, in a later phase. For modern biotechnology, we have already noted that it affects many different sectors, being a first objection to considering it 'one' industry. A second objection is that entry by new firms is sustained over time, in that we observe a series of waves of entry in various fields within, or related to, modern biotechnology. These new firms are often started in relation to the appearance of new technologies, ranging from rDNA and monoclonal antibodies, to so-called platform technologies like combinatorial chemistry and high throughput screening, all the way down to genomics and protenomics and so on.

Moreover, within the overlap of biotechnology and pharmaceuticals, not all incumbents have been swept away by new entrants – indeed, many of the large firms seem to have been strengthening their market position. In the past decades, the large incumbent firms have so far been able to maintain leadership of the pharmaceutical industry, by gradually learning and absorbing the new technologies, by controlling key complementary assets, and by establishing a dense and complex web of both collaborative and competitive relationships with entrants. Similar patterns are evident in the agro-food and chemical industries. Real questions remain, however, about future patterns of collaboration and competition between small and medium biotech firms and large incumbent firms in existing industries.

Sixth, modern biotechnology also sharply illustrates all the difficulties that incumbents in existing industries face when attempting to adapt. Especially in this situation, modern biotechnology is similar to other emerging technological areas where it is common to find a plethora of approaches and methods leading ultimately to the convergence of international standards for statistics that permit some comparability. That convergence is clearly on its way, but future research and policy initiatives will be important. Even with additional work, it will be some time before we can really generalize with any certainty about the quantitative dimensions – and before we can link quantitative analysis to qualitative data.

Finally, modern biotechnology poses difficult and often dramatic dilemmas for government regulation and ethical choices. The history of innovation and the development of science and technology in modern biotechnology suggest that such dilemmas are likely to emerge in different fashions, time and again. Democratic societies still have to learn to cope with them, and thereby, societies and their representatives must make choices about which paths to take.

NOTES

1. This paragraph was originally written by K. Smith as part of Chapter 2.
2. Concepts like dynamic competences (Teece et al 1994) and organizational capabilities (Mowery and Nelson 1999) are key in order to make sense of such processes of corporate and industrial evolution.

REFERENCES

Allansdottir, A., A. Bonaccorsi, A. Gambardella, M. Mariani, L. Orsenigo, F. Pammolli and M. Riccaboni (2002), 'Innovation and competitiveness in European biotechnology', *Enterprise Papers No 7*, European Commission.
Commission of the European Communities (2002), 'Life Sciences and biotechnology – A Strategy for Europe', communication from the commission to the council, the European parliament, the economic and social committee and the committee of the regions, January 23.
Ernst & Young (2002), *Beyond Borders: The Global Biotechnology Report 2002*, Toronto: Ernst & Young.
Mowery, D. and R. Nelson (eds) (1999), *Sources of Industrial Leadership: Studies of Seven Industries*, Cambridge: Cambridge University Press.
Teece, D., R. Rumelt and G. Dosi (1994), 'Understanding corporate coherence: theory and evidence', *Journal of Economic Behaviour and Organisation*, 22, 1–30.

Index